T0331276

Advances in Next Generation Services and Service Architectures

RIVER PUBLISHERS SERIES IN COMMUNICATIONS

Volume 14

Consulting Series Editors

MARINA RUGGIERI
University of Roma "Tor Vergata"
Italy

HOMAYOUN NIKOOKAR
Delft University of Technology
The Netherlands

This series focuses on communications science and technology. This includes the theory and use of systems involving all terminals, computers, and information processors; wired and wireless networks; and network layouts, procontentsols, architectures, and implementations.

Furthermore, developments toward new market demands in systems, products, and technologies such as personal communications services, multimedia systems, enterprise networks, and optical communications systems.

- Wireless Communications
- Networks
- Security
- Antennas & Propagation
- Microwaves
- Software Defined Radio

For a list of other books in this series, see final page.

Advances in Next Generation Services and Service Architectures

Editors

Anand R. Prasad

NEC Corporation, Japan

John F. Buford

Avaya Labs Research, USA

Vijay K. Gurbani

Bell Laboratories, Alcatel-Lucent, USA

River Publishers

Aalborg

ISBN 978-87-92329-55-4 (hardback)

Published, sold and distributed by:
River Publishers
P.O. Box 1657
Algade 42
9000 Aalborg
Denmark

Tel.: +45369953197
www.riverpublishers.com

For my parents Jyoti and Ramjee Prasad – A.R.P.

For my wife Gina and our daughter Jacqueline – J.F.B.

For my family – V.K.G.

Table of Contents

Part 4: Security

Preface

Disruptive technology advances are driving new services and new service architectures. More powerful endpoints and access networks are fueling demand for delivery models that support infotainment, social networking, location-based services, IPTV, real-time streaming, smart grids, remote health care, and a variety of other services. New Internet application platforms such as Web 2.0/3.0/4.0 and peer-to-peer networking, as well as increased support for mobile and virtual world applications, are facilitating the creation and delivery of new services. User sensitive services and user targeted service delivery are being enabled by the use of semantic models and context awareness. Service management and deployment is being enhanced by new capabilities for virtualization.

Due to the parallel developments of these trends there is a need for a comprehensive treatment of these topics. The purpose of this book is to present state-of-the-art results in services and service architectures, identify challenges including business models, technology issues, service management, and security, and to describe important trends and directions.

This material is intended to be accessible to a wide technical audience, and is written for researchers, professionals, and computer science and engineering students at the advanced undergraduate level and higher who are familiar with networking and network protocol concepts and basic ideas about algorithms.

Advances in Next Generation Services and Service Architectures is intended to provide readers with a comprehensive reference for the most current developments in the field. It offers broad coverage of important topics with eighteen chapters covering both technology and applications written by international experts. The 18 chapters of *Advances in Next Generation Services and Service Architectures* are organized into the following four parts:

1. Emerging Services and Service Architectures – This part provides eight chapters which survey many of the important emerging categories of

services, and provides details about architectures, service models, and sample applications.

2. IPTV and Video Services – Video content delivery to a variety of end-points with varying capacities and network connectivity is a fundamental service. In this part, four chapters address enabling technologies including semantic support, context-awareness, QoE optimization, and support for mobile devices.

3. Context Awareness – User sensitive application delivery has long been viewed as an important capability to increase the value of services to users. Context awareness focuses on representing and using the immediate situation and surroundings of the user in the delivery of the service. In this part, four chapters cover recent progress in context awareness and illustrate its use in next generation networks and IPTV.

4. Security – New types of services and service architectures require new security techniques. This part contains two chapters, one an overview of security challenges and the other on the user of reputation in service management.

Advances in Next Generation Services and Service Architectures is complemented by a separate volume, *Future Internet Services and Service Architectures*, which covers future Internet architectures, peer-to-peer service models, event based processing, and VANETs.

Acknowledgements

Thanks are due to the staff at River Publishers.

Finally, we thank our families for their support and understanding while we worked on this book.

Anand R. Prasad
John F. Buford
Vijay K. Gurbani
November 2010

PART 1

EMERGING SERVICES AND SERVICE ARCHITECTURES

1

Networks and Services: A Decade's Perspective

José Soler, Ana Rosselló-Busquet, Lukasz Brewka,
Lars Dittmann and Michael S. Berger

Networks Technology and Service Platforms Group, DTU-Fotonik,
Technical University of Denmark, 2800 Kgs. Lyngby, Denmark;
e-mail: {joss, aros, ljbr, ladit, msbe}@fotonik.dtu.dk

Abstract

This introductory chapter provides a description of the evolution of the work within an academic and research institution in relation to "Next Generation Services and Service Architectures" in the last decade. We will present some of the problems found in Communication Services area and how these problems were approached. In this chapter you will find an overview of the evolution of telephony towards IP telephony, telephony services and hybrid telephony architectures. In addition, a brief presentation on personalized service provisioning and energy efficiency for home environments will be provided. This chapter ends with an overview of end-to-end Quality of Service and its challenges.

Keywords: convergence, interoperation, personalization, software configuration management, ontology, home-grid communication, services for energy efficiency, interdomain QoS, home and access interworking, SIP servlets, GUP, policies.

Anand R. Prasad et al. (Eds.), Advances in Next Generation Services and Service Architectures, 3–19.

1.1 Introduction

For the last decade, emerging telecommunication services and architectures have been an element of the research, industrial collaboration and education activities of the Networks and Service Platforms group. This opening chapter will give a brief description of some of this decade's applied research projects within the group.

Within the chapter we try to outline a logical evolution from topic to topic. The main reason for the variety of the presented scope of projects is to show different sets of activities in the group, supported by specific European projects.

The chapter starts with our early work in the realm of telephony "supplementary services" for converged PSTN/VoIP networks, where we applied concepts and mechanisms from Service Oriented Architectures (SOA) to solve the problem of network-dependency in telephony service invocation.

Thereafter, we present our work in the area of service-data and network neutrality, targeting common service and user profile repositories for heterogeneous mobile public networks (2G/3G/WLAN).

This is followed by a description of our work in the area of home networking, also from a services point of view. Three different lines of work are presented. The first one related to customization of behavior of home services and devices, based on profiling and identity management mechanisms to achieve a "personalized", user-dependent home environment. A second line of work follows, targeting software/firmware update automation of home devices and based on semantic and reasoning technologies to solve software/firmware conflicts resolution in a transparent way for end users in home environments. Finally, a line of work on energy efficiency in home environments, targeting energy usage control, communication with the electrical grid and scalable energy management mechanisms, based on semantic technologies and SOA concepts is also presented within this section.

The chapter ends with the presentation of a different working area, targeting interdomain Quality of Service (QoS) management and the necessary interaction between access and core network elements to achieve coherent and robust mechanisms guaranteeing end-to-end QoS.

1.2 Value Added Services in Hybrid Telephony Networks

In the early days of the 21st century, generalization of broadband access at private household premises at affordable prices, together with adoption of

Session Initiation Protocol (SIP) [22] as the preferred signaling mechanism for Voice over IP (VoIP), paved the way for commercial VoIP services to flourish.

The possibility of replacing traditional telephony systems and services with IP-based ones became a subject of study [35, 43]. Telephony, as we understand it and as users have been led to expect, is more than basic voice related issues. The so called Intelligent Network (IN) was initially developed in the 1980s as a complex architectural system complementing the Public Switched Telephone Network (PSTN) in order to enable creation, deployment and maintenance of Value Added Services (VAS) [40]. Within a VoIP context, different Application Servers (AS) within a SIP architecture could host equivalent IN-like applications, if a differentiation from the call related entities (proxies and redirection servers) was necessary.

But a complete replacement of the PSTN infrastructure was not foreseeable for the immediate future in those days,[1] and therefore interoperation between VoIP and PSTN-IN was seen as a must. Not only at call-control level but at service level as well. There was a need to guarantee user experience homogeneity, despite the complete heterogeneity (principles, architectures, protocols, systems) of the involved interacting networks in the different legs of a call involving PSTN-IN and IP-telephony entities.

With this background, a European project targeting the delivery and extension of PSTN-IN services in hybrid PSTN-IP-telephony environments [32] was initiated. The aim of the GEMINI Project was to demonstrate the feasibility of interoperation of already existing PSTN-IN infrastructure and seamless service access from users in a hybrid telephony setup, considering some target demonstration scenarios and services. Our specific role and task within the GEMINI Project, was to define the infrastructure where IP-based value added telephony services should be hosted, to define, implement and demonstrate the mentioned services and to collaborate in the interoperation definition and integration with the rest of the infrastructure (PSTN-IN and SIP-based VoIP).

Since a SIP environment was the chosen playground for the IP-telephony side of the architecture proposed in the GEMINI Project [33], SIP Servlets were considered as a promising technology to support our main task: creation of SIP-based VAS.

---

[1] Just a few months prior to the writing of these lines, in December 2009, USA's Federal Communications Commission received a filling from AT&T, requesting a complete phase out of the PSTN infrastructure [18].

At that date, late 2002, the Java Community Process had released the JSR116 specification proposal for SIP Servlet Application Programming Interface (API). With that release, a Technology Compatibility Kit was provided [42], containing a basic SIP Servlets Container developed by Dynamicsoft Inc., which provided the needed environment to start our SIP Servlets based experiments, in the so called GEMINI's IP-based Service Control Point [33].

The devised initial architecture proposed in the GEMINI Project for the IP-based infrastructure was a symmetrical implementation of that of PSTN-IN with a SIP Proxy in charge of call related signaling and of communicating to a SIP Servlets-based application server when a service was identified and triggered. These two entities were termed respectively IP-based SSP (Service Switching Point) and IP-based SCP (Service Control Point) as their counterparts in the PSTN/IN.

All the call related protocol conversion was managed by a signaling gateway between the PSTN/IN and the IP-telephony islands, based on SIP for Telephones (SIP-T) [54].

Interworking of Intelligent Network Application Part Protocol (INAP) and SIP as described in [27], was not necessary in the architecture proposed in the GEMINI Project, since none of the involved VAS to be demonstrated implied execution of service logic in PSTN/IN entities. Additionally, the question of "network independent" services was considered as a promising research area. In order to give a feasible reply to the issue within the architecture proposed in the GEMINI Project, a Web-Services based mechanism to invoke telephony VAS was proposed, which treated the request for service (PSTN-IN or IP Telephony) independently of its origin, very similar conceptually to the mechanism standardized by Parlay X around that time [34]. A bridging mechanism for SIP was implemented and demonstrated in the original IP-SCP of the architecture proposed in the GEMINI Project, as illustrated in Figure 1.1.

Finally two sets of services, Televoting and Calendar-dependent Call Forwarding, were demonstrated based on the architecture proposed in the GEMINI Project in a test-bed within the network of a Central European telephony operator in Spring 2003. The results were fully satisfactory and within the targets of the project demonstrative purposes, displaying the need to incorporate adaptation layers or bridging mechanisms in order to absorb the specifics of those different networks involved, abstracting them to the upper service implementation logic.

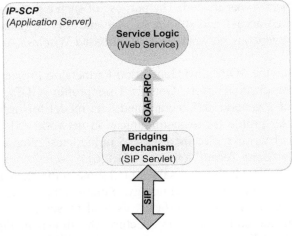

Figure 1.1 GEMINI Project's service invocation decoupled from network signaling.

In the beginning of 2004, it was evident that the research potential regarding services for IP-telephony environments was huge. Issues such as features interaction and service triggering control [44,47,59] appeared as next realms of interest within the subject.

This work can be related to some of the work presented in other chapters, specifically that related to communication and web technologies convergence in Chapter 6, as well as the work related to services in IMS in Chapters 7 and 8.

1.3 Service-Data Centric Networks

By early 2004 deployment of new General Packet Radio Service (GPRS) and Universal Mobile Telecommunication System (UMTS) infrastructures had already started in Europe.

With the deployment of new mobile network elements, a new problem arose: there was a considerable amount of user and service data replicated in different elements of the mobile telephony network(s) infrastructure [19].

As a result, updating and synchronizing these different pieces of replicated information implied an increased volume in the network's signaling traffic, which did not contribute efficiently to the operation.

To demonstrate the feasibility of a likely solution to this problem, a European Project targeting the implementation of an overlay network – the

FlexiNET Project – was created. The purpose of such a network architecture was to centralize access and storage of user and service data, for usage in different heterogeneous networks (i.e. WIFI based Wireless Area Networks) [29].

By that time the 3GPP (3rd Generation Partnership Project) had already started a standardization effort, Generic User profile (GUP), targeting the specification of a generic architecture and data model for user information in a generic user profile for telecommunication networks [5]. 3GPP's GUP is, at the time of writing, also incorporated in the latest release of the 3GPP specification for Long Term Evolution (LTE) [6].

The FlexiNET Project's proposal adopted some of the elements of 3GPP's GUP, mainly the interfacing technology, Simple Object Access Protocol (SOAP)/Web-Services based interfaces, as well as some of the functional splitting within the end overlay architecture. As shown in Figure 1.2, the entry point from different networks to the architecture proposed in the Flex-iNET Project were per-network specialized Access Nodes (AN). Each AN submitted requests and provided responses for user and services data through the web services based generic interface proposed in FlexiNET Project. That interface was the access point to another functional entity, the Data Gateway Node (DGWN), which managed access to the distributed data repository, supported by another specific web-services based interface [16]. Our specific task, within the FlexiNET Project, was to define and implement the integrated generic data profiles for users and services, as well as the DGWN and the specific web-services based data interface [17].

The FlexiNET Project functional results were promising. Nevertheless, its integration and deployment in GPRS/UMTS mobile telephony networks was impractical without improved techniques such as Fast Web Services [53], due to the delay overhead introduced by its double chain of web-services interfaces.

This work is related to Part 3 on context awareness.

1.4 The Intelligent Home: Service Personalization

After participating in the FlexiNET Project, our research line shifted towards profile based customization of applications and services which involved issues related to context aware service configuration and personalization. Within this context, home appliances and their different functionalities (i.e. white-goods, audio/video) can be thought of as generic or user-independent: their functionality and behavior is independent of the user accessing them.

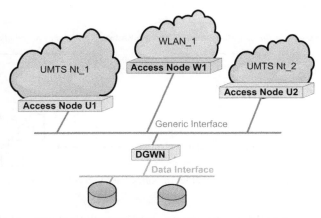

Figure 1.2 Simplified FlexiNET architectural model.

As an example, a TV-set, a video recorder or a washing machine offers the same functionality to whoever accesses their user interface (remote control, button panel), without differentiation or consideration to issues such as ownership, age and administrative rights among others. By 2006, our research group collaborated in another European Project [14], whose aim was to define and demonstrate an architecture, the so called ESTIA architecture, allowing personalized service provision (i.e. management of audio visual contents and white-good functions) within a networked home environment, based on identity management and profiling techniques. Identity management was a hot topic at the moment and multiple identity federation frameworks and applications were proposed by that time [3, 7, 21, 51]. The main idea was to demonstrate the possibility of having identity dependent functionality from home devices, i.e. restricted use of dangerous devices or specific contents to children, possession of devices based on identity and pre-established rights, etc.

Our contribution to this project was the definition of the different user and device profiles, together with initial contributions to the different usage and security policies based on the user's role and identity within the home network [15]. The Extensible Access Control Markup Language (XACML) v1 [2,28] was the technological enabler of our final development whose architectural model was based on the generic architecture proposed by the IETF's Policy Framework Working Group [1,56] and illustrated in Figure 1.3.

A home gateway physical unit integrating different functional elements (profiles and policies, transcoding) and based on a flexible Open Services

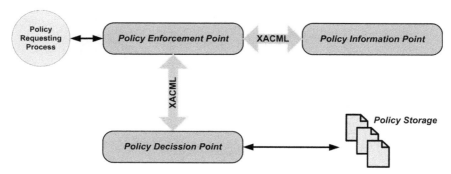

Figure 1.3 XACML as enabler of policy based service architecture.

Gateway Initiative (OSGi) architecture [38], was the outcome of this collaboration, whose results were successfully demonstrated for a set of selected scenarios. Although these results were satisfying, the necessary predefinition and configuration of users and policies, made the approach technically too complex for the "average" home user.

This work is related to Part 3 in this volume, on context aware services.

1.5 The Intelligent Home: Automating Maintenance Services

In order to extend the work related to profile and identity-based service customization, our research group participated in the COMANCHE Project [13] to solve a different challenging problem within home-networked environments: different devices, technologies and services are provided in home environments, which leads to a heterogeneous environment. Managing the maintenance of these different devices and their software can be a confusing and difficult task for "average" end users, especially when dependencies exist. The aim of the COMANCHE Project was to develop a generic framework for Software Configuration Management (SCM) within home environments, which allowed software configuration, conflict detection and error recovery across the multi-vendor, multi-technology environment found in home networks. Very little work had been done on the subject at that date, and mostly for industrial environments [46, 50, 55].

The work was based on the design and implementation of an ontology, termed COMANCHE ontology [20, 24], to represent the knowledge base and enable a context aware solution. The knowledge context considered in this

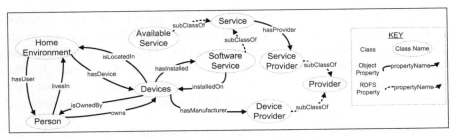

Figure 1.4 Simplified COMANCHE ontology.

project consisted of different devices that compose the home network, their software and their application and services. Based on this, and supported by the necessary communications infrastructure, a home network manager could detect devices connected to the network, their capabilities and their software interdependencies.

To design COMANCHE ontology Protégé [52] was used as the editing tool. Protégé ontologies can be based on a variety of formats including Resource Description Framework (RDF) [57], Web Ontology Language (OWL) [26], and Extensible Markup Language (XML) [4]. The Web Ontology Language is a knowledge representation language for creating ontologies developed by the World Wide Web Consortium. OWL is based on formal semantics and RDF/XML [8] and uses RDF Schema [9]. FACT++ [49] was used as reasoner, to validate the consistency of the ontology and to compute the inferred ontology classification where the dependencies between devices and their software were found.

The ontology shown in Figure 1.4 is a simplified outline of COMANCHE's ontology and consisted of five main classes: *Home Environment, Person, Devices, Services, Provider.*

The *Devices Class* was used to model the devices found in the house. A device is manufactured by a *Device Provider* and has software installed, represented by *Software Service.*

The *Service Class* was further divided into two subclasses: *Available Services* and *Software Services*. These two subclasses represented the software that was available to be installed in devices and the already installed software on devices, respectively.

The *Device Provider Class* and the *Service Provider Class* were subclasses of the *Provider Class* and modeled the different entities that supply the devices or the software.

Using the reasoner over the ontology, the network manager provided control of the devices in the home environment achieving the automatic configuration of the devices' software/firmware. This allowed offering new services to the end user, such as automatic update of software for the individual devices in the network, interoperation and integrity assurance between devices, and finally a holistic service for software optimization of the complete installation.

Although the results matched the initial objectives and success criteria, the created knowledge base (ontology) was too constrained by the chosen demonstration scenarios and therefore limited in applicability.

This work is related to Part 3 in this volume, on context awareness.

1.6 The Intelligent Home: Services for Energy Efficiency

Our work within home-networked environments, presented in the previous sections, was focused on technology enablers of advanced, composed, automated and personalized services. Little attention was paid so far, within our activities, to the impact of those services, or its components to the overall energy consumption within the home or to its aggregated effect on the external energy distribution network.

While metering devices are available at user premises allowing to know the accumulated overall energy consumption at a certain moment, few steps have been taken to provide users with detailed information of per device consumption or the user's own personal energy consumption patterns. This could help them to enhance the efficiency of their energy usage [36, 41, 45]. On the other hand, while energy consumption patterns at home have changed considerably, due to availability of new devices and services, the electrical distribution infrastructure (or grid) has not changed significantly since a century [48]. This electrical distribution grid needs to operate in a more efficient and resilient way to cope with these new demands.

From a communications and services technology perspective, the availability of communication mechanisms allowing a bidirectional data flow between the end users and the grid, would allow the definition of mechanisms to solve the presented challenges. Delivery of punctual and aggregated energy consumption data, from users and geographical areas towards intelligent load distribution decision points within the grid, would allow to balance the load in the distribution grid in an efficient way. At the same time control mechanisms from the grid to the end users, by means of aggregated points of control,

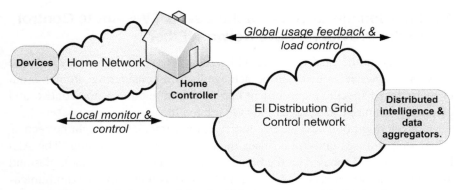

Figure 1.5 In home control and grid data gathering and load balancing.

would allow to minimize the impact of users, or specific consumption patterns on the grid.

With this background, the following two environments, illustrated in Figure 1.5, can be seen as interesting research areas:

- An in-home energy control and monitoring service architecture, based on a flexible OSGi architecture, allowing one to gather consumption information from devices.
- A bidirectional home-grid communication infrastructure, which in a scalable way, can aggregate usage information towards the upstream direction of the electricity distribution grid and provide control mechanisms in the downstream direction.

The expected results may allow electricity operators to create flexible electricity distribution networks enabling the incorporation of new "green" energy sources in a dynamic way. The target is a source and network independent operation of the different distribution sub networks, where control and usage data transmission is defined as an upper information service, independent of the used distribution infrastructure.

The work presented here within our "Intelligent Home" set of activities may be related to that presented in Chapter 11 and Part 3, on semantic technologies support to context aware services, as well as that in Chapter 4 on smart grids.

1.7 Interdomain QoS: From the User's TV Remote Control to the Provider's Core Network Fiber

As presented in the case of our activities related to power efficiency in home networks, synergic effects can be achieved when considering problems out of their closed specific environment and placed in relation to the context and external technological systems they are related to.

Considering home and access domains, their interaction might be seen as a first step towards end-to-end user focused service provisioning. The ALPHA Project [11] focuses on the home and access networks technologies and as such allows research on end-to-end QoS provisioning in telecommunication networks, which is an issue that does not have obvious solutions at the time of writing. The main reasons for this state is the heterogeneity of the telecommunication networks. The reasons for this heterogeneity might vary from political, like ownership, to technical, like scalability.

The result is that the users' data traverse multiple heterogeneous domains on their path from a source to a destination. In such an environment it is a real challenge to provision QoS. This has gathered considerable attention in the ALPHA Project [11]. The activities of the ALPHA Project also cover the control and management of the telecommunication networks where the mapping of the QoS parameters is considered.

The activities around the QoS within ALPHA Project are mainly focused on home and access networks. QoS within home networks is quite a new research area [10, 37]. In the past it might have been considered as not interesting due to very low capacity of the access links compared to the capacity at home.

But nowadays, due to growing access link bandwidth and homes with more and more network end-points, it is becoming important to provide QoS in home networks.

In the scope of the ALPHA Project, after investigating available home QoS mechanisms, provision of the QoS in the home network environment will be handled by Universal Plug and Play Quality of Service Architecture (UPnP QoS Architecture) [12, 31]. The UPnP QoS Architecture [30] defines the services that allow to gain control over the network resources and administer the treatment preferences of particular flows based on their type, priority or users. Three services that are building blocks of UPnP-QoS are: QoS Manager (QM), QoS Policy Holder (QPH) and QoS Device (QD). Additionally the functionality of a Control Point (CP) is defined as the application that is

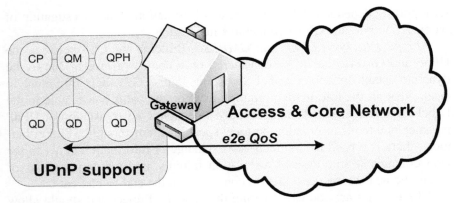

Figure 1.6 In home control and grid data gathering and load balancing.

capable of requesting resources between source and destination of the traffic flow. This is illustrated in Figure 1.6.

The advantage of using UPnP QoS Architecture is that UPnP was designed for architectures where the devices can often leave and join the network. Therefore a Service Oriented Architecture like UPnP will allow automated registration and unregistration of devices and services. Considering the access networks, the way of provisioning the QoS in these networks much depends on their structure, with two distinguishable approaches: Passive Optical Networks (PON) and Active Optical Networks (AON).

- PON – In PON networks a differentiation can be made between architectures with statically allocated resources where the shared part of the media is proportionally distributed among all the users despite the current activity of the end points. The alternative is an architecture that uses the Dynamic Bandwidth Allocation (DBA) methods/algorithms in order to assign the access to the media depending on the users' needs at a particular time.
- AON – Here the attention is mainly focused on the Generalized Multi-Protocol Label Switching (GMPLS) controlled networks. We consider the Resource Reservation Protocol with Traffic Engineering (RSVP-TE) being used for establishment of the Guaranteed Services paths.

Mapping between UPnP and PON (in particular GPON with DBA) – At the time of writing, the project activities are aimed at the mapping between some of the home networks QoS parameters (like reservation level) into buffer reporting dependent DBAs parameters [39] and also the investigation of wireless and wired (PON) architectures integration.

Mapping between UPnP and GMPLS – Building the QoS between the UPnP and GMPLS has to be considered on a couple of levels. Due to the dynamic establishment of Label Switching Path (LSP), it is important to ensure that all the information required to setup the paths is available in the Label Edge Router (LER) and secondly, to enable QoS translation of UPnP parameters into the reservation protocol specific parameters [25,58]. Furthermore, there is a need for bidirectional exchange of information between the QoS domains so in a case of reservation failures, updates, terminations or preemptions, the resources' state is consistent on the end-to-end path.

These issues are addressed within the ALPHA Project and should allow for QoS providing on the connections from home through the access network. The focus on the GMPLS for access networks should ease the extension of the QoS provisioning, looking beyond the access towards the core networks.

Moreover, since the GMPLS is seen as a next generation reliable IP backbone network [23], translation of the QoS signaling between the GMPLS access integrated with UPnP QoS at the home network could be performed building the QoS from the users endpoint to the core servers.

This work is related to Chapter 3 on resource aware networks, as well as Chapter 12 on quality of experience.

1.8 Conclusions

In this chapter, an overview of some concrete European Projects and applied research activities, within the realm of *services* and *service architectures*, in different communication domains and during the last decade, has been provided. With that the evolution of some areas of interest in the Communication Services arena has been partially drafted.

It is not easy to foresee where the focus of Communication Technolgy research will be in the following decade. However, some ongoing and starting trends of this interesting research area are presented in the following chapters.

References

[1] Policy Framework (policy). http://datatracker.ietf.org/wg/policy/charter/. Retrieved on 20 September 2010.
[2] Sun's XACML Implementation. http://sunxacml.sourceforge.net/. Retrieved on 20 September 2010.

[3] Windows Identity Foundation. http://msdn.microsoft.com/en-us/security/ aa570351.aspx. Retrieved on 21 September 2010.

[4] Extensible Markup Language (XML) 1.0 (Fifth Edition), W3C Recommendation. http://www.w3.org/TR/REC-xml/, November 2008. Retrieved on 21 September 2010.

[5] 3GPP. 3rd Generation Partnership Project. 3GPP Generic User Profile. 3GPP TS 29.240 v0.1.0, September 2003.

[6] 3rd Generation Partnership Project. 3GPP Generic User Profile. 3GPP TS 29.240 v9.0.0, December 2009.

[7] Liberty Alliance. http://www.projectliberty.org/. Retrieved on 21 September 2010.

[8] David Beckett. RDF/XML Syntax Specification (Revised). W3C recommendation, W3C, http://www.w3.org/TR/2004/REC-rdf-syntax-grammar-20040210/, February 2004.

[9] Dan Brickley and Ramanatgan V. Guha. RDF Vocabulary Description Language 1.0: RDF Schema. W3C Recommendation, W3C, http://www.w3.org/TR/2004/REC-rdf-schema-20040210/, February 2004.

[10] J. But, G. Armitage, and L. Stewart. Outsourcing automated QoS control of home routers for a better online game experience. *IEEE Communications Magazine*, 46(12):64–70, December 2008.

[11] ALPHA Consortium. EU's Project ICT-2007-212352, Architecture for fLexible Photonic Home and Access networks (ALPHA).

[12] ALPHA Consortium. Requirements and architectural options for broadband in-building networks supporting wired and wireless services, January 2009.

[13] COMANCHE Consortium. EU's project IST-2005-034909. Software configuration management framework for networked services environments and architectures incorporating ambient intelligence features (COMANCHE).

[14] ESTIA Consortium. EU's Project IST-2004-027191. Enhanced networked environment for personalized provision of AV content and appliances control information (ESTIA).

[15] ESTIA Consortium. Deliverable D2.2: Functional Specifications. Technical report, August 2006.

[16] FlexiNet Consortium. Deliverable D2.1: Requirements, Scenarios and Initial Flexinet Architecture. Technical report, June 2004.

[17] FlexiNet Consortium. Deliverable D3.0: Detailed design of Legacy Infrastructure Data & Interworking. Technical report, November 2005.

[18] Cathy Carpino et al. Comments – NBP Public Notice #25. Comments of AT&T Inc. on the transition from the legacy circuit-switched network to broadband, December 2009.

[19] Christoforos Kavadias et al. A novel network architecture for enhanced services and applications in mobile networks. In *Proceedings of 7th International Symposium on Communications Interworking*, November 2004.

[20] E. Meshkova et al. Modeling the home environment using ontology with applications in software configuration management. In *Proceedings of International Conference on Telecommunications 2008 (ICT 2008)*, pages 1–6, 16–19, 2008.

[21] H. Lockhart et al. Web Services Federation Language (WSFederation), v1.1. December 2006.

[22] J. Rosenberg et al. SIP: Session Initiation Protocol. RFC 3261 Proposed Standard, June 2002. Updated by RFCs 3265, 3853, 4320, 4916, 5393, 5621, 5626, 5630.

[23] Kohei Shiomoto et al. *GMPLS Technologies Broadband Backbone Networks and Systems*. CRC Press, 2006.

[24] Nikiforos Ploskas et al. A Knowledge Management Framework for Software Configuration Management. In *Proceedings of Annual International Computer Software and Applications Conference*, pages 593–598, 2008.

[25] S. Shenker et al. RFC 2212: Specification of guaranteed quality of service, Status: Proposed Standard. September 1997.

[26] Sean Bechhofer et al. OWL Web Ontology Language Reference. Technical report, W3C, http://www.w3.org/TR/owl-ref/, February 2004.

[27] V. K. Gurbani et al. Interworking SIP and Intelligent Network (IN) Applications. RFC 3976 (Informational), January 2005.

[28] S. Godik et al. Oasis Open. Extensible Access Control Markup Language. Oasis Open Comitee Specification, August 2003.

[29] FLEXINET. EU's Project IST-2002-507646. Flexible Gateways Architecture for enhanced access network services and applications (FLEXINET).

[30] UPnP Forum. UPnP QoS Architecture: 3 Service template version 1.01 for UPnP, Version 1.0, November 2008.

[31] UPnP Forum. UPnP Forum Standards. http://www.upnp.org/standardizeddcps/, 2009.

[32] GEMINI. EU's Project IST-2001-33465. Generic Architecture for customised IP-based IN services over hybrid Voice over IP and SS7 networks (GEMINI).

[33] GEMINI. GEMINI Comsortium. Deliverable D2.2: Functional & Technical Specifications. Technical report, November 2002.

[34] The Parlay Group. Parlay X Web Services Specification, May 2003.

[35] Vijay Gurbani. SIP enabled IN services – An implementation report, November 2000.

[36] H. Hrasnica, S. Tombros, A. Capone, and M. Barros. A new architecture for reduction of energy consumption of home appliances, March 2009.

[37] Taein Hwang, Hojin Park, and Jin Wook Chung. Home-to-home media streaming system based on Adaptive Fast Replica. In *Proceedings of 11th International Conference on Advanced Communication Technology, 2009 (ICACT 2009)*, volume 3, pages 1665–1666, February 2009.

[38] Open Services Gateway Initiative. OSGi Service Platform. Release 3, March 2003.

[39] Telecommunication Standardization Sector of ITU. *G.984.3 – Gigabit-capable Passive Optical Networks (G-PON): Transmission convergence layer specification*, March 2008.

[40] ITU-T. International Telecommunication Union. ITU-T Series recommendation Q.1200 to 1699.

[41] Koen Kok, Stamatis Karnouskos, David Nestle, Aris Dimeas, Anke Weidlich, Cor Warmer, Philipp Strauss, Britta Buchholz, Stefan Drenkard, Nikos Hatziargyriou, and Vali Lioliou. Smart houses for a smart grid, 2009.

[42] Anders Kristensen. SIP servlet API technology compatibility kit (draft), November 2002.

[43] Jonathan Lennox. Implementing intelligent network services with the Session Initiation Protocol. Technical report, United States, February 1999.

[44] Jonathan Lennox and Henning Schulzrinne. Feature interaction in internet telephony. In *Proceedings 6th Feature Interactions in Telecommunications and Software Systems*, pages 38–50. IOS Press, 2000.

[45] R. Miceli, D. La Cascia, A. Di Stefano, G. Fiscelli, and G.C. Giaconia. Impact of novel energy management actions on household appliances for money savings and CO_2

emissions reduction. In *EVER 09 – Ecologic Vehicles – Renewable Energies*, March 2009.

[46] Eric Endsley Morrison, Morrison R. Lucas, and Dawn M. Tilbury. Software tools for verification of modular FSM based logic control for use in reconfigurable machining systems. In *Proceedings 2000 Japan-USA Symposium on Flexible Automation*, Ann Arbor, pages 565–568, July 2000.

[47] Masahide Nakamura, Pattara Leelaprute, Ken-ichi Matsumoto, and Tohru Kikuno. On detecting feature interactions in the programmable service environment of internet telephony. *Comput. Netw.*, 45(5):605–624, 2004.

[48] U.S. Department of Energy. The smart grid: An introduction, 2009.

[49] University of Manchester. Fast clasification of terminologies ++, FaCT++.

[50] Hamid Haidarian Shahri, James A. Hendler, and Adam A. Porter. Software configuration management using ontologies, 2007.

[51] Shibboleth. http://shibboleth.internet2.edu/. Retrieved on 21 September 2010.

[52] Standford University Stanford Center for Biomedical Informatics Research. Protégé Editor, Platform v3, 2010.

[53] International Telecommunication Union. ITU-T recommendation X.892: IT-generic applications of ASN.1 – Fast Web Services, May 2005.

[54] A. Vemuri and J. Peterson. Session Initiation Protocol for Telephones (SIP-T): Context and architectures. RFC 3372 (Best Current Practice), September 2002.

[55] Shige Wang and Kang G. Shin. Constructing reconfigurable software for machine control systems. *IEEE Trans. Robotics and Automation*, 18(4):475–486, August 2002.

[56] A. Westerinen, J. Schnizlein, J. Strassner, M. Scherling, B. Quinn, S. Herzog, A. Huynh, M. Carlson, J. Perry, and S. Waldbusser. Terminology for policy-based management. RFC 3198 (Informational), November 2001.

[57] RDF Working Group. Resource Description Framework (RDF). http://www.w3.org/RDF/. Retrieved on 21 September 2010.

[58] J. Wroclawski. *RFC 2210: The use of RSVP with IETF integrated services*, Status: Proposed Standard, September 1997.

[59] Xiaotao Wu and Henning Schulzrinne. Feature interactions in internet telephony end systems. Technical report, Department of Computer Science, Columbia University, 2004.

2

Interactive Multimedia Consumer Services: Bridging the Three Islands of the Web, Telecom and Entertainment

Andrea Basso and Gregory W. Bond

AT&T Labs Research, NJ, USA; e-mail: {basso, bond}@research.att.com

Abstract

There is a growing demand for interactive multimedia consumer services both on broadband and wireless, as evidenced by the increasing popularity of social networks, interactive TV, online games, real-time messaging and advanced telecom services. However the infrastructures supporting each of these services largely belong to one of three separate, largely incompatible "islands": the web, telecom or entertainment. A consequence is that the consumer bears the burden of navigating amongst these islands. The evolution and growth of interactive multimedia consumer services imposes a number of requirements on these three technology areas. The focus of this contribution is the discussion, from an architectural and development perspective, of the requirements and possible solutions needed to simplify the consumer experience while supporting a flexible service development model. By establishing bridges amongst these three islands we seek to make them true enablers for future multimedia interactive services.

Keywords: VoIP, multimedia, streaming, telecommunications, entertainment, Internet.

Anand R. Prasad et al. (Eds.), Advances in Next Generation Services and Service Architectures, 21–41.

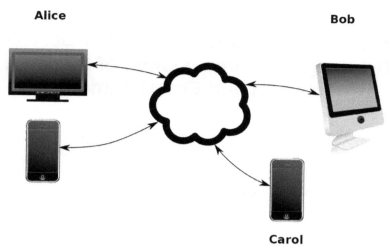

Figure 2.1 Example participants and their devices.

2.1 Introduction

There is a growing demand for interactive multimedia consumer services both on broadband and wireless, as evidenced by the increasing popularity of social networks, interactive TV, online games, real-time messaging and advanced telecom services. However the infrastructures supporting each of these services largely belong to one of three separate, largely incompatible "islands": the web, telecom or entertainment. A consequence is that the consumer bears the burden of navigating amongst these islands.

In order to motivate the discussion we present an example scenario high-lighting the desired consumer experience in the face of using multimedia services that reside on the three islands. At the core of this example is a con-sumer, Alice, who subscribes to a number of different multimedia services, one of which is a co-watching service, e.g. [20]. This particular co-watching service allows participants to collaboratively watch and control TV while maintaining an audio conversation with one another. As shown in Figure 2.1, Alice has a TV with a set-top box and a smartphone. Her friend Bob has a PC and her friend Carol has a smartphone.

Alice is watching TV and wants to invite a remotely located friend Bob to a co-watching experience. To invoke this service Alice uses a co-watching application on her smartphone. Using the application, Alice chooses to view the TV content on the TV she is currently watching and chooses her phone as the device she will use to converse with Bob. Then Alice selects Bob from her address book and the application sends Bob an invitation to the co-watching session. Since Bob is sitting at his PC, Bob receives the invitation on his PC as an instant message. He clicks on a link embedded in the invitation which presents him with a screen where he chooses to view the TV content on his PC and converse with Alice using his PC. At this point the co-watching session is established. Alice and Bob see the same content at the same time, they can control the delivery of the content itself, e.g. rewind or pause, and they can converse with one another. If Alice receives a phone call from Carol while participating in the co-watching session, Carol's call ID is overlaid on Alice's TV display and Alice's phone rings. Alice's phone and TV offer her the ability to accept the call on either device. When the call is accepted the shared TV session is optionally placed "on hold" so that the video stream pauses for both Alice and Bob, and Alice and Bob's conversation is placed on hold while Alice converses with Carol. During the course of the call Alice uses her co-watching application to invite Carol to join. Carol receives an invitation in the form of a pop-up query presented on her phone. Carol accepts the invitation and chooses to use her phone to join the session.

While one could imagine a single monolithic service that provides the user experience given in the example, the reality is that consumer preference is to pick-and-choose from a myriad of primarily standalone services available from their network service providers and from third-party "over-the-top" service providers. In our example, Alice's TV may be offered as an IPTV service from a network provider or it may be provided by an over-the-top provider like Hulu. Bob's instant messaging service is most likely provided by a third-party like AOL. The call alerting and call waiting services may be provided by a network provider or by an over-the-top provider like Skype or Google.

When offered as stand-alone services, the consumer is forced to work with each service separately. Assuming that the services exist stand-alone in our example, then Alice would first have to determine if Bob is online, perhaps by checking his instant messaging presence status using a stand-alone instant messaging service. Then Alice would be responsible for deciding how to invite Bob to the co-watching session. Alice may decide to invite Bob via instant messaging, perhaps by cutting and pasting a co-watching session URL from the stand-alone co-watching service into the stand-alone instant messaging service.

However, Alice should not need to know that Bob can be contacted by instant messaging. It suffices for Alice to know that Bob is online in order to choose to invite him to a co-watching session. Furthermore, Alice's co-watching service should be able to determine how it will contact Bob. Knowing that Bob is using instant messaging, the co-watching service can choose to use instant messaging to deliver its invitation. Ideally, the existing instant messaging session can be used to "bootstrap" the co-watching session in a way that provides a simple user experience.

A similar situation occurs when Alice invites Carol to join the co-watching session. Today, it would be up to Alice to determine how to convey a co-watching invitation to Carol. Most likely Alice would send the invitation using a separate service like SMS or email, or she would simply read an invitation URL over the existing audio connection meaning that Carol would have to transcribe the URL in order to access the co-watching session. An approach that simplifies Carol's experience would be the following: given that Alice and Carol already have an existing session between them it should be possible for the co-watching service to use this session to convey its invitation in a manner compatible with the session; in this case, sending a message over the session that results in presenting a query to Carol. This way, Carol need only respond to the invitation that is put to her.

To provide the consumer with the desirable experience given in the example, the underlying infrastructure must support:

1. Flexibility and independence so that service developers are free to innovate.
2. Management, coordination and discovery of sessions established amongst endpoint devices, network-based application servers and network-based media servers.
3. Coordination of media streams associated with those sessions and low latency synchronization of session signaling and media streams.
4. A compatible representation of media signaling and coding among the different services and devices.

These requirements are only partially satisfied today and in a measure that is still insufficient for their efficient implementation. For example today, not only are content delivery, telecom and internet-access supported by separate infrastructures, APIs and user interfaces, but they are also based on different, often conflicting, media signaling and coding standards.

This chapter explores the nature of these incompatibilities, from an architectural and development perspective, with the goal of identifying the requirements for a system that will manage incompatibilities in a way that simplifies the consumer experience while supporting a flexible service development model. Based on our findings we propose utilizing a number of pre-existing technologies for "building bridges" between the islands in a way that achieves these goals. We call this the Cooperative Context-aware Multimedia Service Architecture (CCMSA).

2.2 Signaling and Media Protocols in the Three Islands

To gain further insight into the differences between the three islands this section presents a brief overview of signaling and media protocols used by them. Each of the protocols we discuss in this section belongs to one of the three islands.

2.2.1 Protocols Overview

SIP. Session Initiation Protocol (SIP) [24] is an IETF standard protocol for establishing bi-directional session-based multimedia communications. SIP is the dominant protocol for IP telecommunications having been adopted by the 3GPP as a signaling protocol used by the IMS (IP Multimedia Subsystem) architecture [1]. SIP is designed to establish sessions between a pair of endpoints. Session establishment between more than two parties (e.g., for conferencing) is achieved by joining multiple sessions at an endpoint or via an intermediary component acting as more than one endpoint (e.g., a back-to-back user agent). The protocol permits endpoints to (re-)negotiate media codecs during and after session establishment. Once a session is established, media is exchanged between endpoints using media-specific protocols agreed upon during codec negotiation. Thus, SIP maintains a clear separation between media delivery and session establishment.

Typically, SIP is used for establishing audio and video sessions, but extensions to the protocol support instant messaging and presence [27] and general purpose event notification [23], and new uses are being discovered for it such as home automation [4].

SIP is not only used for establishing (and tearing down) a session. It is also used to control a session in progress. Since a SIP session persists until the session is torn down, the session can be used to convey in-session signals between endpoints. This makes it possible for endpoints to re-negotiate codecs, exchange information with each other or with intermediary components, and control media flow (directionality, source or destination). All of these attributes make SIP well-suited to support real-time, person-to-person communications.

RTP and RTCP. The Real Time Protocol together with the companion protocol Real Time Control Protocol (RTCP) [21] define a standardized format for delivering media over IP networks. RTP and RTCP have been developed by the Audio Video Transport Group in Internet Engineering Task Force (IETF). While RTP is used for carrying media, RTCP delivers, in the form of sender and receiver reports, information about the quality and the statistics of the media delivery as well as all the information needed for synchronization.

RTP is based on a design principle known as Application Level Framing (ALF). For each application category class (e.g., audio, video), RTP defines a profile and one or more associated payload formats that are media-specific. RTP provides, by means of the generic header, general services for jitter computation via the RTP timestamp, and packet loss detection and computation via the RTP sequence number. The RTP media specific payload format is customized on the basis of the media format transported (i.e., MPEG-2 video or H.264 video) and carries information to better frame the media, based on its characteristics and increase the error resilience to network impairments.

IP multicast. Current IP Multicast protocols are receiving a lot of attention on the Internet island, in particular for applications such IP-based Television (IPTV) and Digital Video Broadcast-Handheld (DVB-H) as well as multi-site corporate videoconferencing, and large scale data distributions. IP multicast is used for the distribution of a variety of data: from metadata and electronic program guides to financial data, as well as database replication and content caching. Several key aspects of multicast are relevant to these types of services including multicast addressing for payload and payload forwarding and routing protocols including PIM-SM, CBT, PIM-DM, DVMRP, and MOSPF as well as Multicast over IPV6. The emergence of this protocol is mainly in the area of managed services [22].

RTSP. Real-Time Streaming Protocol (RTSP) [25,26] is a an IETF standard utilized by services on the Internet island, for delivering multimedia content from a server to clients. The protocol is designed to support delivery of live or stored content.

Like SIP, RTSP supports codec negotiation and it establishes a session between a server and its clients. In contrast to SIP, RTSP is uni-directional since media is intended to flow only from server to clients. Also unlike SIP, an RTSP server is designed to serve more than one client at a time, including support for live content delivery to multiple clients.

By virtue of establishing a session between client and server, RTSP facilitates client control of the server's content delivery by sending in-session signals, e.g. pause, seek, rewind, fast forward.

HTTP. HyperText Transfer Protocol (HTTP) [7] is an IETF standard originally intended to support client access of server-resident hypertext documents. However, while HTTP itself has remained unchanged for over a decade, its applicability has broadened considerably. Of central interest here is its capability for delivering multimedia content on the Internet island.

Most commonly, HTTP is used to deliver stored content by progressively downloading the content to the client allowing the client to play the content before it is completely downloaded. Unlike streaming of stored content with RTSP, stored content provided by HTTP is progressively downloaded and cached on the client.

The recent evolution of HTTP streaming in the form of smooth or dynamic streaming (see Microsoft smooth streaming [28], Apple adaptive video streaming [3] and Adobe dynamic streaming [2] approaches), is gaining a lot of traction. HTTP dynamic streaming works by segmenting the original input video in several chunks of a few seconds each in length. Every chunk is encoded at multiple bitrates. The player is aware of the chunks and concatenates them together during playback again requesting the chunks more suitable for the available bandwidth and resources. Dynamic HTTP streaming in this way implements some form of QoE adapting the delivery bitrate to the available bandwidth and local processing resources [8]. The HTTP streaming approach simplifies the server architecture since a single server can be used to deliver the media and to handle the control. In contrast, RTSP requires in general a server to handle the RTP streams. HTTP streaming combines the advantages that HTTP brings (i.e., firewall traversal) with the ability to seek to non-downloaded parts.

2.2.2 Observations

The use of SIP is growing for the telecom domain as it becomes the dominant protocol underlying mobile telecommunications. While its use is scant for the Internet and entertainment domains, it is the most likely candidate for adoption by these two islands in order to support the interactive, session-oriented services demanded by consumers.

The services that the RTP protocol provides are mainly useful in real time interactive applications and live entertainment but not for video-on-demand (VOD), or over-the-top video. In VOD applications, in fact, inter-media synchronization is less critical as the delivery mechanisms, as HTTP streaming carries multiplexed formats (such as MPEG-2 transport stream, Flash, MPEG-4). An exception is the case of IPTV live media delivery such as linear TV where the real time aspect is important. We expect to see RTP used much more in real-time interactive applications such as telepresence than in non-interactive entertainment services.

IP multicast is the common choice in managed video services such as live IPTV where the network infrastructure is built from scratch. Due to its deployment costs it will have a more limited adoption in existing networks where it is likely that alternatives such as peer-to-peer (P2P) will be considered. The advantage of the P2P approach is more robustness as there is no single point of failure as in the multicast case and much cheaper deployment cost.

HTTP streaming is becoming a key player in the streaming space due to its advantages and simplicity. It is becoming very popular as sites like YouTube support this functionality. We expect that dynamic HTTP streaming will become more common in the player and server streaming implementations.

RTSP is losing momentum in entertainment applications as HTTP streaming delivers a solution to the streaming problem that is cheaper to implement, is simpler and is more flexible.

2.3 Media Coding and Representation on the Three Islands

In the past, the development of video coding technologies has been articulated along two tracks: video coding for entertainment and video coding for teleservices. This brought an inevitable distance between these two classes of services that instead now are converging. In this section we want to highlight the differences among the two approaches, how the convergence is taking place and the issues that this implies. ITU-T has been the key organization that drove the standardization effort in the area of video coding for telecommunications. It started with the recommendation H.261 followed by H.263 and its several revisions (H.263+ and H.263++) while ISO/MPEG focused more on video codecs that were more suitable for entertainment services and digital TV. More recently the efforts of these two organization came together in the standardization of the much more performant codec H.264. In this section we will briefly overview the evolution of the coding technologies in these key application areas and pinpoint their differences and commonalities. We will finally discuss H.264 and how it is used today for both entertainment and video services.

2.3.1 Video Coding Standards Overview

H.261 and H.263. The H-261 [12] standard was developed by the ITU-T study group XV in the nineties. The group was given the task to standardize a video coding algorithm to enable audio-visual communications such as videotelephony and videoconferencing over ISDN. A Hybrid block based DCT/DPCM coding scheme with motion compensation was selected to boost compression performance by means of inter-frame coding. All the subsequent coding schemes will be based on the same principle. In H.261 the first frame of an input video sequence is encoded in Intra-frame mode (I-picture). Each subsequent frame is coded using Inter-frame prediction (P-pictures) where the prediction is computed on the data from the nearest previously coded frame. An essential feature of such coding scheme is the "conditional macroblock replenishment" algorithm that allows the update of macroblock information at the decoder end only if the content of the macroblock has changed with respect to the content of the corresponding macroblock in the previous frame. H.263 [13] is similar to H.261, however with some coding performance and error resilience improvements. The differences between the H.261 and H.263 coding algorithms include a more accurate motion estimation and compensation with half pixel precision, better coding of the symbols to be transmitted (syntax-based arithmetic coding) and more articulated prediction modes (advanced prediction, forward and backward frame prediction similar to MPEG called P-B frames). H.263 had several revisions and improvements (H.263+ and H.263++).

MPEG-1. In parallel MPEG-1 [9] was standardized by ISO-IEC. MPEG-1 focused primarily on CD-ROM multimedia applications. It was designed to support only non-interlaced video, with flexible picture size and frame rate. Essential features for entertainment applications include frame based random access, fast forward/fast reverse (FF/FR) reverse playback and editability in compressed domain. In order to incorporate the requirements for storage media the notion of I-pictures and B-pictures was introduced in MPEG-1. I-pictures allow access points for random access and FF/FR but are not efficient from a compression perspective. Inter-frame predicted pictures (P-pictures) are coded with reference to the nearest previously coded frame (either I-picture or P-picture), usually incorporating motion compensation. They do not provide suitable access points for random access or editability as they depend on I frames for decoding. Bi-directional predicted/interpolated pictures (B-pictures) require both past and future frames as references. As a general rule, a video sequence coded using I pictures only allows the highest degree of random access and editability but achieves only low compression. A sequence coded without B pictures achieves moderate compression and a certain degree of random access and FF/FR functionality. Incorporation of all three pictures types brings a good compromise between random access, FF/FR functionality and compression ratio but on the other side increases the coding delay significantly. This delay may not be tolerable for video services such as videotelephony or videoconferencing.

MPEG-2. MPEG continued its standardization efforts with a second phase called MPEG-2 [10] in order to support emerging applications, such as digital TV and satellite and terrestrial digital broadcasting. New coding features were added to achieve sufficient functionality and quality, thus prediction modes were developed to support efficient coding of interlaced video. Furthermore, scalable video coding extensions were introduced to provide additional functionality such as graceful degradation in the presence of network impairments. The concept of I-pictures, P-pictures and B-pictures is kept in MPEG-2. In addition MPEG-2 has introduced the concept of frame pictures and field pictures along with particular frame prediction and field prediction modes to accommodate coding of progressive and interlaced video. New motion compensated field prediction modes were introduced by MPEG-2 to efficiently encode field pictures and frame pictures.

H.120. The H.120 was an attempt to standardize coding schemes aimed to videoconferencing applications using TV quality interlaced video. The coding schemes adopted were more focused in keeping the delay low than the video quality. The H.120 standard was not successful because it could not be established as a unique world-wide standard and also because the price-performance ratio was not adequate: the costs for the codec equipment and the costs to deliver the video with the relatively high data rate required for transmission were too high to find acceptance.

MPEG-4. ISO MPEG-4 [11] started its standardization activities in July 1993 with the goal to develop a low bitrate video coding algorithm. It evolved into a standard focused on interactivity and content composition providing the capability to manipulate and edit the visual objects part of a given scene in the compressed domain without transcoding. The standard provides tools to support content analysis and indexing, content-based multimedia data access, support coding of multiple views or soundtracks of a scene efficiently. In addition it provides efficient methods for combining synthetic components with natural scenes at the bitstream level (i.e., text and graphics overlays). Finally it includes erro resilience schemes for error-prone network environments. To achieve such target a significant change in the video coding scheme was needed. Many of the algorithms used are based on so-called "second generation coding techniques", e.g. object based coding schemes, model based coding techniques and segmentation based coding schemes. Many of the schemes diverge significantly from the successful hybrid DCT/DPCM coding concept employed in H.261, H.263, MPEG-1 and MPEG-2 standards.

H.264/AVC/MPEG-4 Part 10. H.264/MPEG-4 AVC [19] codec developed by a join partnership known as JVT of the ITU-T Video Coding Experts Group (VCEG) in conjunction with the ISO/IEC Moving Picture Experts Group (MPEG). The resulting spec H.264/AVC/ MPEG-4 Part 10 is widely used today in a variety of applications that span from Blu-ray Disc, DVB broadcast, to real-time videoconferencing and telepresence. Due to the advances in VLSI technology the H.264/AVC/MPEG-4 Part 10 contains a number of new algorithms and methods that were considered too complex and expensive in previous standardization efforts. As a result its compression efficiency has improved considerably, and in general almost doubled with respect to MPEG-2. In addition the range of applications supported has been also greatly extended. From a coding efficiency perspective the new features include more sophisticated algorithms that allow multi-picture inter-picture prediction, variable block-size motion compensation, the ability to use multiple motion vectors per macroblock, the ability to use any macroblock type in B-frames, quarter-pixel precision for motion compensation, enabling precise description of the displacements of moving areas, as well as new transform and quantization schemes. From a network support perspective, H.264 specifies a very flexible Network Abstraction Layer (NAL) that allows an efficient packetization of the coded video in RTP. In summary, H.264 increases the coding efficiency, is able to satisfy the requirements of low delay applications as teleconference but it can boost coding performance and insure random access, precise frame rate playback and editability for entertainment applications.

2.3.2 Observations

The success of coding standards has always been deeply related to the evolution of technology and in particular VLSI design has been a major factor in this respect. In particular in the area of video coding a large set of sophisticated coding methods exist but may be too complex for implementation. The H.261 standard was intended to give freedom to the user to find his trade-off between video quality and transmission costs. However, at the time the standard was issued only a poor spread of ISDN connections was available and the costs for implementation of the video coding equipment was high due to the limited performance of the VLSI technology. In this respect the MPEG-1 standard tried unsuccessfully to provide considerable advantages for a variety of multimedia terminals, with the additional flexibility provided for random access of video from storage media and the diverse image source formats supported. MPEG-2 became a success because there was a strong commitment from industries, cable and satellite operators and broadcasters to use this standard. The standard is today still very popular in Digital TV broadcasting, pay TV, pay-per-view, video-on-demand, interactive TV. H.264 offers the highest range of capabilities to cover th needs of the three islands but the coding modes still differ considerably.

2.4 Building Bridges

In order to achieve the desired consumer experience described in the introductory example from Section 2.1, an infrastructure must exist that:

- allows services to manipulate sessions and media channels without introducing cross-service conflicts;
- supports delivering content compatible with different endpoint devices;
- does not dictate the technology used by a service developer.

Today, no single existing technology meets all these requirements. However, a number of technologies exist that, if combined, would satisfy the requirements. In particular, we propose combining *service composition, distributed media control*, and traditional *protocol/codec bridging* to serve as the constituent parts of an infrastructure we call the Cooperative Context-aware Multimedia Service Architecture (CCMSA) that will provide bridges between the three islands. Service composition provides the means for services to share sessions. Once services are composed, distributed media control provides a means for services to independently manipulate individual media streams in a consistent fashion. Protocol and codec bridging provide a means for a service to deliver content to an endpoint without concern for endpoint or media server capabilities. The bridge formed by these constituent technologies is a framework with an API for service developers. A service developer is free to choose how much of CCMSA to use depending on the nature of a service and the desired level of integration with other services.

In the following discussion, we adopt the view that sessions are stateful, persistent associations between system elements, whether those elements are endpoint devices or network elements. Sessions enable the management of media channels: their creation, destruction and modification. We require the existence of a session between elements prior to establishing a media channel between elements. We also allow multiple media channels to be associated with a given session. This concept of a session is consistent with the one supported by the SIP protocol.

2.4.1 Service Composition

To provide the consumer with a simplified experience, the network needs to provide a means for services from third parties and network providers to utilize pre-existing sessions associated with other services. By doing this it becomes possible to simplify the consumer experience and to unify service offerings so that the unified whole is greater than the sum of its individual parts. To perform this task we propose using the *application routing* mechanism which is part of the SIP Servlet 1.1 standard [16]. Application routing is the realization of the Distributed Feature Composition (DFC) virtual architecture [14, 29] in the SIP realm. Application routing provides a means for independent applications (services) to be inserted into sessions established between endpoints. It effectively allows services to share sessions while providing a framework for managing interactions between them.

Application routing dynamically assembles services into arbitrarily complex session graphs interconnecting services with endpoint devices. For example, Figure 2.2 reflects the final state of the co-watching example when all three partipants are taking part in the co-watching session. In this figure:

- IPTV represents the IPTV media server that is providing the audio/video TV content.
- MIX represents the media mixer responsible for mixing the participant's conversations.
- IM represents the instant messaging server for Bob's instant messaging client.
- CW represents Alice's mobile phone call waiting service.
- COW represents Alice's TV co-watching service.

The graphs may evolve over time as sessions are added and destroyed. Each arc in the graph represents a SIP session and each leaf node in the graph represents a SIP user agent. The interior nodes are typically SIP back-to-back user agents since they serve to link SIP sessions.

Each service intercepts signaling messages passed to them along the sessions they are connected to. A service is free to absorb, modify, propagate

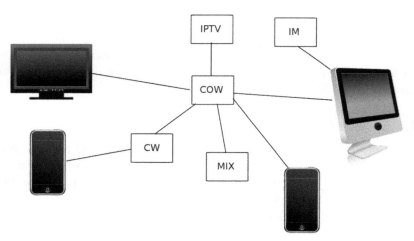

Figure 2.2 A session graph resulting from application routing.

or generate a message. For example, by generating an initial SIP INVITE request, a service is able to create a new session, and by generating a SIP BYE message a service can terminate a session. With this flexibility comes the challenge of managing interactions that may occur amongst composed services. An application router does not automatically prevent bad interactions, but the constrained way in which services are permitted to interoperate makes the analysis of possible interactions practical [29].

Application routing is provided by standard SIP servlet containers, of which there are a number of commercial and open source implementations. SIP servlet containers play the role of IMS application servers, but they are capable of operating independent of IMS infrastructure. The nodes representing services in the graph are SIP servlet applications running in a SIP servlet container. All such containers also provide standardized HTTP servlet support [17] which means that a service may also include an HTTP interface for the purpose of interacting with the service. A service with both SIP and HTTP interfaces is called a *converged service*. A converged service gives the service developer flexibility in how their service will be controlled. For example, the co-watching application that Alice uses on her smartphone can use HTTP instead of SIP to control the network-resident co-watching service. Note that converged services use SIP to establish and maintain sessions, and HTTP to control the session, e.g. via a REST-ful interface. The existence of an HTTP interface makes it possible to unify call control logic and web-based services using a service composition mechanism [5].

2.4.2 Distributed Media Control

Given the existence of a composition mechanism, it is inevitable that multimedia services will compete for use of the same endpoint devices and network-based media resources for content delivery. Therefore, any composition mechanism must also provide support for mediating amongst potentially conflicting media requests so that the consumer is not faced with performing the mediation themselves. At the same time, we require that services can be developed independent of one another.

The solution to the problem is the adoption of a SIP-based distributed media control algorithm [6]. This algorithm provides formal guarantees with regards to the correctness of the media connections that will result from different services making potentially conflicting requests for media resources. Given the complexity of the algorithm, it is necessary to embed it in a programming abstraction [30] for it to be of practical use for services.

2.4.3 Service Developer API

It must be possible to develop a service independent of other services it may be composed with. This is the freedom that third party developers expect and it is a requirement for innovation. On the other hand, if the composition mechanism provides a superior consumer experience then developers are likely to use it. The challenge, then, is to provide service developers with an API that offers compelling advantages for their service while not technologically constraining them. For example, it should be possible for an pre-existing service, like the third-party instant messaging service that Bob uses, to utilize the API without requiring a major redesign.

Our proposal for solving this problem consists of an endpoint-based API and a network-based API. The endpoint-based API is light-weight and suitable for resource-constrained mobile devices. The network-based API provides developers with the high level abstractions required to specify session and media control logic without being required to know SIP. It also provides support for receiving and delivering content without concern for the specific protocols and codecs required for endpoint devices. To provide some intuition for how these APIs would work, consider Figure 2.3. The figure shows the progression of our example, beginning in a state where a TV media stream is being delivered to Alice's TV. Note that each endpoint device possesses a client API. Some functions supported by this API are: ascertaining or updating current session state, and creating or terminating sessions associated with the endpoint.

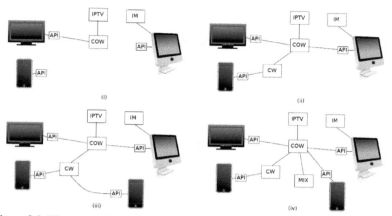

Figure 2.3 The co-watching example implemented using the service developer API.

In the first diagram, Bob's third-party instant messaging program (most likely a non-SIP-based) takes advantage of the API by notifying it that it has established a third-party instant messaging session and provides it with the coordinates for accessing the IM session. Apart from the modification to perform the notification, Bob's instant messaging client remains unchanged. The co-watching application on Alice's smartphone first obtains a reference to the network-resident co-watching service for inclusion in the invitation to Bob. When Alice invites Bob to the co-watching session, the co-watching application uses the client API to query if Bob has any existing sessions to send the invitation on and then uses the API to convey the invitation to Bob's instant messaging session via a gateway to the third-party instant messaging server.

In the second diagram, Bob accepts the invitation by clicking on a link embedded in the instant message. This has the effect of initiating a HTTP request to the network-resident co-watching service, which responds by sending a SIP INVITE message to the client API on Bob's PC. By accepting the invitation, a session is established between COW and Bob's PC. Similarly, COW establishes a session with Alice's smartphone. At this point Alice and Bob are participating in a co-watching session.

In the third diagram Carol calls Alice and the call is handled by Alice's call waiting service. To invite Carol to the co-watching session, the co-watching application on Alice's smartphone accesses the client API to discover the session between Alice and Carol and then sends a message to Carol's client API instructing it to initiate an invitation to the co-watching service. Carol's client API responds to the request by notifying Carol of the request and, when Carol accepts the invitation, the API initiates a session with COW.

In the final diagram, Carol has joined the co-watching session. COW has established a session with a network-resident media mixer in order to mix the three participant's conversations. Alice has instructed her call waiting service to switch back to her COW session and Carol has dropped her session with Alice's call waiting service.

From this description of the API it can be seen that the client-based API sometimes behaves as a service and sometimes as a traditional API. For example Alice's smartphone API behaves like a telecom service when the co-watching application uses the API to send a co-watching invitation to Carol's smartphone API via the session that exists between them. Alice's API acts like a traditional API when the co-watching application uses it to query the session status of Bob or Carol.

Work comparable to what we propose here is the JSR 281 IMS services API for client devices [15] and the IMS services API for Java EE [18]. These APIs aim to simplify development of IMS-based services. For example, they provide developers with the ability to create or tear down sessions, and access IMS core services like presence and messaging. While we share the goal of simplifying development, one important difference is that our work is not dependent on IMS infrastructure. Our proposal can be deployed in an IMS environment, or it can be deployed in a lighter-weight environment consisting of registrars, application servers, and media resources. Another important difference is that our proposal aims to simplify life for the consumer. This broadens the scope of our work to encompass session sharing across services, distributed media control, protocol bridging and codec bridging.

2.4.4 Protocol Bridging

Bridging protocols is an important part of CCMSA as it supports integrating and translating from one island's protocol to another's. The goal is to transparently provide bridging support for services utilizing CCMSA. Protocol bridging is relatively straightforward. For example, assume that a consumer requests media content by invoking a media service using SIP but the media source is an HTTP server. In this case a SIP to HTTP adapter acts as a SIP user agent on the consumer side and as a HTTP user agent on the HTTP side. The adapter requests content from the HTTP server and then encapsulates the media content received from the server in RTP packets.

The type of bridging to perform is dependent on the protocols supported by the media source and sinks, as well as on the nature of the service providing the media. For example, if we assume that the IPTV media server in our co-watching example provides content in the form of an MPEG-2 IP multicast transport stream then it will be necessary to add logic to support low latency delivery of the stream to the co-watching participants with minimal skew across participants, a feature that IPTV systems do not currently support.

2.4.5 Codec Bridging

The current trend on the media codec side sees H.264 as the main player in this space. H.264 should be regarded as a flexible collection of standards that allow adapting to the different application requirements. Thus, the fact of using a single coding standard does not mean that a video encoder for real time applications, such as video conferencing and telepresence, will be readily usable for entertainment applications, as the first will minimize delay while the second will have inherent larger delay and random access. A single adopted standard means some ease on the implementation side and an easier way to switch from one coding mode to another depending on the application specifics.

2.5 Conclusions

The proposed Cooperative Context-aware Multimedia Service Architecture (CCMSA) addresses the requirements enumerated in the introduction by being agreeable to developers, addressing protocol and codec incompatibilies, and managing and coordinating sessions and media streams utilized by current and future consumer services. The biggest impediment to realizing CCMSA is the neccessity that endpoints must include support for it. While most device vendors are rapidly adopting support for entertainment media, they have been unable to provide interactive media services that harmoniously integrate existing media sources and media service components. This severely limits the composition of complex multimedia services. To do this requires adopting an API like the endpoint API we describe here.

Interactive video services and entertainment have inherently different characteristics. The first is characterized by low delay at the price of inconsistent video quality and variable frame rate. The second is instead characterized by initial delay for buffering, the need for random access, editability, high quality and constant frame rate. Such differences can be handled today by an standard such as H.264. The issue is how to move efficiently from one island to the other as, from the consumer perspective, the video experience is substantially the same on the same display surface.

The social aspect is introducing new constraints such as the inter-session synchronization across participants needed for the co-watching service. This needs to be handled by existing protocols such as RTP/RTCP or it may need new extensions. From a media protocol perspective, HTTP streaming with its adaptive capabilities will likely supersede RTSP as it delivers, from a consumer perspective, all the needed features at a lower price point.

From a signaling protocol perspective SIP will play a larger role due to the popularity of IMS and its extensibility. From a service side things look bright as the protocol and the media 'jungle' is in the process converging towards a few key protocols, simplifying the design and deployments of such services. Naturally, it must be possible to integrate session-based services based on SIP with existing and future web infrastructure. Fortunately, this is possible due to the existence of standards-compliant application servers supporting converged (web and telecom) applications.

Our introductory example shows that, to build complex services with a simplified consumer experience, a consumer's services must be able to query the state of a consumer's sessions and be able to interact with existing sessions. The use of service composition and distributed media control in conjunction with a suitably abstract API layer accomplish this while still providing the developer with the freedom to innovate.

CCMSA need not be deployed by a traditional network service provider. Given the light-weight, standardized nature of the constituent technologies and given that many of the technology solutions are available as open source software, we can envision a future where an over-the-top service provider can deploy a system like the one we propose here.

We think that CCMSA offers advantages in terms of the usability of a service for a consumer will be attractive to developers. It is important though to ensure that the architecture does not make onerous technological demands. Developers should be able to use as much or as little of the architecture as they please. The goal is accommodate existing services in a way that only minor modifications are necessary in order to gain some advantage from CCMSA. Services which take advantage of all that CCMSA has to offer will provide an even better consumer experience.

References

[1] 3GPP IP Multimedia System (IMS). `http://www.3gpp.org/article/ims`.

[2] Adobe dynamic streaming. `http://www.adobe.com/devnet/flashmediaserver/articles/dynamic_stream_switching.html`.

[3] Apple adaptive video streaming. `http://developer.apple.com/iphone/library/documentation/networkinginternet/conceptual/streamingmediaguide/StreamingMediaGuide.pdf`.

[4] Bertran Benjamin, Consel Charles, Kadionik Patrice, and Lamer Bastien. A SIP-based home automation platform: An experimental study. In *Proceedings 13th International Conference on Intelligence in Next Generation Networks*, pages 1–6, October 2009.

[5] Gregory Bond, Eric Cheung, Ioannis Fikouras, and Roman Levenshteyn. Unified telecom and web services composition: Problem definition and future directions. In *IPTComm '09: Proceedings of the 3rd International Conference on Principles, Systems and Applications of IP Telecommunications*, pages 1–12, ACM, New York, 2009.

[6] Eric Cheung and Pamela Zave. Generalized third-party call control in sip networks. In *Proceedings of Second International Conference on Principles, Systems and Applications of IP Telecommunications (IPTComm 2008)*, pages 45–68. Springer-Verlag, July 2008.

[7] R. Fielding, J. Gettys, J. Mogul, H. Frystyk, L. Masinter, P. Leach, and T. Berners-Lee. Hypertext transfer protocol – HTTP/1.1. IETF RFC 2616, June 1999.

[8] HTTP live streaming. IETF Internet Draft draft-pantos-http-live-streaming-04, June 2010.

[9] ISO/IEC. ISO/IEC 11172 part 1-5 MPEG-1, coding of moving pictures and associated audio for digital storage media at up to about 1.5 mbit/s. Technical report, ISO/IEC, 1993.

[10] ISO/IEC. ISO/IEC 13818 part 1-11 MPEG-2 generic coding of moving pictures and associated audio information. Technical report, ISO/IEC, 1993.

[11] ISO/IEC. ISO/IEC 14496-1:2004 part 1-27 – Information technology – Coding of audio-visual objects. Technical report, ISO/IEC, 1993.

[12] ITU-T. H.261: Video codec for audiovisual services at p × 64 kbit/s. Technical report, 1993.

[13] ITU-T. H.263: Video coding for low bit rate communication. Technical report, 2005.

[14] Michael Jackson and Pamela Zave. Distributed feature composition: A virtual architecture for telecommunications services. *IEEE Transactions on Software Engineering*, 24(10):831–847, October 1998.

[15] JSR 281: IMS services API v1.1. Java Community Process, 2009.

[16] JSR 289: SIP Servlet API version 1.1. Java Community Process, 2008.

[17] JSR 315: Java servlet 3.0 specification. Java Community Process, 2009.

[18] Salvatore Loreto, Tomas Mecklin, Miljenko Opsenica, and Heidi-Maria Rissanen. IMS service development API and testbed. *IEEE Communicationns*, 48(4):26–32, April 2010.

[19] D. Marpe, T. Wiegand, and G.J. Sullivan. The H.264/MPEG4 advanced video coding standard and its applications. *IEEE Communications Magazine*, 44(8):134–143, August 2006.

[20] M. Nathan, C. Harrison, S. Yarosh, L. Terveen, L. Stead, and B. Amento. CollaboraTV: Making television viewing social again. In *Proceedings of the 1st International Conference on Designing Interactive User Experiences for TV and Video (UXTV'08)*, pages 85–94. ACM, 2008.

[21] Colin Perkins. *RTP: Audio and Video for the Internet*. Addison-Wesley Professional, 2003.

[22] B. Quinn and K. Almeroth. IP multicast applications: Challenges and solutions. IETF RFC 3170, 2001.

[23] A.B. Roach. Session initiation protocol (SIP)-specific event notification. IETF RFC 3265, June 2002.

[24] J. Rosenberg, H. Schulzrinne, G. Camarillo, A. Johnston, J. Peterson, R. Sparks, M. Handley, and E. Schooler. SIP: Session initiation protocol. IETF RFC 3261, June 2002.

[25] H. Schulzrinne, A. Rao, and R. Lanphier. Real-time streaming protocol (RTSP). IETF RFC 2326, April 1998.

[26] H. Schulzrinne, A. Rao, R. Lanphier, M. Westerlund, and M. Stiemerling. Real time streaming protocol 2.0 (RTSP). IETF Internet Draft Music 2326bis Version 23, March 2010.

[27] SIP for instant messaging and presence leveraging extensions (SIMPLE). `http://datatracker.ietf.org/wg/simple/charter/`.

[28] Microsoft smooth streaming. `http://www.microsoft.com/silverlight/smoothstreaming`.

[29] Pamela Zave. Modularity in feature composition. In Bashar Nuseibeh and Pamela Zave, editors, *Software Requirements and Design: The Work of Michael Jackson*, chapter 12, pages 267–292. Good Friends Publishing Company, 2009.

[30] Pamela Zave, Eric Cheung, Gregory W. Bond, and Thomas M. Smith. Abstractions for programming SIP back-to-back user agents. In *IPTComm '09: Proceedings of the 3rd International Conference on Principles, Systems and Applications of IP Telecommunications*, pages 1–12. ACM, 2009.

3

Towards an Expressive, Adaptive and Resource Aware Network Platform

Antonio Manzalini, Roberto Minerva and Corrado Moiso

Strategy and Innovation, Telecom Italia, I-10148 Turin, Italy;
e-mail: {antonio.manzalini, roberto.minerva, corrado.moiso}@telecomitalia.it

Abstract

This chapter compares Next Generation Networks (NGNs), service control patterns and platforms against requirements like pervasiveness, mobility; programmability and de-perimeterization. The analysis suggests to provide services through a platform made out of overlays of virtualized resources. This platform, enhanced with autonomic capabilities, should optimize the usage of resources, allow the Service Providers to select the best control pattern, and enable the integration of Operators and external Providers resources.

Keywords: client-server, network intelligence, peer-to-peer, resource awareness, virtualization, autonomics, self-management.

3.1 A View of the Future Communication Context

The pervasive distribution of capable devices (e.g., computer, mobile phone, sensors, smart objects, etc.), interconnected through a variety of network technologies, will create a large pervasive infrastructure, spanning several administrative domains. Users will constantly be peers in many networks (personal, local, and public ones), and will have an active role in the provision of contents, functions and resources.

Anand R. Prasad et al. (Eds.), Advances in Next Generation Services and Service Architectures, 43–63.

This evolution will propel the creation of ecosystems where individuals, enterprises, service and network providers will create environments in which ecosystems share contents, resources, functions and applications, provided both by servers, personal devices and smart objects. The borders of each single network will blur into a "cloud of resources", whose cooperative usage must be optimized. Services of this infrastructure have to supply to users personalized and situation-aware services, even for a limited time and region of space. Service environments should dynamically adapt their behavior to accommodate users' needs and preferences, environmental context, resource availability, and their continuous changes, by discovering, selecting, and composing suitable resources. In this context, person-to-person services will be one of the many classes of possible services.

Telecommunication Operators (TelCos) need to exploit the recently deployed Next Generation Networks (NGNs) in order to generate new revenue streams, and they must face this emerging scenario. There is the need to identify (from a TelCo point of view) the merits, weaknesses and applicability of major technologies and paradigms usable to build a platform able to support a variety of services. Such an analysis focuses on the expressive power and the capabilities offered by three major paradigms: client/server (C-S), network intelligence (NI), and overlay networking (ON). In this chapter ON refers mainly to peer-to-peer overlay networks.

Expressive power is the ability of a paradigm or a technology to describe, support, and enable classes of services. A formal treatment of expressive power is outside the scope of this chapter, more information can be found in [11, 28, 33]. The scope of this chapter is: to compare different service control paradigms currently adopted for providing services on Internet and NGN against requirements posed by new network and service scenarios; and to advocate and provide guidelines for a novel architecture designed to satisfy those needs.

Section 3.2 focuses on new service platform requirements. Section 3.3 discusses three major service control paradigms and related protocols. Section 3.4 analyses the merits of service architectures adopting the considered control paradigms. Section 3.5 discusses their inadequacies with respect to new requirements. In Section 3.6 we introduce the need for and the characterizing elements of a novel programmable and dynamic service platform. Finally, in Section 3.7 some conclusions are drawn.

3.2 Service Platform Requirements for NGNs

For competing in the service arena, a TelCo should have a suitable Service Platform that accommodates new features and classes of services that go beyond telephony added value services. Some classes of services that introduce new requirements are:

- *Ambient Intelligence*, i.e., services that allow users to interact with intelligent objects in the surroundings by creating temporary ad hoc (personalized) service networks.
- *Extended Reality*, i.e., services that expand the perception of users of the real world by mixing, augmenting or recreating virtual and real data.
- *Social Media*, i.e., services that allow users to create, manage, share, and extend their data within communities.

These classes of services set requirements to NGNs in terms of: *seamless pervasiveness, enhanced mobility and resource awareness*. Requirements on privacy, security and identity will not be considered in this chapter. Instead, two other needs, important for TelCos, hold for NGN service platforms:

- *Programmability*, the possibility to increase the compose-ability of network resources, functions and data for enabling a rapid service creation, the "real-time" and automatic service creation according to resources and needs of a dynamic context; the composition of services by mashing up elements provided by different networks and providers;.
- *De-perimeterization of services*, i.e., the decoupling of telecom services from networks. TelCos provide services in regional markets by strongly coupling functions and networks. There is a need to overcome this limitation. The de-perimeterization of services leads also to the integration of loosely coupled networks, i.e., a Network of Networks, NoNs where public, private community, mesh networks are seamless integrated for creating a reliable system in which services can effectively execute.

3.3 Service Control Paradigms and Expressive Power

Three main service control paradigms are widely used for the provision of services: client-server, network intelligence [26] and overlay (or peer-to-peer, P2P [20]) models (Figure 3.1).

In the C-S paradigm, clients request a function to a Server that processes the request and returns a response (Figure 3.1a). Many classes of services, especially in the Internet, have adopted this control method. The model can

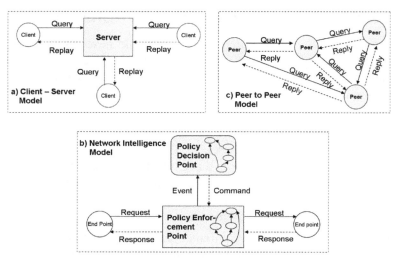

Figure 3.1 Service control paradigms.

be used in a stateless or stateful fashion depending on the design choice. A stateless implementation of this paradigm is REST [12], while SOAP, Simple Object Access Protocol [19], is a stateful implementation. The distinction is a different balance between number of functions and simplicity. The paradigm implementation is simple and its applicability is general, because the request-response pattern fits many services. Developers can focus on functions they want to get/provide from the servers with a minimum comprehension of networking mechanisms. This is an extremely crucial practical aspect: the paradigm implementation on the Internet clearly decouples communication and service control. However, the implementation of other control patterns (needed for implementing new classes of pervasive services of Section 3.2), as Publish Subscribe, PubSub, Model [10] or generic event/notification mechanisms, is clumsy: either clients continuously poll the server for notifications, or clients have to expose call back interfaces (transforming clients into servers and adding complexity). The "minimalist" C-S approach is complicated and inefficiencies are introduced.

In the NI paradigm (Figure 3.1b), the service execution adopts an event/command model between functionally rich entities in the upper layers (Policy Decision Points) and communication related nodes (Policy Enforcement

Points). Nodes are stateful and governed by Finite States Machines related to session and communication establishment.[1]

Service execution mixes connectivity control with added value service functions. Programming services is cumbersome because implementation requires deep knowledge of network mechanisms. It is questionable whether this paradigm enables an efficient way to implement non-session related services [25].

An event/command model has a great expressive power, especially if any entity can emit or receive events. Events can occur asynchronously and be propagated to the right recipient. All the services implemented with client/server mechanisms can be fully implemented by this model, because each request and reply primitive can be represented by asynchronous events. Adopting the approach suggested in [13], it can be assumed that the expressive power of the event/command model is equal or greater of the client/server model because the former can fully describe the capabilities and features of the latter.

Service implementation according to this paradigm has to manage queuing or prioritization of events. This complexity that has led, from a programming point of view, to the adoption of the simpler C-S paradigm. In addition, in the case of NI, the coupling of connection related and generic control functions greatly limits the expressive power of the paradigm. This integration is noticeable also at the protocols level. Albeit SIP, the Session Initiation Protocol [32] (the replacement of old signalling protocols in NGNs), can be used in a stateless fashion and without reference to connectivity services, its inner design targets a stateful session establishment and control of connectivity. SIP can implement many control patterns (a stateless client-server approach like REST, a NI one, the PubSub model, device-to-device), but, in the implementation of a specific pattern, the protocol results over specified or complicated.

Overlay networks offer abstraction, flexibility and reliability [2]. The power of ON [31] relies on its characteristics of symmetry, scalability and robustness: any involved party, a peer, adds resources and capabilities to the overlay network and can equally be a requester or a servant of functions. Overlay networks control mechanisms (Figure 3.1c) are required for managing the complexity of the internal organization of the overlay network and

---

[1] Not all the telecom service solutions adopt an event-command model, but in Intelligent Network and its evolution, IMS [35], the control layer triggers towards the service layer.

the dynamics of single nodes. They are a framework that enables the creation of distributed services (e.g., flie sharing).

The basic control mechanisms is message passing: peers are linked and exchange messages. This mechanism guarantees reliability in controlling the network and flexibility in service control because it allows the adoption of several patterns, like event/command and client/server for executing the specific services. Message passing allows the full implementation of client/server and event/command patterns. Since message passing is such a powerful basis on top of which to implement other control patterns, it could be stated (going along with [13]) that the ON model has an expressive power equal or greater than the C-S model. As a consequence, all the C-S services can be realized according to this paradigm. The synchronization between peers and their orchestration require governance (e.g., by means of gossiping [17]) and, in general, the ON paradigm is more complex from a programming point of view than a C-S model. The definition of protocols for overlay networking reflects the needs to provide generic mechanisms for messages exchange between peers plus functions for peer logical addressing, discovery, and logical binding.

The Invisible Internet Client Protocol (I2CP) [27] specifies a network model based on few entities: Routers, Destinations, Tunnels, and a Network DataBase. The protocol is session based and stateful. Routers act as store and forward boxes with respect to messages sent and received by Destinations.

For its generality, the control model for ON is powerful and its flexibility enables the implementation of many service control paradigm. These two features, together with the high expressive power of this paradigm, set ON as a good pattern for the service provision in pervasive environments. The additional cost is the need to tackle the complexity of programming distributed systems in a stateful manner.

3.4 Service Control Architectures

Various service architectures adopt the analyzed paradigms: Application Server model, Service Oriented Architecture (SOA [9]), or data center architectures advocate the C-S paradigm. The NI control applies to IP Multimedia SubSystems (IMS [30]) and Service Delivery Platform (SDP) [34] frameworks. The ON paradigm is adopted in several P2P application specific or general purpose platforms and architectures.

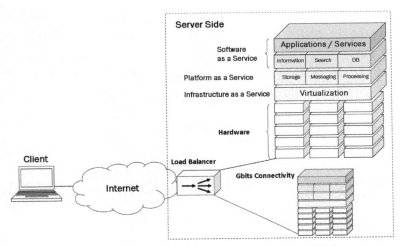

Figure 3.2 Client-server paradigm supported by data centers.

3.4.1 From Application Servers to Massive Data Centers

Client-server architectures have progressed from the N-tier definition of application servers to new models for the distributed execution of services, Cloud Computing. It enables new ways of designing, providing and using virtualized services, infrastructures, computing platforms, storage and processing capabilities within the Internet. The ability to provide functionalities or even computing and storage capability on demand by means of the C-S paradigm is termed "XaaS" (X as a Service, where X stands for Service, Platform or Infrastructure).

Figure 3.2 is a generalization of a data center architecture based on N-Tier and XaaS approaches. For achieving high availability, servers, functions and data are distributed and replicated in networked environments. The layering of functions eases the programming of these complex infrastructures. The bottom layer deals with basic functions as virtualization, replication, management and allocation of storage and processing resources; the middle one supports data management with approaches like map-reduce [7] or Hadoop [36]; the upper ones offer specific service functions and exposure of interfaces.

These platforms come with Application Programming Interfaces (APIs) for facilitating the creation of services. They are compliant with the newest technologies offered by the web and so programmers are encouraged to develop services in these environments. Solutions are usually proprietary

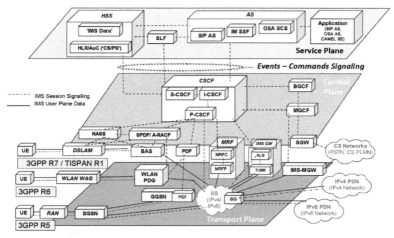

Figure 3.3   The IP Multimedia SubSystem (IMS) functional architecture (source: Wikipedia).

and open to maintain a competitive edge and differentiate from other Providers. These platforms are not designed to interoperate and they provide services, functions and APIs in closed administrative domains. This creates a dependency of developers to a particular infrastructure and to a specific set of APIs.

Service architectures aligned to SOA principles show similar features due to the centralization of the service logic and related middleware functions. Data centers do not exploit the pervasiveness of end-points, and force the processing and storage of local information in remote data centers, dissipating processing and bandwidth resources.

3.4.2 IMS and Service Delivery Platforms

IMS is designed as an enabler platform for NGN services. Its architecture (Figure 3.3) stems from GPRS [3]. It adopts an "all-IP" approach decoupling transport, control and service planes.

The transport plane ensures the efficient transport over an All-IP infrastructure; the control plane provides call and session control, and quality of service enforcement; and the service plane accommodates for services that require the intervention of specialized Application Servers. The arrangement of Call Session Control Functions, CSCFs, is devoted to the session establishment with quality of service and roaming functions across different administrative domains. For specific treatment, CSCFs send events (triggers) to logics in the Service Plane. All IMS systems are governed by Session State Models. The communication between nodes occurs by means of protocols such as SIP or Diameter.

At the Service Plane, different Application Servers, sometimes structured as Service Delivery Platforms, SDP, provide services. They adopt triggering mechanisms for interaction with the control plane and a SOA approach for interactions with applications. They are designed in such a way to facilitate service development for the Operator and Third Parties by means of simple APIs, like ParlayX [8] that allow to map abstract functions onto SIP primitives. SDPs can be considered as a means to mediate between control patterns: an event based one for interaction with underlying control plane and a C-S one for interaction between applications and the platform. Programming in this context implies knowledge of the network and specific Basic State Models. The programming of services can be a daunting effort and Web programmers are usually reluctant to use these features. Even the introduction of high level APIs dos not help: abstracting too much from protocol functionalities certainly simplifies the programming task, but it reduces the granularity of control. This protocol dependence of the middleware [25] is an obstacle to the programming of reasonably rich services on top of these platforms.

Also the IMS functional design has impact on programmability: each system can be fully compliant to standards but still be a non-programmable blackbox.

3.4.3 Building Services on an Overlay Platform

In spite of the lack of standardization, overlay platforms follow similar architectural principles. For example, JXTA [14] (Figure 3.4) adopts a layered architecture orientated towards programmability of nodes and services. Its basic elements are edge and super peers. Edge peers have a dynamic nature and are less stable and trusted. Super peers have specific roles and support functions within a JXTA overlay. Each peer has a unique Identity, has a Membership within groups, and is connected to the overlay by means of a peer transport. The software architecture of JXTA is organized in layers: at the bottom the JXTA core comprises basic functionalities needed for supporting the aggregation of peers. Another feature is Peer Advertisement, i.e., an XML description of a peer. It can be used for integrating in the system new resources coming from external environments. At the intermediate layer there are general services used to manage activities and properties of groups

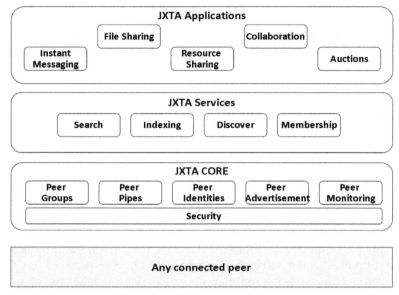

Figure 3.4 JXTA overlay platform.

of peers. At the top level there are applications like Instant Messaging or newer ones like distributed auctions.

The platform is flexible and the definition of general and reusable functionalities alleviates the complexity in programming distributed services. One of the drawbacks of P2P overlay networks is the lack of coordination with the underlying networks causing the misuse of physical resources. Some initiatives [15] aim at coordinating overlay and physical networks for optimizing the usage of scarce resources.

3.5 NGNs and Viable Service Control Paradigms

Requirements arising from future services have an impact on service control paradigms and the supporting service architectures. Some of their consequences are discussed in this section.

3.5.1 Seamless Pervasiveness and Native Mobility

Extended reality, Social media and Ambient Intelligence services operate on large amounts of data with constraints on real-time availability. Pervasive resources will be service enablers that need data for correctly and timely executing functions. There are at least two important consequences in future data distribution: the network needs to become symmetrical and the service platforms need to be extremely efficient in gathering the produced information and in distributing it. Data produced locally can have a double value: to be used/consumed/aggregated locally and/or to be produced locally and aggregated/consumed remotely. Both models should be supported in service platforms. They should also take care of data replication: copies of data could be scattered in many places and there will be the need to simultaneously update them maintaining consistency.

Variety in data distribution is a key aspect that has pushed for the adoption of different paradigms like PubSub, tuple space, or even hadoop and others. Richness in data distribution patterns is not easily supported by the C-S paradigm and, in addition, distributed data management is a relevant issue that could have a major impact on networking [16] transforming the network from a bit pipe into an information network. ON is a viable solution for coping with these issues. The increasing mobility of nodes will challenge the new service environment. A mobile node will instantly attach to several local networks and environments: the ability to quickly create, compose and update a "usage context" acquiring (and discarding) the right information will be fundamental for the new classes of services (Ambient Intelligence and Extended Reality in primis). From a local point of view, nodes will be characterized by a very dynamic behavior and environments will have to cope with it by integrating nodes, but not assuming them as permanent resources.

A reflective organization of the software infrastructure, i.e., a middleware organization able to accommodate to real-time changes of resources will be needed [4]. There is the need to create and consolidate dynamic aggregation of resources mechanisms at a local level. These resources could be granted and allocated to a specific user for a short period of time after a negotiation or a bid for them. At the same time a fast response at local level based on a granular knowledge of the context could be preferable for services for better accommodating user actions and requirements.

Under a mobility perspective the C-S approach seems to rely on solutions at the IP level. The advantage is that programmers will not cope directly with networking issues, but at the same time they will not be able to take advantages of local optimization of resources for better supporting services. These optimization will be at a network level and not at a service level. Mobility will challenge the NI paradigm: the value of SIMcards and roaming mechanisms will decrease and the entire mobility infrastructure should be revised to support a fast change of network provider comprising also small networks and ad hoc networking. The NI infrastructure is the one more prone to problems and issues. Under the mobility perspective, ON can be complementary to a Self-Organizing access Network [1]. The ON properties of flexibility and reorganization extended with a reflective platform and autonomic networking can be used for supporting higher levels of mobility of nodes, systems and especially services. The lack of standardization is a limitation of this approach requiring the construction of ad hoc solutions for each overlay architecture.

3.5.2 Resource Awareness

Resource awareness refers to two general properties: the ability to fulfill a single service requirements by minimizing the set of resources allocated and their usage; and the ability to maximize the number of services supported by means of the pool of available resources satisfying the requirements of each single service and providing an optimized and fair allocation of resources to the requesting services.

In the client/server approach the optimization of networked resources is detached from the service control. The infrastructure optimization problems are demanded to other layers, protocols and mechanisms. This has lead to ad hoc solutions for load balancing and correctly dimensioning Data Centers and consequently to a lack of standardization. Interworking between heterogeneous data centers is out of scope because they tend to be self-contained and centralized. The integration of services and resources pertaining to different environments is left to the users. Sometimes this task is relatively simple (by means of mashup techniques), but in the future, due to the complexity of pervasive systems, it could become much more complex. In a C-S paradigm, services are not able to dynamically request (or release) resources according to their effective usage.

The NI paradigm is suitable to orchestrate and optimize resources usage within a specific administrative domain. The management of resources is exerted by means of standard interfaces and protocols that pursue a rigid and static policy enforcement model. Compared to the data center approach, the possibility to manage the infrastructure in order to accomodate dynamic service needs is more effective and clear. From a global capacity point of view, the NI approach is similar to the Data Center, i.e., in order to support a large number of users and service requests, the centralized systems devoted to service provision have to grow to a suitable size.

Overlay platforms support internal dynamic management of peers. Aggregation of processing, communication, sensing and storage resources can be exerted according to dynamic needs of sevices. ON allows an optimization of the usage resources within the overlay itself. The optimization of the underlying physical network resources is still disregarded, however there are projects that are trying to allow a negotiation of resources between the network and the overlay themselves [15]. An overlay system is more complex to manage, but its global capacity (if a consistent number of nodes shares capabilities) could compare with the one offered by gigantic Data Center, with the advantage of being able to aggregate capabilities close to the users requesting them [5]. Conversely there is the stringent need to organize the entire system for optimize the use of resources.

3.5.3 Network Programmability

The client/server model facilitates the service creation through a simple control pattern and by strictly decoupling the service and the communications functions. Its drawbacks are mainly related to:

- The difficulty to support different control models that are required by new classes of services: for instance Internet of Things, Ambient Intelligence and Extended Reality applications will likely require the adoption of different data or interaction models (e.g., PubSub, Tuple Space [6], Message Queuing or others) that do not fit in the C-S paradigm.
- The increasing complexity of data center architectures and their closed and proprietary development environments create too many dependencies on a specific provider. For instance Social Media applications will need to compose information flows stemming from different sources, and this capability is not easy to realize with closed data centers.

The NI paradigm is adequate for creating and providing connectivity services, but it is less effective in supporting generic services. Even if the SIP protocol shows flexibilities, the IMS architecture in conjunction with SDPs limits the programmability by introducing constrained Call Models, too abstract APIs and a blackbox approach. The functional approach and the closeness of the platform create additional dependencies of the TelCos and programmers towards vendors. Scalability issues of the service layer are similar to those of data centers.

The three envisaged classes of service could benefit from a highly distributed implementation of these functionalities by means of P2P overlay networks. Overlay platforms are not standardized, but many of them are open source and can be extended for specific implementations. In principle, this creates fewer dependencies to a specific vendor. The major drawback is that overlay networks introduce mechanisms in the nodes for facilitating the management of the system. This increases the complexity and force programmers to take care of some networking issues.

3.5.4 De-Perimeterization

The de-perimeterization of services with respect to the underlying supporting network is a fundamental feature that has determined the success of Web2.0 and P2P applications. Two facets are comprised under the concept of de-perimeterization: on one side the need to decouple services from the underlying networks; on the other side the need to provide services that span over several networks. The problem of perimeterization of service is becoming a general issue: e.g., services provided by enterprise networks should be made generally accessible and usable. Services should be made available over adaptive or opportunistic infrastructures and over local ones. There is a requirement to provide services over compound networks, i.e., on a network of networks. The cooperation between different infrastructures should be enabled and promoted. In the long run, TelCos should be able to offer their services on other Operators infrastructures independently from their location.

The client/server paradigm is the epitome of the de-perimeterization. Users, by standardized and widespread Web mechanisms, can easily access services independent of location. However, large data centers are striving to compose to divergent requirements: the possibility to offer services globally from remote data centers, and the need to move services closer to users for a better quality of experience and responsiveness. The drawback is related to investments, costs and the need to locally create and manage massive infrastructures.

IMS is an implementation of the "ownership of customers" concept, i.e., a customer is tied to a specific network, and so services are only those offered by that network. They will be even activated and used by means of roaming mechanisms. The chain of controlling CSCFs seeks for specific Application Servers in the home network allowed to serve the customer. Overlay platforms seem a promising solution for providers because of their openness, flexibility and scalability. It is possible to use overlay technologies for integrating different networks into a cooperating system or to aggregate local resources for coping with peaks in service requests, reaching one of the major goal of de-perimeterization.

De-perimeterization is better supported by the C-S and the ON paradigms. In particular, development and deployment costs of ON are marginal with respect to those of a NI infrastructure.

3.5.5 On the Best Service Control Paradigm

The analysis in the previous sections highlighted limitations of currently adopted control paradigms and related service architectures for new scenarios. However, the different control paradigms have been profitably used in service implementations and in several specific architectures. One important lesson is that there is not a single paradigm that fits all possible services and usages: the expressive power of a control paradigm has to be traded with the easiness in implementation. New services based on pervasiveness of nodes will stress even more the need for flexibility: programmers should be free to select the best service control paradigm that matches the most with the service functions and their programming skills and style.

This requirement moves the problem from a control paradigm to the supporting platforms, services architectures, and even programming languages. In fact, new programming languages [29] are evolving considering networking environments and the distribution of nodes. This has an impact on their structure and also on the way applications will be developed in the future. The stringent requirement is then the ability to support as many as possible control paradigms and even to mix them for the construction of a single service. Under this profile overlay networks are the most flexible ones, but they need to introduce new functions and architectural patterns in order to satisfy the envisaged requirements and should also move to standardization in order to benefit from shared solutions.

3.6 Towards an Expressive Service Platform for NGNs

Overlay technology is not sufficient to fulfill the discussed requirements. Autonomic and Cognitive networking can reinforce the capabilities of overlay technologies by smoothing and coping with the dynamic behavior of peers and networks. They allow peers to be able to react to changes in the usage context and to modify their behavior according to local policies. To this extent, gossiping protocols act locally but converge to a global effect [18]. The need for a novel service platform for NGNs based on these guidelines is advocated. It aims at the optimization of resource usage, the selection of the best combination of control patterns for each service, and the integration of Operators resources with those of Users and external Providers. It is made out of layered overlays comprising virtualized resources, enhanced with autonomic and cognitive [21, 22, 24] capabilities to achieve self-awareness and self-organization.

There is a strong requirement to decouple the network functions from the service control, however this sharp separation should not lead (as it is today) to two separate and incommunicable worlds: services will perform better if they are able to adapt to the available network capabilities. In other terms, a clear interface between the network layer and the service layer has to be defined. The proposed architecture relies on three layers (Figure 3.5):

- The lower layer comprises pervasive heterogeneous resources spanning from network nodes and servers, to users' devices and "smart things" (e.g., sensors, actuators); they are interconnected through several types of (wired and wireless) connections and networks.
- The intermediate layer comprises clouds of software components [23] which provide virtualization of services and features of underneath resources and networks; they are grouped in overlays, and have autonomic capabilities to provide self adaptation, management, and organization.
- The upper layer is an ecosystem of reusable functions leveraging features of the underlying clouds of autonomic components, offered and used by several actors (e.g., Providers, LEs, SMEs, Prosumers).

The following subsections provide a high level description of functional features and enabling technologies of the first two layers.

3.6.1 Pervasive Resources in a Network of Networks

The considered network and services scenario (a Network of Networks) encompasses pervasive heterogeneous resources spanning from network element (transport nodes, switches, routers and even home gateways) to servers, to users' devices (e.g. laptop, PDA, etc.) and "smart things" (e.g., sensors, actuators); these resources are interconnected through several types of (wired and wireless) networks. At the connectivity level (specifically Layers 3, 2 and 1), it is advisable to consider technologies and architectures (such as Generalized Multi Protocol Switching) capable of extending the control of packet switching to time division (e.g., legacy circuits), to wavelength (e.g.

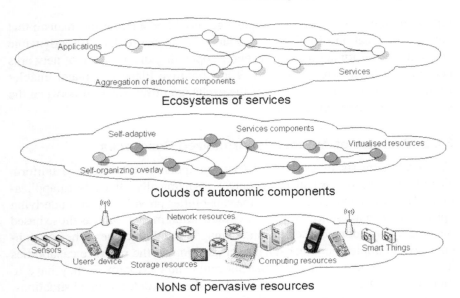

Figure 3.5 The proposed layered service platform.

DWDM) and to spatial switching (e.g., for incoming port or fiber to outgoing port or fiber).

The introduction of autonomic and cognitive capabilities in this context allows the creation and maintenance of optimized overlay networks of storage and processing virtual resources. In particular, autonomic and cognitive capabilities will complement the Network Control Plane (performing the connection management functions for the data plane) and the Service Control. The basic motivation is that if a network node is enhanced with some knowledge of the overall status of the network, autonomic operations decisions should be at least as good, if not better (in terms of global effectiveness and optimization) than those made in absence of data and knowledge.

In order to gather and represent the knowledge of the Network of Networks, entities have to be described and represented. Properties of nodes like intended usage, linkage and aggregations with other nodes should be represented as well. Finally relationships between the different nodes, aggregates of nodes and other resources and users should be expressed. This approach recalls the semantic web: the description and aggregation of information will have a paramount importance for policy decision aiming at the optimization of resources. There is a need to develop a language for resources and services representation and request. This language should be a major mechanism that enables the creation of a Cognitive Plane (in this particular case a dynamic collection of information about resources operating in a network of networks and their respective relationships). The Cognitive Plane is a major enabler for the resource awareness and for the exchange of information between the service and the network layers.

3.6.2 Overlays of Autonomic Virtualized Resources

This layer consists of sets of software components that create a uniform shared substrate of virtualized resources, dynamically allocable to applications. The virtual resources rely on the capabilities provided by the underlying "physical" resources, according to a virtualization model, such as those based on hypervisors adopted to create and manage virtual machines. The components have to implement mechanisms for partitioning the resource capabilities in "isolated" slices, which are the allocation unit. Moreover, they have to monitor the state of the resources and of its slices, to detect critical situations, and trigger corrective actions. Finally, they include functions for negotiating, (de)allocating and, possibly, migrating virtual resources according to needs.

Virtual resources are enhanced with autonomic capabilities. The autonomic logic is exerted by decision modules that determine the actions needed for self-CHOP (i.e., configuration, healing, optimization, protection) behaviour and for tuning slices allocation. The single autonomic logic is expected to cooperate at the local level with those executed by other nearby components in order to implement decentralized algorithms able to exert supervision at global level. In order to deal with scalability, each component communicates only with a few nearby neighbours, interconnected through a (self-managed) overlay network. They aggregate resources sharing similar functions (e.g., resources which provide storage in a given geographical area) and support the spreading of cooperation messages among the local autonomic components.

This allows achieving distributed and global management policies. Moreover, overlays by interacting with the Cognitive Plane (e.g., by means of information hashing in a DHT like fashion) allow interconnecting virtual resources allocated to a given application/user. These application-specific overlays create a computational and networking environment able to decouple applications from actual physical resources and networks. Self-organization mechanisms can be used to optimize and maintain these overlays. The features of this layer of autonomic virtual resources enable the applications to achieve a suitable level of resource awareness: applications are able to self-adapt and self-organize by taking into account availability of resources, applications' requirements, and their continuous changes.

3.7 Concluding Remarks

The analysis of service control mechanisms for NGNs has focused on major requirements tending to the conclusion that all the control paradigms have drawbacks that will hinder a smooth evolution of next generation service platforms. Their capabilities should be carefully reassessed by TelCos that want to compete in the service arena. The evolution of overlay networking together with autonomic capabilities fosters the creation of a framework that will support many pervasive services and an optimized usage of available resources in a de-perimeterized environment.

Before the deployment of similar architectures, there is the need to solve a number of technical issues: the standardization of overlay technology; the integration of autonomics functions within unreliable nodes; the virtualization of physical resources; the definition of a new control layer that encompasses mobility, identity (of people, things and resources); and more. This approach could be so disruptive as to require enhancements of network lower layers in favor of autonomic and cognitive capabilities as well as deep changes in current approach to service layer. This novel architecture will be able to re-balance the relationship between clients and the gigantic Data Centers, and to introduce the concept and mechanisms for resource aware networking and service provision.

References

[1] 3rd Generation Partnership Project. Telecommunication management; Self-organizing networks (son); Concepts and requirements (release 8), 2008.

[2] D. Andersen, H. Balakrishnan, M.F. Kaashoek, and R. Morris. Resilient pverlay networks. In *Proceedings of 18th ACM Symposium on Operating Systems Principles (SOSP)*, Banff, Canada, pages 131–145, October 2001.

[3] C. Bettstetter, H.-J. Vogel, and J. Eberspacher. Gsm phase 2+ general packet radio service gprs: Architecture, protocol, and air interfacer. Technical Report 3, 1999.

[4] L. Capra, W. Emmerich, and C. Mascolo. Carisma: Context-aware reflective middleware system for mobile applications. *IEEE Transactions on Software Engineering*, 29:929–945, 2003.

[5] K. Church, A. Greenberg, and J. Hamilton. On delivering embarrassingly distributed cloud services. In *HotNets*, 2008.

[6] P. Ciancarini. Distributed programming with logic tuple spaces. *New Gen. Comput.*, 12(3):251–284, 1994.

[7] J. Dean and S. Ghemawat. Mapreduce: Simplified data processing on large clusters. In *Proceedings of the 6th conference on Symposium on Opearting Systems Design & Implementation*, Berkeley, CA, USA, pages 10–10, USENIX Association, 2004.

[8] G. Di Caprio and C. Moiso. Web services and parlay: An architectural comparison. In *Proceedings of International Conference on Intelligence in Networks – ICIN2003*, Bordeaux, France, 2003.

[9] T. Erl. *SOA – Principles of Service Design*, 2nd edition. Prentice Hall, August 2007.

[10] P. Eugster, P. Felber, R. Guerraoui, and A.M. Kermarrec. The many faces of publish/subscribe. *ACM Computing Surveys (CSUR)*, 35(2):144–131, 2003.

[11] M. Felleisen. On the expressive power of programming languages. In *Science of Computer Programming*, pages 134–151. Springer-Verlag, 1990.

[12] R. Fielding, D. Software, and R. Taylor. Principled design of the modern web architecture. *ACM Transactions on Internet Technology*, 2:115–150, 2002.

[13] D. Gorla. On the relative expressive power of asynchronous communication primitives. In *Proceedings of 9th International Conference on Foundations of Software Science and Computation Structures (FoSSaCS'06)*, Lecture Notes in Computer Science, Vol. 3921, pages 47–62. Springer-Verlag, 2006.

[14] J. Gradecki. *Mastering JXTA: Building Java Peer-to-Peer Applications*. John Wiley & Sons, 2002.

[15] V.K. Gurbani, V. Hilt, I. Rimac, M. Tomsu, and E. Marocco. A survey of research on the application-layer traffic optimization problem and the need for layer cooperation. *Comm. Mag.*, 47(8):107–112, 2009.

[16] V. Jacobson, D. Smetters, N. Briggs, M. Plass, P Stewart, J. Thornton, and R. Braynard. VoCCN: Voice-over content-centric networks. In *ReArch'09: Proceedings of the 2009 workshop on Re-architecting the internet*, New York, USA, pages 1–6. ACM, 2009.

[17] M. Jelasity, A. Montresor, and O. Babaoglu. Gossip-based aggregation in large dynamic networks. *ACM Trans. Comput. Syst.*, 23(3):219–252, 2005.

[18] M. Jelasity, A. Montresor, and O. Babaoglu. Gossip-based aggregation in large dynamic networks. *ACM Trans. Comput. Syst.*, 23(3):219–252, 2005.

[19] J. Kangasharju, S. Tarkoma, and K. Raatikainen. Comparing SOAP performance for various encodings, protocols, and connections. In *Proceedings of IFIP-TC6 8th International Conference, Personal Wireless Communications (PWC 2003)*, Venice, Italy, September 23–25, 2003. Lecture Notes in Computer Science, Vol. 2775, pages 397–406. Springer, 2003.

[20] Keong Lua, J. Crowcroft, M. Pias, R. Sharma, and S. Lim. A survey and comparison of peer-to-peer overlay network schemes. *Communications Surveys & Tutorials, IEEE*, 7(2):72–93, 2005.

[21] A. Manzalini, P.H. Deussen, S. Nechifor, M. Mamei, R. Minerva, C. Moiso, A. Salden, T. Wauters, and F. Zambonelli. Self-optimized cognitive network of networks. *The Computer Journal*, online http://comjnl.oxfordjournals.org/content/early/2010/03/23/comjnl.bxq032.abstract, Oxford Journals, 2010.

[22] A. Manzalini, R. Minerva, and C. Moiso. Bio-inspired autonomic structures: A middleware for telecommunications ecosystems. In A. Vasilakos, M. Parashar, S. Karnouskos, and W. Pedrycz (Eds.), *Autonomic Communication*, pages 3–30. Springer, 2009.

[23] A. Manzalini, R. Minerva, and C. Moiso. Autonomic clouds of components for self-managed service ecosystems. *Journal of Telecommunications Management*, 3(2), 2010.

[24] A. Manzalini and F. Zambonelli. Towards autonomic and situation-aware communication services: the cascadas vision. In *IEEE Workshop on Distributed Intelligent Systems: Collective Intelligence and Its Applications*, pages 383–388, 2010.

[25] R. Minerva. On some myths about network intelligence. In *Proceedings of International Conference on Intelligence in Networks – ICIN2008*, Bordeaux, France, 2008.

[26] R. Minerva and C. Moiso. The death of network intelligence? In *Proceedings of International Symposium on Services and Local Access (ISSLS)*, Edinburgh, UK, 2004.

[27] I2P Anonymous Network. Invisible internet client protocol (I2CP). August 2003.

[28] C. Palamidessi. Comparing the expressive power of the synchronous and the asynchronous pi-calculus. *Mathematical Structures in Computer Science*, 13:685–719, 2003.

[29] J. Paulus. New languages and why we need them. *MIT Technology Review*, on line edition, 2010.

[30] M. Poikselka, A. Niemi, H. Khartabil, and G. Mayer. *The IMS Second Edition: IP Multimedia Concepts and Services*. John Wiley & Sons, 2006.

[31] M. Ripeanu, I. Foster, and A. Iamnitchi. Mapping the gnutella network: Properties of large-scale peer-to-peer systems and implications for system design. *IEEE Internet Computing Journal*, 6(1), January/February 2002.

[32] H. Rosenberg, J. Schulzrinne, G. Camarillo, A. Johnsto, J. Peterson, R. Sparks, M. Handley, and E. Schooler. SIP: Session Initiation Protocol. RFC 3261 (Proposed Standard), June 2002.

[33] N. Santoro. *Design and Analysis of Distributed Algorithms, Wiley Series on Parallel and Distributed Computing*. Wiley-Interscience, 2006.

[34] A. Schuelke, D. Abbadessa, and F. Winkler. Service delivery platform: Critical enabler to service providers' new revenue streams. Presented at World Telecommunications Congress, April 2006.

[35] J. Waclawsky. A critique of the grand plan. *Business Communications Review*, 35(10):54–58, October 2005.

[36] T. White. *Hadoop: The Definitive Guide*, 1st ed. O'Reilly, June 2009.

4

Services in Smart Grids

Thomas M. Chen

*School of Engineering, Swansea University, Swansea SA2 8PP, Wales, UK;
e-mail: t.m.chen@swansea.ac.uk*

Abstract

The smart grid is a vision to modernize the existing power system with
extensive use of information and communication technologies. For utility
companies, the smart grid promises improvements in reliability, efficiency,
resilience, distributed power generation, and ability to meet peak demand. For
consumers, the smart grid will offer the means to intelligently manage their
energy consumption and the opportunity to sell consumer-generated energy
to the grid. As the smart grid evolves, it will likely create new opportunities
for service providers and third parties to offer a variety of new services related
to power generation, smart appliances, and home power management.

Keywords: smart grid, electricity, demand response.

4.1 Introduction

The existing power system was designed for one-way distribution of electricity from central power plants. It has not changed much in the past few
decades, and many believe it is becoming increasingly outdated, inefficient,
and unreliable. Energy losses in the transmission and distribution system
have been increasing steadily. The electrical grid has also suffered costly
outages [21].

The smart grid is a vision to modernize the electrical power infrastructure
and the services it will support [9,12,14,28]. Smart meters will allow two-way

*Anand R. Prasad et al. (Eds.), Advances in Next Generation Services and Service
Architectures,* 65–79.

communications between consumers and service providers. Smart appliances within the home may interface with the smart meter to better manage electricity usage and reduce energy costs. Opportunities will be created for service providers or third parties to offer new services taking advantage of the smart grid infrastructure.

The emerging smart grid will have far reaching implications for service providers and consumers. At the same time, there are considerable risks and many uncertainties. The future of the smart grid is not just a question of new technologies. It will depend on the power industry, government regulations, national and international standards, and the public.

4.2 Current Power Grid

The existing power system was created mainly for one-way distribution of electricity from central power plants [8]. In the US, there are approximately 10,000 central power plants using mostly coal and natural gas to turn steam turbines (about 70% of electricity). From power plants, high voltage electricity is transmitted along 200,000 miles of transmission lines to neighborhood substations, where it is stepped down to medium voltages and distributed to consumers.

At customer premises, most power meters do not have communication capabilities and depend on monthly field visits to make readings. In North America, approximately 86 million meters are equipped with communications capabilities. Among these, 90% are automatic meter reading (AMR) meters with one-way communications to allow remote readings, and the other 10% are advanced metering infrastructure (AMI) or smart meters capable of two-way communications [16].

Efficiency and reliability are two major issues in the existing power system. Energy losses in the transmission and distribution system have increased from 5% in 1970 to 9.5% in 2001. Power outages are estimated to cost the US economy about 100 billion dollars every year [21]. The 2003 blackout in the northeastern US and Canada incurred a loss of 6 billion dollars. Protection systems are limited only to specific components, and responses to outages are carried out centrally through the supervisory control and data acquisition (SCADA) system which is relatively slow [3].

Another major issue is a mismatch between consumer tariffs and actual energy generation costs. Consumers typically see flat pricing and pay according to the total monthly usage. However, power generation during peak demand is more costly to utility companies. Baseload generating units (mostly coal and nuclear) are used to meet the minimum customer demand and operate continuously at a relatively constant rate. Generally they are the largest and most efficient power generators but lack flexibility to quickly adjust their output to meet changing demand. Fluctuations in energy demand are met by peakload generating units. Peakload generators usually running on natural gas are more flexible in their output but are relatively costly to operate.

As a consequence of dynamic demand, utilities are required to maintain a certain "spinning reserve" capacity that can be brought online within minutes to meet sudden surges in demand. The spinning reserve represents costly spare capacity.

4.3 Smart Grid Goals

As a broad concept, the idea of the smart grid is to leverage information and communication technologies to add "intelligence" to the existing power system for the benefit of both consumers and service providers [20]. The intelligence is enabled by better sensing, communications, and distributed automation throughout the entire system. At a minimum, consumers should see a higher quality service (e.g., fewer outages). Even more, the smart grid aims to empower the consumer with more information and more flexibility. For example, consumers will receive dynamic pricing signals which can be used to optimize their electricity usage. They will be able to use third party services to help manage their electricity usage to meet financial goals. The smart grid can be viewed as an information gateway between energy service providers and "smart" devices in the home.

At this time, the smart grid is more of a vision than a specification. The US Department of Energy (DoE) outlines a few desirable characteristics for the smart grid related to consumer participation, efficiency, resilience, and flexiblity [17, 22]. In summary, the smart grid should:

- encourage consumer participation (e.g. by demand response);
- operate more efficiently;
- accommodate all generation and storage options;
- enable new products, services, and markets;
- provide higher quality power (e.g., avoiding outages);
- be resilient (e.g., self healing) against attacks and natural disasters.

The Federal Energy Regulatory Commission (FERC) has identified several high priorities for the smart grid: system security; demand response; wide-area situational awareness; energy storage; communication and coordination across inter-system interfaces; electric vehicles; AMI; and distribution grid management [10].

Consumers are encouraged to participate through financial incentives made possible by demand response [22]. In demand response, a smart meter provides historical usage data and time-based pricing signals to the consumer. Consumers can realize financial benefits by modifying their usage behavior, e.g., using less electricity during peak periods or voluntarily allowing high-usage appliances to be turned off remotely.

In the DoE vision, smart (AMI) meters are only the first step towards a smart grid. Over the long term, it is expected that the smart grid will also implement [16]:

- advanced distribution operations (ADO) to use AMI communications to improve distribution and self healing;
- advanced transmission operations (ATO) to use ADO to better manage transmission congestion and voltage;
- advanced asset management (AAM) to use AMI, ADO, and ATO to improve operating efficiency and asset utilization.

4.4 Smart Grid Architecture

The NIST conceptual reference model identifies seven domains: bulk generation, transmission, distribution, markets, operations, service provider, and customer [24]. Each domain contains actors and networks, and communicates with other domains through well defined gateways. Here we describe a slightly simpler view of power generation, distribution and transmission, smart meters, networking, and controls.

4.4.1 Power Generation

The smart grid recognizes the need for more flexibility to accommodate energy sources that may be geographically distributed and diverse. Renewable energy sources will be important but may be inherently intermittent (e.g., depending on the sun or wind). Consumers may also generate and contribute energy into the grid, for example by electric vehicles.

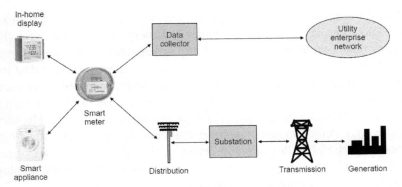

Figure 4.1 Smart meter located at customer premises.

4.4.2 Power Distribution and Transmission

With distributed energy sources, the distribution and transmission networks will need to be more complicated and flexible than the existing network for centralized power generation. The existing network is configured in a simple tree-like topology to distribute power in one direction from a central source. In the smart grid, there will likely be more choices and competition among multiple energy providers, which implies a more interconnected network topology. Furthermore, the goal of self healing implies a mesh topology with multiple redundant paths.

4.4.3 Smart Meters

As mentioned earlier, smart (AMI) meters enable two-way communications between the consumer and service provider. In one direction, the smart meter receives time-based pricing signals and historical usage data. Examples of time-based pricing include:

- time-of-use (TOU) pricing varying by set intervals during the day;
- critical peak pricing (CPP) reflecting costs during peak periods;
- real-time pricing (RTP) varying regularly according to supply and demand.

In the other direction, the smart meter measures time-of-day power consumption and sends that data to the service provider for monitoring and billing purposes [5]. Communications and electrical flows through the smart meter are illustrated in Figure 4.1.

NIST's conceptual reference model separates the smart meter into two logical functions, referred to as the meter and energy services interface (ESI), although both functions are likely to be implemented in the same physical device [24]. The meter function measures energy usage; transmits data to the utility; communicates with the utility for outage management; and enables automated provisioning and maintenance functions. The ESI is an information gateway between the consumer and energy service providers. The service provider may offer demand response aggregation or energy management services. The ESI handles demand response signaling (time-based price information) and provides energy usage information to an in-home display or residential energy management system.

4.4.4 Networking

Smart meters will communicate with a local data collector or access point within a neighborhood area network (NAN) using wireless (radio mesh, 3G cellular, WiFi, WiMax, ZigBee) or wired (power line carrier, broadband over power line, fiber optics, DSL) technologies [1, 25]. Regardless of physical layer technology, communications are widely expected to be based on TCP/IP protocols due to the maturity of IP standards and availability of IP-based equipment, tools and applications [19]. The access point is connected to the utility company.

On the other side, the smart meter serves as a gateway to the home area network (HAN) interconnecting appliances, computers, and energy control devices within the home.

4.4.5 Controls

The smart grid is envisioned to be self healing against faults and resilient against active attacks, which implies more sophisticated control than exists in the current power grid. In the existing power system, control is centralized through the supervisory control and data acquisition (SCADA) system. Centralized control will be too slow for self healing which needs distributed monitoring for faults and automated rerouting around detected faults. Centralized control is also inappropriate for resilience against attacks because the SCADA system represents a single point of vulnerability. Mechanisms to detect and respond to malicious attacks will need to be distributed throughout the smart grid.

4.5 Current Activities

The US federal government initiated support for the smart grid with the Energy Independence and Security Act of 2007. It set out 100 million dollars in annual funding from 2008 to 2012; established a matching program to states and utilities to build smart grid capabilities; and created a Grid Modernization Commission to assess potential benefits and recommend protocol standards. NIST (National Institute for Standards and Technology) was given responsibility to coordinate the development of smart grid interoperability standards. In 2009, the US government invested 3.4 billion dollars in 100 smart grid related projects through the American Recovery and Reinvestment Act, with an additional 4.7 billion in corporate matching.

In Europe, the European Commission wrote the European Technology Platform (ETP) SmartGrids in 2005 as a long term vision for European electric grids. The Strategic Energy Technologies Plan (SET-PLAN) is a strategic plan to accelerate innovations in cost-effective low carbon technologies. SET-PLAN includes the European Electricity Grid Initiative (EEGI), a nine-year strategic plan to research futuristic electricity networks in Europe. The EEGI is supported by two multinational industry groups, the European Network of Transmission System Operators for Electricity (ENTSO-E) and European Distribution System Operators Association for SmartGrids (EDSO-SG).

Asia is also actively pursuing smart grid projects. South Korea established the Korea Smart Grid Institute in August 2009 to coordinate the nation's smart grid projects. China announced a "Strengthened Smart Grid" plan in May 2009 with an aggressive timetable: planning and trials by 2010; technology development and mass deployment of a basic operational smart grid by 2015; and completion of advanced technologies in the smart grid by 2020. However, the existing electrical power system in China is quite different than the US one, implying that the smart grid will likely evolve differently with more emphasis initially on the transmission infrastructure rather than distribution and smart meters.

4.5.1 Standards

Standards primarily for interoperability are universally understood to be important for the smart grid, but standards are currently incomplete. The standardization situation is challenging because the smart grid involves a broad range of technologies and many organizations. There is a great deal of uncertainty about how the smart grid will evolve in the next couple of decades. A number of standardization efforts are underway but new standards will be very likely in the near future.

In the US, NIST is coordinating interoperability standards for utilities, equipment manufacturers, and service providers to work together. The current NIST interoperability report presents a conceptual reference model; identifies relevant standards; and identifies important gaps in standards [24]. It examines the need for cyber security but draws mostly from the NIST-led Smart Grid Interoperability Panel - Cyber Security Working Group (SGIP-CSWG), formerly the Cyber Security Coordination Task Group [26].

Jointly with NIST, IEEE P2030 seeks to define interoperability between various types of power grids. Three task forces have been formed to address distribution systems, information technology, and communications technology.

IEC (International Electrical Congress) Strategic Group 3 is aiming to provide a comprehensive set of global standards. Standards related to the smart grid are being covered by 24 IEC technical committees. For instance, TC57 has published a number of standards related to power systems management, control, and information exchange.

The ANSI C12.19 standard provides a common data structure for communicating data to and from utility end devices, including smart meters. Specifically, C12.19 defines the standard data structure as sets of tables. ANSI C12.22 is a recent standard to transport ANSI C12.19 table data over existing communication networks. As an application layer protocol, C12.22 can work over any type of wired or wireless networks including TCP/IP (used in the Internet).

4.5.2 Trials

Many field trials are active in various locations around the world. In the US, more than 8.3 million smart meters have already been installed, and the number is projected to reach 52 million by 2012. Examples include:

- Austin Energy has built a smart grid in Austin, Texas that encompassed 500,000 devices servicing 1 million consumers and 43,000 businesses in 2009.
- Xcel Energy is building a smart grid infrastructure called SmartGridCity in Boulder, Colorado involving hundreds of miles of optic fibers for broadband-over-power-line (BPL) communications to tens of thousands of homes.

- Dominion Virginia Power is carrying out a smart meter field test before proceeding with a plan to replace all 2.4 million meters in its service area with smart meters.
- Southern California Edison Company (SCE) will install smart electric meters throughout the eastern part of Los Angeles county as part of the Edison SmartConnect program.
- PG&E is reportedly installing 12,000–15,000 smart meters daily in central and northern California.
- Hydro One is installing a smart grid that will reach million customers in Ontario, Canada by the end of 2010.
- One of the earliest smart meter trials, the Telegestore project in Italy was completed in 2006 involving more than 30 million meters.

The field trials are important for a number of reasons: demonstrating feasibility; experimenting with new equipment; evaluating real costs; and observing real consumer responses. Early implementers have an opportunity to lead standards and serve as models for other projects.

4.6 Challenges

Service providers are motivated to embrace the smart grid for a number of practical reasons mentioned earlier, namely reliability, efficiency, resilience, and better management of peak demand. The smart grid also offers benefits to consumers who want to subscribe to its services. However, the smart grid requires an enormous investment that would ultimately have to be passed onto consumers. It also presents risks associated with new untried technologies, e.g., new vulnerabilities and exposure to new malicious attacks. Public acceptance of the smart grid is not guaranteed. Some field trials have demonstrated serious public concerns about privacy and energy prices.

4.6.1 Consumer Participation

From a service perspective, the smart grid sounds quite appealing. Smart meters will offer detailed energy usage information to consumers that they never had before. Consumers will also informed about dynamic pricing that can be used to optimize energy usage. The smart grid will create opportunities for new products like smart appliances and new services for helping consumers manage their appliances. Basically, the smart grid will empower the consumer with more information and choices.

On the other hand, no one quite knows how consumers will respond, what type of feedback signals will be most helpful, and exactly how much consumers will benefit. Traditional flat pricing is well understood and ingrained in the consumer's mind. The consumer knows the strategy to reduce electrical bills - simply use less electricity. Demand response enabled by the smart grid is more complicated. It assumes that consumers will strategically adjust their energy usage based on pricing signals and past energy usage information [11].

Dynamic pricing can take different forms, for example: time-of-use pricing; critical peak pricing; or real-time pricing. It is uncertain what type of dynamic pricing will be more effective at motivating consumer participation. Also, it is uncertain if additional feedback would be useful to the consumer. For example, perhaps environmental benefits would motivate some consumers. It is uncertain if consumers want to deal with the additional complexity of dynamic pricing signals, and if they are willing to strategize, how much benefit could be achieved. A few studies with real data have suggested that consumers reduce energy usage by an average of 10% just by having information readily available.

More research is needed to understand and model consumer behavior in demand response. Some empirical understanding is being gained from smart grid field trials. A better understanding of the consumer can lead to better estimates of potential benefits, which can strengthen the case for smart grid acceptance.

4.6.2 Privacy

Privacy is one of the major concerns of consumers and potential obstacles to public acceptance [15, 18]. Privacy may be defined as the control of individuals and organizations over how information about them is collected, shared, and used. The main concern is about the capability of smart meters to collect detailed energy usage information which could lead to an unprecedented invasion of privacy. Energy usage information could allow inferences about activities occurring inside the home. There are additional concerns that even anonymized data might be combined with other pieces of information available from other sources to reveal personally identifiable information (PII). That is, information collected from the smart grid may not be that revealing by itself, but people leave other information about themselves online. The information aggregated from various sources could be much more revealing than any single piece of information.

The Privacy Sub-group of the SGIP-CSWG found a notable lack of consistent and comprehensive privacy policies, standards, and supporting procedures throughout the states, government agencies, and utility companies [26]. There is no consensus among state public utility commissions (PUCs) on how to address the specific privacy implications of the smart grid. Furthermore, enforcement of state privacy-related laws is often delegated to agencies other than PUCs. Comprehensive and consistent definitions of privacy-related information with respect to the smart grid typically do not exist at state or federal regulatory levels, or within the utility industry. The Privacy Sub-group concluded that a substantial privacy risk is created by new types of information created and collected by the smart grid, and the privacy risks are not addressed or mitigated by existing laws and regulations. Also, the smart grid may create new privacy risks and concerns beyond the existing practices and policies of the organizations that have been historically responsible for protecting energy usage data collected from the traditional electrical system.

Public acceptance of the smart grid will depend on trust, and a major trust issue is assurance that privacy protection (such as verifiable enforcement of privacy policies) is designed directly into the system. Some recommendations have been proposed including [23]:

- only the minimal amount of information should be provided to third parties given the nature of the relevant service;
- identities should be pseudonomyzed where possible;
- consumers should be able to maintain control over the type of information disclosed to third parties;
- communication channels should be secured (e.g., encrypted), commensurate with the type of data transmitted;
- third parties should agree not to correlate data with data obtained from other sources or the individual, without the consent of the individual.

It is widely understood that encryption of data messages will be necessary in the smart grid. However, the other recommendations are far from certain because they will depend on regulations. At the present time, privacy seems to be underestimated as an important issue, hence it is not clear that it will be properly addressed [6].

4.6.3 Smart Meters

Clearly, smart meters are a new component that represents a substantial investment for utility companies. Smart meters must be designed with required processing, storage, monitoring, and communications functions. Many smart meters must be built and installed, and then managed remotely by the service provider.

At the same time, smart meters are exposed to risks because they reside at the untrusted customer premises [2,4]. They must be physically tamper proof but it will be impossible to protect meters completely from an attacker with sufficient resources and effort. Meters should at least be capable of detecting tampering if it happens [7].

In addition to physical attacks, there are concerns about cyber attacks. The smart meter will be part of the NAN and interface to a consumer's HAN, which potentially expose it to remote attacks. Physical and cyber risks are mainly speculative at the current time and need to be understood better before any real deployments of smart meters.

4.6.4 Self Healing

Self healing is a property of the smart grid referring to automatic recovery from unexpected faults. Self healing should be faster than current centralized responses through the SCADA system. Self healing is well known in other contexts such as communication networks but has not been studied much yet for the smart grid.

Generally, self healing has four requirements: spare backup resources; distributed fault detection; fault notification; and automated reconfiguration. If a component fails, service can be recovered if a backup component is available to put into operation. A fault event should trigger self healing but generally faults can occur at any time and place. Hence fault detection must be distributed so that detection can be done as close to the fault as possible (for speed). Once detected, a notification must be sent around the system to other affected components that will be involved in reconfiguration. The notification will trigger the components to act together to reconfigure resources to restore service.

4.6.5 Attack Prevention and Defense

As critical infrastructure, the smart grid is recognized as an attractive target for malicious attacks [13]. The smart grid vision identifies the need for resilience against attacks. Unlike random faults, a malicious attacker is assumed to be intelligent enough to seek out and exploit vulnerabilities in the system. Hence, it is important to deter and prevent attacks by minimizing vulnerabilities. Unfortunately, the smart grid will be more complex than the existing power system, and the increased complexity will likely create new vulnerabilities which may be more difficult to understand and identify.

Practical defenses will use multiple layers of protection [29]. Intrusion detection is always part of any system defense [27]. By itself, intrusion detection does not protect a system from attacks but alerts the service provider to investigate and respond. Traditional intrusion detection depends on signature-based or anomaly-based approaches. Signature-based detection depends on prior knowledge of possible attacks. Given the lack of experience with the smart grid, signature-based detection will likely be inadequate.

Anomaly-based detection can potentially identify unknown attacks without a signature but relies on knowledge of "normal baseline" activities. A baseline can be learned by observing normal activities over a length of time. However, anomaly-based detection has its challenges as well. Even normal activities have natural variations and can change over time. It is usually difficult to define what are significant deviations. Moreover, deviations from normal activities might be suspicious but not necessarily malicious. Anomaly-based detection does not identify the specific nature of anomalous activities; alerts require investigation which take time and manpower.

Intrusion tolerance recognizes that intrusions can not be totally prevented. It tries to maintain normal services in the face of successful intrusions. Somewhat similar to self healing, intrusion tolerance involves reconfiguration of spare resources. However, the disruption is not a random fault but more likely the most vulnerable or critical components that would be the main targets of an intelligent attacker.

4.7 Conclusions

The emerging smart grid promises to be a vast improvement to critical infrastructure which in principle should open up a range of new service opportunities. Field trials and new standards are progressing, but at the present time, the future is uncertain. The smart grid encompasses many technologies and involves many entities. Several major issues need to be understood better and resolved.

References

[1] A. Aggarwal, S. Kunta, and P. Verma. A proposed communications infrastructure for the smart grid. In *2010 Innovative Smart Grid Technologies (ISGT)*, 2010.

[2] AMI Security Acceleration Project AMI-SEC Task Force (AMI-SEC-ASAP). Ami system security requirements v1.01. www.oe.energy.gov/DocumentsandMedia/, 2008.

[3] S. Amin and B. Wollenberg. Toward a smart grid. *IEEE Power and Energy Mag.*, 3(5):34–41, September 2005.

[4] Advanced Security Acceleration Project (ASAP-SG). Security profile for advanced metering infrastructure. osgug.ucaiug.org/, 2009.

[5] C. Bennett and D. Highfil. Networking ami smart meters. In *Proceedings of IEEE Energy 2030 Conference*, 2008.

[6] T. Chen. Survey of cyber security issues in smart grids. In *Cyber Security, Situation Management, and Impact Assessment II; and Visual Analytics for Homeland Defense and Security II (part of SPIE DSS 2010)*, 2010.

[7] F. Cleveland. Cyber security issues for advanced metering infrastructure (ami). In *2008 IEEE Power and Energy Society General Meeting – Conversion and Delivery of Electrical Energy in the 21st Century*, 2008.

[8] J. Fan and S. Borlase. The evolution of distribution. *IEEE Power and Energy Mag.*, 7(2):63–68, March 2009.

[9] H. Farhangi. The path of the smart grid. *IEEE Power and Energy Mag.*, 8(1):18–28, January 2010.

[10] US Federal Energy Regulatory Commission (FERC). Smart grid policy, 128 ferc 61,060, docket pl09-4-000. hwww.ferc.gov/whats-new/comm-meet/2009/071609/E-3.pdf, 2009.

[11] US Federal Energy Regulatory Commission (FERC). National action plan on demand response, docket ad09-10. www.ferc.gov/legal/staff-reports/06-17-10-demand-response.pdf, 2010.

[12] T. Garrity. Getting smart. *IEEE Power and Energy Mag.*, 6(2):38–45, March 2008.

[13] C. Hauser et al. Security, trust, and qos in next generation control and communication for large power systems. *Int. J. Critical Infrastructures*, 4(1):3–16, 2008.

[14] A. Ipakchi and F. Albuyeh. Grid of the future. *IEEE Power and Energy Mag.*, 7(2):52–62, March 2009.

[15] H. Khurana, M. Hadley, L. Ning, and D. Frincke. Smart-grid security issues. *IEEE Security and Privacy*, 8(1):81–85, January 2010.

[16] National Energy Technology Laboratory. Advanced metering infrastructure. www.netl.doe.gov/smartgrid/referenceshelf/whitepapers/, 2008.

[17] National Energy Technology Laboratory. A vision for the smart grid. www.netl.doe.gov/smartgrid/referenceshelf/whitepapers/, 2009.

[18] P. McDaniel and S. McLaughlin. Security and privacy challenges in the smart grid. *IEEE Security and Privacy*, 7(3):75–77, May 2009.

[19] A. Moise and J. Brodkin. ANSI C12.22, IEEE 1703 and MC1222 transport over IP. IETF draft-c1222-transport-over-ip-03.txt, 2009, work in progress.

[20] US Dept. of Energy. Grid 2030 a national vision for electricity's second 100 years. www.ferc.gov/eventcalendar/files/20050608125055-grid-2030.pdf, 2003.

[21] US Dept. of Energy. The smart grid: An introduction. www.oe.energy.gov/, 2008.

[22] US Dept. of Energy. Smart grid system report. www.oe.energy.gov/, 2009.

[23] The Future of Privacy Forum. Privacy for the smart grid: Embedding privacy into the design of electricity conservation. www.privacybydesign.ca, 2009.

[24] Office of the National Coordinator for Smart Grid Interoperability. Nist framework and roadmap for smart grid interoperability standards release 1.0. Technical Report special publication 1108, NIST, Gaithersburg, MD, 2010.

[25] G. Prasanna et al. Data communication over the smart grid. In *Proceedings of IEEE International Symposium on Power Line Communications and Its Applicications*, 2009.

[26] Smart Grid Interoperability Panel Cyber Security Working Group. Smart grid cyber security strategy and requirements. Technical Report draft NISTIR 7628, NIST, Gaithersburg, MD, 2010.

[27] J. Verba and M. Milvich. Idaho national laboratory supervisory control and data acquisition intrusion detection system (SCADA IDS). In *Proceedings of 2008 IEEE Conference on Technologies for Homeland Security*, 2008.

[28] J. Wang, A. Huang, W. Sung, Y. Liu, and B. Baliga. Smart grid technologies. *IEEE Industrial Electronics Mag.*, 3(2):16–23, June 2009.

[29] D. Wei, Y. Lu, M. Jafari, P. Skare, and K. Rohde. An integrated security system of protecting smart grid against cyber attacks. In *Proceedings of 2010 Innovative Smart Grid Technologies (ISGT)*, 2010.

5

Connections, Conversations and Awareness: Themes of Social Networking Services

Ajita John

Avaya Labs Research, 233 Mt. Airy Road, Basking Ridge, NJ 07920, USA;
e-mail: ajita@avaya.com

Abstract

In recent years, social networking services such as Facebook, YouTube, Flickr and Twitter have radically changed people's online experience. People connect with others, search for content, share content, thoughts and ideas differently and on a scale much larger than before. This chapter presents an overview of social networking services with a view towards understanding the inherent models, themes and architecture. These aspects will drive the next generation services as they seek to offer users new experiences and extract revenue for the providers. The chapter discusses three unifying themes for these services – users connect with each other, users have conversations with each other, and users are made aware of other users and their conversations. The chapter categorizes social networking services into three classes based on the focus of the services - contacts, content, and real-time interaction. It discusses how the services in these classes support the different themes and how users can be at odds with social networking providers and enterprises on the theme of awareness with respect to privacy.

Keywords: social networking, services, social software, online conversations, Facebook, YouTube, online awareness, online activity.

Anand R. Prasad et al. (Eds.), Advances in Next Generation Services and Service Architectures, 81–100.

5.1 Introduction

The last few years have seen a proliferation of online social networking services such as MySpace, Facebook, Bebo, Flickr, YouTube and Twitter. These services have radically changed user interactions on the World Wide Web from a static, one-way, consumption model where web content was created by owners and relevance of content was determined by the link structure of webpages. Today, user interactions follow a dynamic, multi-way, participation model where users connect and converse with others in a public space, upload, comment on and rate content, thus impacting the relevance of content and determining what other users may be seeing. Such broad user power and flexibility have changed how people engage in and experience their interconnections, interests, and collaborations.

From services to connect with friends and professional colleagues to sharing videos and photos to collaboratively creating content repositories such as Wikipedia, engaging in collective political action and playing online games, social networking services have changed the way we live and engage with the world. They have brought great benefits for users, enterprises, and institutions. We check online reviews before we buy products, uncover popular and relevant articles on bookmarking websites and chance upon interesting links through our friends' postings. Enterprises gauge reactions to products and services by monitoring online forums and celebrities keep in touch with their fans through Twitter. Grassroots movements spread the word and organize through online social networking tools. The list is endless.

Social networking services will mature significantly over the next few years to offer new kinds of services and interactions, improvements in design, leveraging of networks for marketing, stronger revenue models, and policies over data. Improvements in design will include better recommendations for content and connections, and visualizations for data. Algorithms will attempt to better understand the networks in these services to extract value for businesses for advertising which in turn will provide revenue for the providers. Better integration between different services will place focus on their architecture. The chapter presents an overview of social networking services from perspectives that will drive the growth of next generation social networking services: how they may be defined, what is the network model and a unified architecture. Additionally, the chapter presents a classification of the services: contacts-based (e.g. Facebook), content-based (e.g. YouTube), and real-time interaction-based (e.g. Warcraft). The chapter picks one representative example from each class to discuss characteristic features in each class. The classification will highlight the focus of each class and how they bring value to the users.

While the services may seem to be disparate and disjoint, there are common dimensions that they build upon. Overall, the chapter discusses the services from the perspective of three dimensions: connections, conversations, and awareness. The services enable connections and conversations between users, the conversations strengthen the connections, and the services are designed to increase awareness of user conversations which leads to more conversations. Users, social networking service providers and businesses benefit from growth along these dimensions. As next generation services evolve, users are eager for new experiences especially those that bring value and service providers are eager to generate revenue and business are ready to capitalize on the human networks in these services. While, in general, users seem to be in agreement with regards to growth, privacy is an issue that tempers it. With the availability of large amounts of user data around connections and conversations, the debate is about the level of control users have over the how this data gets shared and the ethics of using complicated user settings which, in many cases, is configured to share data as a default. The need for privacy and the human limitations to managing attention may indeed provide bounds to the growth of these services.

5.2 What Are Social Networking Services?

Social networking services are services provided by social software and there has been some debate about what social software means and encompasses. In 2003, Clay Shirky defined social software to be "software that supports group communications, includes everything from the simple CC: line in email to vast 3D game worlds like EverQuest, and it can be as undirected as a chat room, or as task-oriented as a wiki [34]. A later definition from Tom Coates was "Social Software can be loosely defined as software which supports, extends, or derives added value from, human social behaviour - message-boards, musical taste-sharing, photo-sharing, instant messaging, mailing lists, social networking" [13]. Boyd and Ellison [9] define a narrower class of services, namely, social networking sites to be web-based services that allow individuals to (1) construct a public or semi-public profile within a bounded system, (2) articulate a list of other users with whom they share a connection, and (3) view and traverse their list of connections and those made by others within the system. Essentially, the debate about a definition for social networking services was about the boundaries of social software and what it extends to. Is it just the new technologies under the umbrella of Web 2.0 [30] or do they encompass earlier ones as well?

While a precise definition is hard to come by due to the nature of these services, at a fundamental level, it is clear that they enable connections between people leading to the formation of communities or networks. These connections may be made explicit by the user (e.g. Facebook, LinkedIn) or they may be implicit, i.e. formed because the users are linked, perhaps, through a tag, a shared interest in a topic or product (e.g. delicious, YouTube, Amazon). These connections may be strengthened by conversations among the users and users may use objects such as videos, photos and blog posts to have such conversations or the objects themselves may be the conversation. It is important to note that these conversations span a wide range: from point-to-point messages and comments/ratings on a video to reviews on a product, and the uploading of a photo and tweets to the world. These are all conversations – expressions of thoughts and sharing of content either directed at a group of people, small or large, or directed at no one in particular. Growth in the connections and conversations among people leads to the growth of communities or networks around them. A facet to these services that makes them different from earlier computer-mediated services like email and IM is that users and applications are made aware of the existence of these communities and the activities in them, leading to their growth at a speed and on a scale much larger than before.

5.3 Network Model

Social networking services operate essentially on a network model where nodes represent people and links represent the strength of connections between them. The services promote the growth of these social networks both in terms of the number of nodes and the strength of connections. The facet of awareness between users is primarily to enable and strengthen these connections; the more aware people are of activities and other people, the more connections there will be and the stronger they will be.

The networks that get formed through these services are important because their structure may determine, for instance, how an idea or influence may spread through a community, how communities are forming, or how advertising may be strategically targeted. Social networks exhibit small world network characteristics where most nodes are not neighbors of each other, but can be reached from each other in a small number of hops. This concept was made famous by Stanley Milgram's experiment in the 1960s where 300 people in Boston and Omaha were asked to send a message to a target in Boston by sending it to someone they knew was closer to the target. Each person receiving the message got the same instructions and 64 messages reached the target. The average length of successful chains was six, leading to the popular phrase "six degrees of separation between us and everyone else on the planet".

Random graphs, where connections between nodes are generated at random, exhibit short path lengths. Research shows that random social connections are not a reality due to properties such as homophily (the tendency to associate with others who are similar to oneself) [29] and triadic closure (if two people have strong connections with a third person, then they have a strong/weak connection between them) [35]. Social networks, typically, exhibit clustering at various levels such as in groups, organizations, etc. Watts [36] showed that, in large networks, a small number of "shortcuts" will reduce the global path length just as in random graphs and yet, preserve local clustering. These graphs are termed small world networks and are not limited to just social networks. So, while social networks are locally clustered to reflect tightly knit communities such as a circle of friends, it has connections or ties to outside the clusters. These ties, typically weak, may be the ones that produce new ideas and connections to members within the clusters [17].

Social networking services provide the opportunity to gather data regarding these human networks and the social behavior in them using online activity on these services. Aspects such as community formation and evolution have been studied using such data [6]. Social scientists have gathered data regarding human networks for decades using other means such as interviewing people. Of course, data gathered from online activity logs cannot provide the nuances and depth of understanding that face-to-face interviews with people can provide, however the data is there and is driving the growth of a field called computational social science [14] which is facilitated by technologies such as Hadoop [1] which enables analysis of large scale data. Computational social science has the power to drive the design of these services as it can inform about facets of user behavior.

5.4 Architectural Model

Despite the diversity of social networking services existing today and the organic growth of these services, an architectural model has emerged for them and is presented in Figure 5.1. The model is an approximation and may evolve as new services emerge. The model consists of four layers (starting

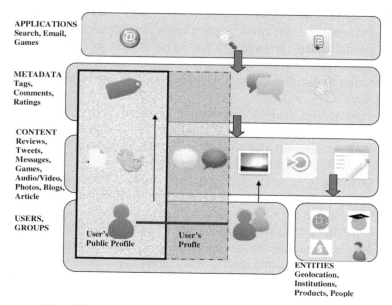

Figure 5.1   Architectural model for social networking services.

from the bottom): (1) profiles for users and entities, (2) content, (3) metadata, and (4) applications. Each layer is discussed in more detail below.

5.4.1 Profiles for Users and Entities

The bottom layer consists of user profiles that contain information about users such as their username, demographic information, and their activities on the service. Users typically have control over how some of this information gets exposed to other users and applications. The user profiles are used by the services to build more connections among the users. So, the presentation of this profile is a key aspect in many services.

This layer also contains profiles for entities that have a digital represent-ation – products on Amazon (e.g. a book or the iPhone), geolocation (e.g. location on Gowalla or Foursquare), people (e.g. a celebrity in a movie), institutions (e.g. a university or political party, The White House).

The FOAF (Friend of a Friend) project is an effort to create a web of machine-readable profile pages describing people, the links between them and the things they create and do [2]. It uses RDF and OWL and is part of the wider Semantic Web project. Essentially, it stores the connections and conversations data from multiple sources (social networking services) and allows it to be used in new ways.

5.4.2 Content

At the next layer is content created and/or uploaded by users. This may consist of textual content such as reviews, tweets, blogs and rich media such as audio, video, photos, etc. Also included in this layer are messages and game instances between users. This layer contains the objects, such as a video or a song, which form the context for the user conversations as well as the conversations around these objects. Sometimes, the object may the conversation itself such as a video created by a group of users that reflects something about them. Additionally, content in one service may get shared with other services. For example, twitter posts may be posted to Facebook. There are several issues that arise in this layer such as those of data ownership, copyright, and liability. Typically, the service provider owns the data once users have uploaded or created content on the service. Some sites monitor the content through editors (Wikipedia) or by having the users report inappropriate content.

5.4.3 Metadata

At the next layer is metadata that may be added by users or may be inferred by the system. These include user-generated tags (public/private), comments, ratings, and other meta-information such as the time a photo was taken, the length of a video, etc. User-generated meta-data such as comments and tags form conversations in a community. It should be noted that comments and tags are also content that is created and/or uploaded by the user. In this sense, this layer is a sub-layer of the content layer, however it is helpful to distinguish this layer because this kind of information helps to drive the design of many of the services.

Collaborative tagging [16] indexes the content and hence, helps to search and link them. Tags are especially useful for indexing rich media content [23, 33]. Content and users get linked through tags enabling, for instance, a user A to step from a news article tagged with "politics" by another user B to user B's bookmarks discovering another set of resources, potentially useful to user A. With so much data being available online, the challenge has been to figure out what may be relevant to a user and how to present the data. Metadata can be used in computational models to infer properties such as expertise in a community [21] or interestingness of videos [12]. Most commonly, they are used to infer models of relevance to construct visualizations such as tag clouds (Figure 5.4a) or to promote content on sites such as digg. For example, news articles and videos that are rated higher are typically presented to the user before those that are rated lower. Tags that are applied more frequently appear larger drawing more users to content that is linked to them. Recommendation algorithms [24] can be used to offer suggestions for new friends/contacts and social data may be used to infer relevance [22] for search. Relevance and recommendations drive the awareness dimension as they help to make the user aware of connections and conversations that the community has deemed relevant. There are, however, limitations to what can be done computationally as the systems are often missing key offline data that is very relevant in social relationships.

5.4.4 Applications and Technologies

At the top layer are applications such as search, email, chat, games, and virtual reality worlds. These applications support users to have conversations. The design of these applications is key to their success. While services such as Facebook offer their own set of Application Programming Interfaces for developing these applications, efforts such as OpenSocial [3] move towards a common interface for social applications.

Visualization technologies will play a key role as social networking services evolve and the data related to these services expands. With respect to data presentations, tools such as Vizster [18] allow exploration of the social networks of connections. However, many of the social networking services lag behind in terms of offering users a means to explore the data and make sense of the content and relationships. Many such as Twitter and associated tools offer list-based representations of search results from which it is hard to extract results that are relevant to the user or for the user to be able to explore the data to extract interesting facets or gain a comprehensive viewpoint. This aspect may be quite important in next generation services as the design of interactions systems plays a key role in determining user interactions with these systems [5].

A number of technologies have played a major role in the design and success of social networking services. Search Engines, Collaborative Tagging, RSS feeds have helped to drive the scale and speed at which people have access to information. Search engines index and bring a trove of data to the user's fingertips. RSS feeds update users instantly when content related to certain topics have been posted. Mobile devices too have played a powerful role in driving social networking services. With the advent of smartphones such as the iPhone and social networking services being accessible on these devices, users can connect to other users and access content nearly all the time and everywhere. Technologies such as touchscreen with tactile feedback help users to input data in a manner closer to a physical keyboard [20]. GPS enables services such as FourSquare to log a user's location from different places.

5.5 Classes of Social Networking Services

Current social networking services may be categorized into three broad classes based on the types of services they offer and the kinds of user interactions they enable. While the features in the services of each class vary widely, they still have unifying characteristics and the classification helps to understand the differing focus in each class. The classes are contacts-based, content-based, and real-time interaction-based. Note that these classes are only representative; they are not meant to be exhaustive and are likely to change somewhat as these services evolve over the next few years. Also, the services have been classified based on the predominant focus in the services. Some services like Facebook while being predominantly contact-based offer services such as Farmville which are real-time interaction-based.

5.5.1 Contacts-Based

This class of services allows users to create profiles and can form explicit connections with contacts. For example, users connect with other users as their friends on Facebook and professional contacts on LinkedIn. Other examples in this class include MySpace and Friendster. This class matches with the group Social Networking Sites or SNS as defined in [9] where users (1) construct a public or semi-public profile within a bounded system, (2) articulate a list of other users with whom they share a connection, and (3) view and traverse their list of connections and those made by others within the system. These services have become very popular in the last few years and Facebook has emerged as a leader in this class [32].

Facebook is discussed in more detail in this section to provide an illustrative example in this class. Facebook was launched in February 2004 and its design primarily focuses on the user, enabling each user to develop his profile and connect to online friends. He can search for other users and invite them to his friends list. To be searchable and ensure growth of the network, a user has to keep a minimal aspect of his profile public. Facebook relies on the user's tendency for self-presentation to his friends to ensure network growth. A user's profile on Facebook is a dynamic entity that consists of some static information, but mostly dynamic information based on the user's activity in Facebook. Some examples of a user's profile information are as follows:

- work and education, likes and interests, and contact information,
- a "wall" that displays

  1. what she has shared with others (photos, video, events, etc.),
  2. her activities (who she became friends with, what she commented on),
  3. photos posted by a friend in which she got tagged,
  4. things sent by friends through applications, e.g. birthday wishes,
  5. groups she has joined,
  6. things posted by her friends on her wall,

- photos,
- videos,
- applications.

Figure 5.2 shows the "wall" part of a user profile. It shows information such as the friends' list on the left pane, the user's activity and friends' postings on the wall in the middle pane and advertisements on the right pane.

The user can share a variety of content with other users such as text, photos, videos, and news articles. She can import content from sites such Flickr, digg, Picasa, delicious, yelp, and google reader or she can export content from these sites and share them on Facebook. She has various options to control who sees this, e.g. everyone, friends and networks, friends of friends, friends only. She can customize this to include specific people (enter a name or list), only she or one or more networks. She can even hide the content from a particular name or list. She can do this per posting or set this as a default setting. The sharing of content is key to generating conversations. When other users see a posting, such as a news article, they can comment on it. The author and other users can reply to comments generating a discussion. Additional control options that a user has include some of the following:

- can control whether friends can post to her wall,
- who can see posts made by friends. Options here are "everyone, friends and network". It is evident that this is fairly tricky because what I post on a friend's wall may be exposed to everyone and I may not know it,

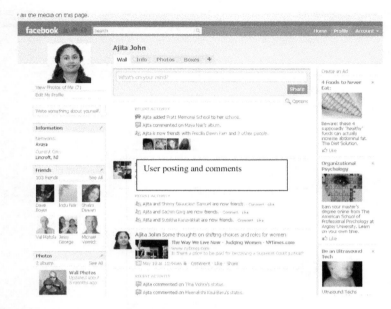

Figure 5.2 A snapshot of the Facebook profile page listing the user's activity and friends' postings. The right pane shows ads and the left pane lists information such as the friends list.

- control which applications can post on my wall.

It must be evident from this description that the types of information that can end up on a user's profile are diverse and can be changing rapidly since users may post comments on content shared on the wall. Boyd and Heer [10] discuss how profiles on such systems are not just about one user, but rather are representations of others too and how profiles are communicative bodies in conversation with other represented bodies.

The options available to the user are complex, making it hard for an average user to keep track of what information is getting shared and to whom [28]. Complex privacy models make it difficult or impossible to discern who can see some of the information one shares with others. For example, if a user posts something on a friend's wall, there is no way to know who can see that information. Is it only the friend's friends or is it everyone? This is determined by the user whose wall it is. User A's posting, say, a video is broadcast to all users on A's friends list, unless A has set the broadcast option to more people. A's commenting on a video posted by B, however, gets broadcasted to all of A's friends and B's friends (and possibly even broader if B has set the option

Figure 5.3 A snapshot of Facebook's NewsFeed feature.

to be so). The same is true if A shares something on B's wall. Research has shown that users have misconceptions about the visibility of profiles [4].

In 2006, without any notice, Facebook rolled out its NewsFeeds feature (see Figure 5.3b) which lists the activities of a user's friends when the user logs in. The right pane in Figure 5.3b shows how users are presented with various options to connect with more people on Facebook. Newsfeeds list all kinds of information about people such as who they befriend, what their scores are on applications, when they commented on their friends' posts, etc. [8]. The NewsFeeds feature was able to push up the level of awareness on Facebook. Prior to this, users had to seek out much of this information. By pushing this information to users when they login, Facebook reduced barriers to awareness of others' activities, much to user consternation. Earlier this year, Facebook exposed user activity to external websites such as CNN. Users can turn this feature off, but it is turned on by default. As Facebook pushes the limits on the awareness dimension of social networking services, it has had to deal with user frustration.

5.5.2 Content-Based

The main theme of services in this class is around content. Typically, each service focuses on a specific type of content – videos on YouTube, photos on Flickr, tweets on Twitter, bookmarks on delicious.com and digg.com, and blogs. Customer product reviews on Amazon, Wikipedia, or topic-based services also fall in this class. Twitter is a service where a user can have "followers" who get updated when the user posts something. Posts are short (140 characters) text messages which may contain URLs which are either private to followers (this has be to set in the account settings, else it is public) or public to everyone. Wikipedia is an interesting example where users collaboratively contribute and edit content leading to a collaboration structure [11]. Recent services such as Gowalla and FourSquare may also be categorized into this class. Here, users use an application on a mobile device to check in and post tips about locations which are identified using GPS technology, and they earn points based on their visits and what they uncover about that location. The application can be integrated with Facebook and Twitter, so users on those networks can see your posts. A user can see the tips from other users about different locations and can get notified of offers from local businesses.

While this is a broad class of services, they have distinguishing features from contacts-based services such as Facebook. By focusing on a specialized kind of content, the services are able to grow communities dedicated to that content. Users go to the specific service when they are interested in the associated type of content. Users may search, tag and rate the content. The content gets linked through the tags to other content and users. For example, the contacts' photostream in Flickr enables social browsing [26] where users can see the photos uploaded by them facilitating new ways of interacting with information. Users may post comments on the content leading to conversations around the content. Users rely on the community of users to provide guidance with respect to relevance through features such as tag clouds (see Figure 5.4a).[1]

YouTube is discussed in more detail as an illustrative example in this class. YouTube allows users to post videos which can be tagged with keywords facilitating search. Additionally, users can comment on the videos generating conversations. Users can be subscribed to their contacts on Gmail, Facebook, Yahoo, etc., and YouTube extracts these contacts from these services. Users are kept updated of the videos uploaded by their subscribed

---

[1] This file is licensed under the Creative Commons Attribution-Share Alike 2.5 Generic license.

Figure 5.4 (a) Tag cloud depicting relevance of difference tags through the size of tags. (b) A snapshot of a delicious page showing how tags and user information (numbers) can be used to navigate to content.

contacts, their ratings, comments and favorites. YouTube can match the email addresses of a user's contacts on other sites such as Facebook and Gmail with the email addresses of those YouTube users who have chosen to allow themselves to be found by their email address or Facebook friends who have also connected their Facebook accounts to YouTube. If a user decides to import from Facebook, all of the user's information on Facebook (including friends lists) is imported into YouTube (see Figure 5.5). Even though, YouTube will only show the accounts of those users who have connected their Facebook accounts to YouTube, YouTube now has all the data from the user's activity on Facebook, which exposes the user's friends' activities too without their permission. While it is great to be able to share data across services, concerns about privacy are an issue here.

Services like Twitter and FourSquare are examples where people have been comfortable sharing posts with, potentially, anyone online. Twitter started off as a public medium, so people understood what they posted could be shared with anyone in the world. While Twitter does offer private tweets (shared with only one's followers), this was a later feature.

5.5.3 Real-Time Interaction-Based

This class of services offers users a simulated environment creating a virtual world where users may have an online graphical representation called

Figure 5.5 (a) YouTube can import all of a user's information including friends lists. (b) A user's activity on Facebook can be shared with other services.

"avatar" and can participate in a collaborative activity. Second Life, Farmville, and Warcraft are examples in this class. While some services offer virtual worlds based on real-life environments, others create fantasy worlds. The Massively Multiplayer Online Role-Playing Games (MMORPG) have dominated this class. Users play in games that typically feature a fantasy world where the characters may collaborate in teams to achieve goals. Some key features of these games are character development in the context of a fantasy world, role-playing, and collaboratively working towards a goal. World of Warcraft is a popular game in this category.

What distinguishes this class of social networking services from the contacts-based and the content-based services is that the primary focus is real-time interactions and collaborations between users leading to real-time conversations. Real-time here means that users interact with others while the users are logged on at the same time, so user responses, typically, are made within a short period of time without the user logging off.

Since users play different characters and roles in these games, the presentation of self is different from other social networking services. Often, the self is presented through a character whose physical characteristics such as looks and strengths can be configured. Character development is complex where progression is quick early on, but slows down as the game advances and often requires cooperation and dependency on others, often beneficial in a mutual manner [37]. Social connections are strongly emphasized in this way. Connections between users consist of temporary groups of a few users or persistent user-created membership organizations, and larger ideological alliances [37] and the adventure towards the goal and stories forms conversations in these games.

5.6 Growth of Social Networking Services

What will the next generation of social networking services bring? While technological advances will drive much of push towards next generation services. the push will be shaped by the different entities who are involved and there are different incentives for users, providers, enterprises, and institutions in social networking services. Why do users spend time on social networking services? The motivations may be varied. Studies on people using Facebook have pointed to the formation and maintenance of social capital [15]. Enterprises and institutions employ social networks for viral marketing purposes as they provide a medium for propagating recommendations through people with similar interests [27]. Motivations for play in Massively Multiplayer Online Role-Playing Games (MMORPG) have been studied in [38].

Service providers are looking for advertising revenue, which has been gaining momentum on social networking sites. A recent study in 2009 showed that social networking sites accounted for more than 20% of all display advertisements viewed online, with MySpace and Facebook combining to deliver more than 80% of advertisements among sites in the social networking category [31].

The primary goal of providers seems to be

- To build out the network in size (increase number of nodes). This increases the target for potential advertising and hence revenue. This explains why Facebook and LinkedIn offer suggestions for new friends.
- To increase activity among users using the service, which in turn, builds out the network in terms of links (increase links and link strength). The more time users spend using the service, the more advertisements they can be targeted with. They spend more time using the service if they connect with more people, "connect more with people", and they are presented with engaging activities.

There are factors such as bounds on connections and the user demand for control that may temper the growth. Research suggests that people use these networks to learn more about people they meet offline, rather than initiate new connections [25] and people interact with a few that matter and reciprocate their attention [7]. There is research is sociology to support this. Robin Dunbar showed that people could maintain at most 150 active relationships at any time. This is known as the Dunbar number [19]. So, while users may have hundreds and even thousands of people on their friends list on Facebook, these may not be active relationships in the conventional sense.

As services move towards more viable business models (and this may drive much of the shift) and as businesses seek to leverage these social networks for advertising, users will demand more control over the data. While the conversations data is owned by the providers, users will demand better control over how that data gets shared. Next generation services will have to provide simpler and better privacy models. Public forums such as Twitter may make easier transitions than private forums such as Facebook.

5.7 Conclusions

This chapter provides a perspective on myriad domain of social networking services. While seemingly diverse, there are some unifying principles around these services. At a core level, these services enable connections between users resulting in what are known as social networks. These social networks provide value to the users in the form of conversations that occur in the networks. These conversations take the form of comments, tags, ratings, games, etc., and provide a platform for users to exchange ideas and thoughts, interact and collaborate. The conversations serve to strengthen the connections between the users.

Users, providers, enterprises, and non-profit organizations stand to benefit from the growth of these networks and growth is spurred by greater awareness of the community of users. Hence, social networking services are pushing the boundaries on features that promote awareness of user activity. While users, to a large extent, stand to benefit from such awareness and growth, the critical concern for users is privacy. In addition to the user's demand for new experiences, businesses targeting the social networks for marketing, and the provider's push for revenue, privacy is likely to shape the evolution of these services in the future.

References

[1] Apache Hadoop. http://hadoop.apache.org/.

[2] The friend of a friend (FOAF) project. http://www.foaf-project.org.

[3] Open social. http://code.google.com/apis/opensocial.

[4] A. Acquisti and R. Gross. Imagined communities: Awareness, information sharing, and privacy on the facebook. In *Proceedings of 6th Workshop on Privacy Enhancing Technologies*, Cambridge, UK, 2006.

[5] A. D. Angeli, A. Sutcliffe, and J. Hartmann. Interaction, usability and aesthetics: what influences users' preferences? In *Proceedings of the 6th Conference on Designing Interactive systems*, pages 271–280. ACM, 2006.

[6] L. Backstrom, D. Huttenlocher, J. Kleinberg, and Xiangyang Lan. Group formation in large social networks: Membership, growth, and evolution. In *KDD'06: Proceedings of the 12th ACM SIGKDD International Conference on Knowledge Discovery and Data Mining*, pages 44–54, 2006.

[7] A. Bernardo, Huberman, Daniel M. Romero, and Fang Wu. Social networks that matter: Twitter under the microscope. http://arxiv.org/abs/0812.1045, 2008.

[8] D. Boyd. Facebook's privacy trainwreck: Exposure, invasion, and social convergence. *Convergence: The International Journal of Research into New Media Technologies*, 14(1):13–20, 2008.

[9] D. Boyd and N. Ellison. Social network sites: Definition, history, and scholarship. *Journal of Computer-Mediated Communication*, 13(1), 2007.

[10] D. Boyd and J. Heer. Profiles as conversation: Networked identity performance on friendster. In *Proceedings of the Hawaii International Conference on System Sciences (HICSS-39), Persistent Conversation Track*. IEEE Computer Society, 2006.

[11] U. Brandes, P. Kenis, J. Lerner, and D.V. Raaij. Network analysis of collaboration structure in Wikipedia. In *Proceedings of the 18th International World Wide Web Conference (WWW)*, 2009.

[12] M.D. Choudhury, H. Sundaram, A. John, and D.D. Seligmann. What makes conversations interesting? Themes, participants and consequences of conversations in online social media. In *Proceedings of the 18th International World Wide Web Conference (WWW)*, 2009.

[13] T. Coates. An addendum to a definition of social software. http://www. plasticbag.org/archives/2005/01/an_addendum_to_a_definition_of_ social_software, 2005.

[14] D. Lazer, A. Pentland, L. Adamic et al. Social science: Computational social science. *Science*, 323(5915):721–723, 2009.

[15] N. B. Ellison, C. Steinfield, and C. Lampe. The benefits of Facebook "friends": Social capital and college students' use of online social network sites. *Journal of Computer-Mediated Communication*, 12(4):1143–1168, 2007.

[16] S.A. Golder and B.A. Huberman. The structure of collaborative tagging systems. CoRR, Volume ABS/CS/0508082, http://arxiv.org/abs/cs/0508082, 2005.

[17] M.S. Granovetter. The strength of weak ties. *The American Journal of Sociology*, 78(6):1360–1380, 1973.

[18] J. Heer and D. Boyd. Vizster: Visualizing online social networks. In *Proceedings of IEEE Symposium on Information Visualization (InfoVis 2005)*, pages 23–25, 2005.

[19] R.A. Hill and R.I.M. Dunbar. Social network size in humans. *Human Nature*, 14(1):53–72, 2003.

[20] E. Hoggan, S.A. Brewster, and J. Johnston. Investigating the effectiveness of tactile feedback for mobile touchscreens. In *CHI'08: Proceedings of the Twenty-Sixth Annual SIGCHI Conference on Human Factors in Ccomputing Systems*, 2008.

[21] A. John and D.D. Seligmann. Collaborative tagging and expertise in the enterprise. In *Proceedings of the Collaborative Web Tagging Workshop at the World Wide Web (WWW06) Conference*, 2006.

[22] S. Kelkar, A. John, and D.D. Seligmann. Visualizing search results as web conversations. In *Proceedings of the Workshop on Web Search Result Summarization and Presentation, the 18th International World Wide Web Conference*, 2009.

[23] S. Kelkar, A. John, and D.D. Seligmann. Some observations on the "live" tagging of audio conferences in the enterprise. In *Proceedings of International Conference on Human Factors in Computing Systems (CHI)*, pages 10–15, 2010.

[24] L. Terveen and W. Hill. Beyond recommender systems: Helping people help each other. In Jack Carroll (Ed.), *HCI in The New Millennium*, Addison-Wesley, 2001.

[25] C. Lampe, N.B. Ellison, and C. Steinfield. A Face(book) in the crowd: Social searching vs. social browsing. In *Proceedings of the 2006 20th conference on Computer Supported Cooperative Work (CSCW)*, pages 167– 170, 2006.

[26] K. Lerman and L. Jones. Social browsing on Flickr. In *Proceedings of the International Conference on Weblogs and Social Media*, 2007.

[27] Jure Leskovec, Lada A. Adamic, and Bernardo A. Huberman. The dynamics of viral marketing. *ACM Transactions on the Web (TWEB)*, 1(1), 2007.

[28] H. R. Lipford, A. Besmer, and J. Watson. Imagined communities: Awareness, information sharing, and privacy on the facebook – Understanding privacy settings in Facebook with an audience view. In *Proceedings of the 1st Conference on Usability, Psychology, and Security*, 2008.

[29] M. McPherson, L. Smith-Lovin, and J. Cook. Birds of a feather: Homophily in social networks. *Annual Review of Sociology*, 27:415–444, 2001.

[30] T. Oreilly. What is web 2.0: Design patterns and business models for the next generation of software. *Communications and Strategies*, First Quarter(1):17, 2007. Available at SSRN: http://ssrn.com/abstract=1008839.

[31] ComScore Press Release. Social networking sites account for more than 20 percent of all U.S. online display ad impressions. `http://tiny.cc/wp781`.

[32] ComScore Press Release. Social networking explodes worldwide as sites increase their focus on cultural relevance. `http://www.comscore.com/Press_Events/Press_Releases/2008/08/Social_Networking_World_Wide`, 2008.

[33] A. Renduchintala, S. Kelkar, A. John, and D.D. Seligmann. Designing for persistent audio conversations in the enterprise. *Proceedings Designing for User Experience Conference, DUX'07*, 2007.

[34] C. Shirky. Social software and the politics of groups. March 9, 2003 on the "Networks, Economics, and Culture" mailing list, 2003

[35] D. Watts. Six degrees: The science of a connected age, 2003.

[36] D. Watts and S. H. Strogatz. Collective dynamics of 'small-world' networks. *Nature*, 3934:440–442, 1998.

[37] N. Yee. The psychology of massively multi-user online role-playing games: Motivations, emotional investment, relationships and problematic usage. In *Avatars at Work and Play: Collaboration and Interaction in Shared Virtual Environments*, pages 187–207, Springer-Verlag, 2006.

[38] N Yee. Motivations of play in mmorpgs. In *Proceedings of DiGRA 2005 Conference: Changing Views – Worlds in Play*, 2005.

6

Communication Services and Web 2.0

Xiaotao Wu

Avaya Labs Research, 233 Mt. Airy Road, Basking Ridge, NJ 07920, USA;
e-mail: xwu@avaya.com

Abstract

The interactive nature of Web 2.0 makes communication functions must-have features for Web 2.0 sites. On the other hand, Web 2.0 applications, such as wikis, blogs, social networking, and mashups, can also enrich and enhance people's communication experiences. The integration of Web 2.0 and communication functions introduces new services, new challenges, as well as new opportunities to service developers. This chapter provides an overview of these new services, and illustrates the new challenges and some existing solutions to the challenges. The first five sections of this chapter present an overview and different aspects of integrating Web 2.0 and communication services. Section 6.6 discusses the enabling technologies for the integration and Section 6.7 discusses the challenges and some existing solutions.

Keywords: Web 2.0, communication services, Voice 2.0, feature interactions, convergent networks, VoIP, mashup.

6.1 Overview

The term 'Web 2.0' is commonly associated with web applications that facilitate interactive information sharing, interoperability, user-centered design, and collaboration on the World Wide Web. [43]

Anand R. Prasad et al. (Eds.), Advances in Next Generation Services and Service Architectures, 101–122.

Web 2.0 sites, such as wikis, blogs, social networking, and mashups, become important parts of people's online life. Statistics show that social networking now accounts for 11% of all time spent online in the US in 2009 [11]. Among all the Web 2.0 sites, one of the common characteristics is their interactive nature, which allows users to actively participate into online community activities and share information.

On the one hand, the interactive nature of Web 2.0 makes communication functions must-have features for Web 2.0 sites. On the other hand, Web 2.0 applications, such as wikis, blogs, social networking, and mashups, can also enrich and enhance people's communication experiences. The integration of Web 2.0 and communication functions seems to be a natural match.

Nowadays, most of the Web 2.0 sites only support end-to-end or group-based text chatting, and have very limited support on voice and video communication functions. As popularity of Voice over Internet Protocol (VoIP) grows, and the telecommunication networks and data networks converge in enterprises as well as in residential environment, it is not difficult to introduce voice and video communication functions into Web 2.0 sites. However, the questions are "do we really need voice and video communication functions in Web 2.0 sites" and "what new services can we provide for Web 2.0 sites". If we consider that adding voice and video communication into Web 2.0 sites is simply to add another modality of communication, this integration is not very attractive because the real-time and high fidelity nature of voice and video communication is not critical for social networks, considering that many users like to hide their real identity and real emotion on social networks.

This chapter will discuss different aspects of integrating Web 2.0 and communication services and try to answer the above two questions. Our focus is on introducing new services, exploring the new challenges and investigating existing approaches to meet those new challenges.

The chapter is organized as follows: Sections 6.2 to 6.5 present different aspects of the integration by using existing service examples. Section 6.6 and Section 6.7 then summarize the enabling technologies and present the challenges.

Since VoIP technologies, especially Session Initiation Protocol (SIP), are the keystones for the convergence of data and telecommunication networks, in the following sections, we base our discussion on SIP for voice and video communications.

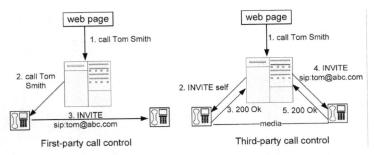

Figure 6.1 First-party call control vs. third-party call control.

6.2 Embed Communication Functions into Web 2.0

This section presents three different service examples to integrate voice communication services into Web 2.0 sites, including click-to-call, Facephone, and embedding voice communication into Google Wave.

6.2.1 Click-to-Call in Web 2.0

The most commonly supported voice communication function in Web 2.0 sites is the click-to-call feature, which makes phone numbers clickable on web pages.

There are two different ways to achieve the click-to-call feature, one uses first-party call control and the other uses third-party call control. Figure 6.1 shows the difference between these two approaches. First-party call control requires a control channel to directly operate users' phones, while third-party call control does not require the control channel. Therefore, third-party call control can work for any phone. From a user's point of view, first-party call control is a more natural way of making calls because it dials out as users normally do, while third-party call control requires users to answer an incoming call prior to making an outgoing call. However, third-party call control may hide callee's information from the caller so it may protect people's privacy on social networks.

6.2.2 Avaya Facephone

The following two sections will introduce two more sophisticated examples of embedding voice communication services into Web 2.0 sites. One is Avaya's Facephone project [21] and the other is to embed voice communica-

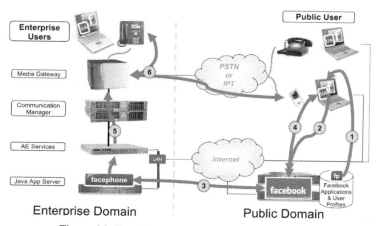

Figure 6.2 Facephone architecture and call flow [21].

tion capabilities into Google Wave. Both projects nicely integrate enterprise communication features with social networking sites.

At the fore-end, Facephone has a user agent running on Facebook [2] as a Facebook application so it can extract user contacts from Facebook. While at the back-end, Facephone runs applications within enterprise communication networks so it can invoke enterprise communication features, such as call transfer or call converage features. From a user's point of view, Facephone can perform numberless click-to-call and have richer information of the called party.

Figure 6.2 shows the architecture and call flow of Facephone. In the figure, calls are initiated on Facebook, which is outside of enterprise communication networks. The call information is brought into enterprise environment by Facephone application running on an enterprise application server. Facephone application then uses Application Enablement (AE) services to connect to the Communication Manager (CM) and build the call control logic, then finally establish the call.

6.2.3 Embedding Voice Communication Capabilities into Google Wave (Wavecall)

Google Wave [3] is a tool developed by Google for communication and collaboration on the web. At the moment of starting this chapter, Google Wave was still actively discussed in communication industry. But communication technologies are changing so quickly that prior to the book getting published, Google had discontinued this product in August 2010. Even though Google Wave is a discontinued product, it can still be considered a good Web 2.0 practice in managing people's communication activities. Google is planning to release Google Wave's code in a project called "Wave in a box", which intends to allow developers and enterprising users to run wave servers on their own hardware. Google Wave combines different communication means, such as email, instant messaging, wikis, web chat, social networking, and project management in one in-browser communication platform. It is open sourced and has many good features such as embeddability, extensibility, and drag and drop file sharing. But among all its features, it lacks voice communication functions.

Google Wave provides two ways to extend its functionalities: gadgets and robots. A gadget is an application which users can participate with, for example, a chess game can be a Google Wave gadget. A robot is an automated participant within a wave. It can talk to wave users, check the content of a wave, and perform actions to alter a wave, such as inviting another user to join the wave. The idea of using robots allows developers to add arbitrary features to a particular wave. There are several existing Google Wave extensions that enable voice communications in Google Wave.

Twiliobot [8] is a Robot extension that provides click-to-call function in Google Wave. The Twiliobot can translate a click-to-call number into a URL pointing to Twilio server, which then handles call control logic. Once the user clicks the URL, Twilio server will perform the third-party call control logic as we introduced in Figure 6.1.

Ribbit [6] is a Gadget extension that provides conferencing capabilities in Google Wave. The call control logic of Ribbit is similar to Twilio that uses third-party call control logic to establish a conference. Both Twiliobot and Ribbit cannot get incoming call events.

Researchers from Avaya Labs Research proposed a different Robot extension that can listen to incoming call events and making outgoing calls [26]. It monitors call events by subscribing to SIP dialog events [34] and uses first-party call control to handle calls. Figure 6.3 shows the architecture. This architecture can bring enterprise voice communication services into Google Wave.

Though Facephone and Wavecall use different ways to provide voice communication services in different Web 2.0 sites, the rationale behind these two services are the same. Both consider that introducing voice communication to Web 2.0 sites is not to add another media channel, rather, it is to utilize

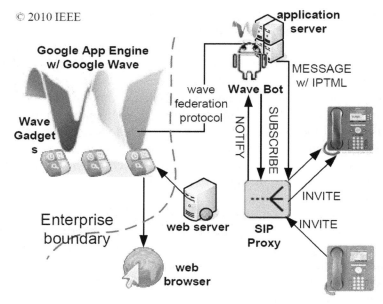

Figure 6.3 Integrate voice communication functions into Google Wave [26].

the existing voice communication services so as to maximize users' capabilities and efficiency. Generally, enterprise communication networks have much more features than residential communication networks, both service examples integrate Web 2.0 sites with enterprise communication networks.

For both integration, firewall traversal is a critical problem to solve. Figure 6.4 shows an example implementation that bring enterprise communication events into Google Wave by using a DMZ (a DMZ is a subnet in an enterprise that can expose the enterprise's services to a untrusted network, such as the Internet) server sitting on the boundary of an enterprise.

6.3 Bring Web 2.0 Capabilities into Communication Sessions

The previous section introduces several services that bring voice communication functions into Web 2.0 sites. This section will show examples that bring Web 2.0 concepts into voice communication services.

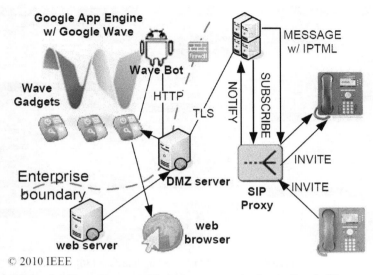

© 2010 IEEE

Figure 6.4 Firewall traversal for voice communication in Google Wave.

Considering a communication session intrinsically builds a relationship among the participants for a period of time. During this period of time, the more information each participant can get, the more value this communication session will be. The means to acquire information during the session should not be limited to voice or video communication itself. Rather, Web 2.0 technologies, such as the social networking sites, mashup concepts, and collaboration features can all help information sharing. Meanwhile, the relationship initiated by the communication session can guard sharing policies, such as ensuring appropriate privacy protection.

To experiment this concept of using Web 2.0 technologies for information sharing during a communication session, researchers from Avaya Labs Research have developed a context-based communication project called PhoneMash [46]. The goal of the project is to build a mashup page that incorporate information retrieved from different sources, such as corporate directories, social networking sites, email archieves, call histories, and instant messages. The information will be filtered by communication session attributes, and the acquired information will be easily shared to other session participants. Figure 6.5 shows the PhoneMash page for an incoming call.

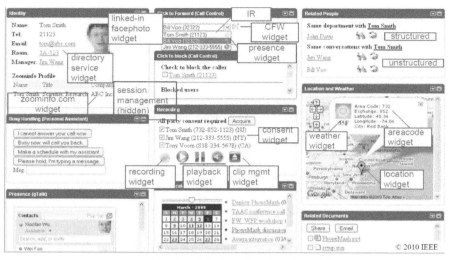

Figure 6.5 PhoneMash mashup interface [46].

To achieve the functions presented in Figure 6.5, we need to overcome the following three challenges to integrate communication sessions with Web 2.0 technologies:

- Bring call events and call control functions from telecommunication domain into web domain.
- Widgets may use different resources, such as phones, desktops, and communication servers and there must be an architecture to access and integrate different resources.
- Widgets may conflict to each other and the confliction must be detected and properly handled.

These challenges will be fully addressed in Section 6.7.

6.4 Web 2.0 and Voice 2.0

There is no clear definition of Voice 2.0. In December 2006, Paul Kretkowski gave the definition of Voice 2.0 on the VoIP News Web site as "an umbrella term for a loosely defined set of technologies and ideas that let people transmit voice, data, video and instant messages via IP, anytime, from anywhere." This definition seems to limit Voice 2.0 to what unified communication can offer. However, Voice 2.0 can be more powerful by embracing Web 2.0 con-

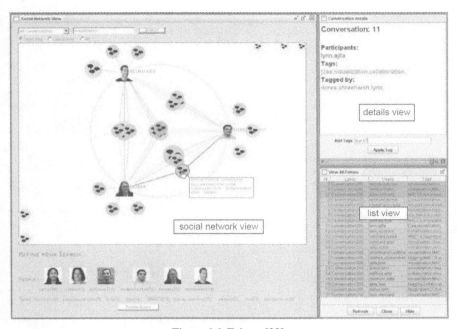

Figure 6.6 Echoes [22].

cepts to make it easier for users to share and collaborate. Below we use Avaya Echoes [22] and Connote [45] as examples to show the integration of Web 2.0 and Voice 2.0.

Echoes is a system that can help to enhance enterprise communication by the use of Web 2.0 technique of tagging. The analytic applications behind Echoes can then benefit from the tags in many different ways: searching conversations, identifying experts, building social networks, and easily retrieving specific audio clips. Figure 6.6 shows the user interface of Echoes.

As shown in the figure, Echoes uses different views to present tag information. The details view provides more detailed information about a people, conversation, or tag. The list view shows a list of conversations. In Echoes, multiple users may tag the same conversation, which is called "collaborative tagging" in Web 2.0. The most interesting view is the visualized view, which uses different color scheme, size, position, connections, and pictures, to show the relationship among people, conversations, and tags. For example, the variations in the color tint can indicate the age of a conversation; if one

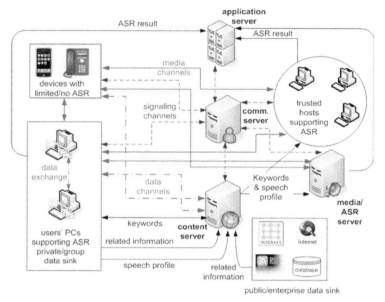

Figure 6.7 Connote architecture [45].

user tags many of another user's conversations, it indicates that they may have common interests.

The conversation data in Echoes are usually recorded audio clips of meeting sessions. Users usually perform post-call tagging in Echoes. While another project called Connote [45] allows users to handle mid-call tagging, either manually or by speech-to-text (STT). Figure 6.7 shows the architecture of Connote.

Both Echoes and Connote utilized Web 2.0 tagging or annotation concept and provide collaborative platforms for users to share the tags.

6.5 Web 2.0 and Mobile Communications

The Web 2.0's SLATES (Search, Links, Authoring, Tags, Extensions, and Signals) features and techniques can also be applied on mobile devices. Many wireless service providers have already deployed their mobile web services. From a user's point of view, there are two major factors that differentiate Web 2.0 for mobile communications from its desktop version: portability and mobility. Both portability and mobility means on-the-spot interaction with users and stronger time, event and location dependence of the information. Portability also means limited resources (reduced screen, less bandwidth, and limited computation capability).

The most commonly used Web 2.0 features for mobile communications is location-based features. According to a report published by Juniper Research [5], revenues from mobile location-based services could be more than $12.7 billion by 2014. Based on a survey from compete.com [4], right now, people are mainly using LBS for navigation and weather services, but people are increasingly interested in local alerts, such as traffic jam, gas sale, and special offers. Many people also constantly use location information to look for local stores, restaurants, and movie show times.

For mobile voice communications, location information can be very useful, especially for emergency services. Section 6.6.6.1 will discuss how to convey location information in IMS in detail.

6.6 Enabling Technologies

This section briefly introduces some technologies that enable the integration of Web 2.0 services and communication services. From Web domain, the enabling technologies include Ajax, RIA, HTML 5, and Web services. From communication domain, the enabling technologies include SIP, IMS, VoiceXML, and XML-based event notification and call control technologies.

6.6.1 Asynchronous Javascript and XML (Ajax)

Ajax (shorthand for Asynchronous JavaScript and XML [16]) allows web pages showing on users' web browsers to retrieve data from servers asynchronously in the background and then update foreground display accordingly. It is not a new technology by itself, instead, it is a group of existing technologies including HTML/XHTML/CSS [31, 33], Document Object Model (DOM) [29], XML, XMLHttpRequest [13], and Javascript. The biggest advantage of Ajax is that it is based existing technologies, so it is widely supported on many popularly used browsers.

When using Ajax for crossing telecommunication domain and web domain services, the same origin policy [12] of Javascript may block third-party web applications to access telecommunication domain information, which limits the abilities to use Web 2.0 technologies, such as mashup, to integrate web applications and telecommunication applications. Some client-side technologies, such as browser plug-ins, Rich Internet Applications, and HTML 5, or server technologies, such as Web Services for Remote Portlets (WSRP) [39], can help to solve the problem.

6.6.2 Rich Internet Application Platforms

Rich Internet Applications (RIAs) are like desktop applications but have been designed specifically for web services. They usually have greater flexibility for web programming than Ajax-based technologies, while have more restrictions than native desktop applications. For example, RIA applications may use cross domain policy file to access sites from different domains. It may also access local devices. This flexibility makes the integration of telecommunication and web domain services easier by using RIA.

Adobe Flash, Java and Microsoft Silverlight are currently the three top platforms. Users generally need to install a software framework or a browser plug-in to enable RIAs. For example, Adobe Integrated Runtime (AIR) platform or Flash player plug-in is required to run Adobe Flash applications.

6.6.3 HTML 5

HTML 5 is the next major revision of the HTML standard. It incorporates many features presented in popularly used RIAs such as Adobe Flash and Microsoft Silverlight. There are many new APIs specified in HTML 5, such as canvas element, drag-and-drop capability, video support, and new MIME type support. Among these new APIs, the following APIs are specifically useful for integrating telecommunication services into Web 2.0 sites.

6.6.3.1 Server-Sent Events, WebSockets, and Web Messaging

Server-Sent Events [19] defines an API for receiving push notifications from event sources on a web server in the form of DOM events. It uses HTTP as transportation protocol. For a client that cannot maintain a persistent connection, such as a smartphone, it also suggests of using a "push proxy" as delegate to receive events, then uses technologies such as OMA push to convey events from "push proxy" to the client.

WebSockets [20] is designed to provide a bi-directional, full-duplex communications channel between a web browser and a web server over a single TCP socket. This technology provides a mechanism for browser-based applications of using a single HTTP connection to handle two-way communication with servers.

"Web browsers, for security and privacy reasons, prevent documents in different domains from affecting each other; that is, cross-site scripting is disallowed. While this is an important security feature, it prevents pages from different domains from communicating even when those pages are not hostile" [18]. Web Messaging [18] allows documents from different domains to communicate with each other in a secure way.

For integrating Web 2.0 and telecommunication services, the above mentioned three technologies can be used to convey telecommunication events to a web widget as well as sending telecommunication control actions, such as call transferring, from a web widget.

6.6.3.2 Microdata

The design goal of Microdata [17] is to provide a simple way to embed semantic markup into HTML documents. It can annotate HTML content with machine-readable labels. For example, it can define the type, e.g., "book", of a string item, e.g., the title of a book, on a web page. This kind of semantic information can be very useful for the integration of communication features. For example, we can enable click-to-call feature only if the numbers that have the type *"phone-number"*.

6.6.4 Web Services

Web services refer to the Application Programming Interfaces (APIs) that can be accessed via HTTP. The service logic of web services usually run on a web server. Web clients usually use one of two approaches to access the web services, one heavier and stricter way of following Simple Object Access Protocol (SOAP) [30] standard, and the other lighter way of using Representational State Transfer (REST) [14] based communications. There are many articles addressing the differences between these two approaches. For integrating Web 2.0 and communication services, Telecommunication entities can provide call control function as Web services so that Web 2.0 widgets can access and control calls.

6.6.5 Session Initiation Protocol

The Session Initiation Protocol (SIP) is a signaling protocol used for creating, controlling, and terminating multimedia communication sessions (two-party or multiparty) over Internet Protocol (IP). "SIP is based on an HTTP-like request/response transaction model. Each transaction consists of a request that invokes a particular method, or function, on the server and at least one response for that request" [35].

One of the design goals of SIP is to make it easier to integrate telecommunication services with other Internet services, such as presence, web, and instant messaging. Traditional telephony services, such as call forwarding, transfer, and call screening can be enhanced by this integration. For example, a user can easily convey additional information about himself by putting his personal web page's URL or Facebook link in the Call-Info header of a SIP message. Web site designers can also put a SIP URI on a page to enable click-to-call feature.

There are many SIP extensions, such as SIP event notification mechanism [34] and SIP extension for instant messaging [10], which further facilitate the integration of SIP and other Internet services.

6.6.6 3rd Generation Partnership Project (3GPP) and Internet Multimedia Sub-system

The 3rd Generation Partnership Project (3GPP) [1] is a collaboration between groups of telecommunications associations to make a globally applicable third-generation (3G) mobile phone system specification. The IP Multimedia Subsystem (IMS) [38] is an architectural framework originally designed by 3GPP for delivering IP-based multimedia services. One of the design goals of IMS is to ease the integration with Internet. IMS uses SIP as its call signaling protocol. The IMS framework can greatly facilitate the integration of Web 2.0 with mobile communications, especially for location-based features.

6.6.6.1 Integrating Location Services into IMS

In 3rd Generation Partnership Project (3GPP) architecture, the central component of providing location services (LCS) is the Gateway Mobile Location Center (GMLC). In the architecture, location information are transferred by Open Mobile Alliance (OMA) Mobile Location Protocol (MLP) between a LCS client (watcher) and a GMLC as well as between different GMLCs. The Requesting GMLC (R-GMLC) communicates with the LCS client. The Home GMLC (H-GMLC) resides on the User Equipment's (UE) home domain and talks with Home Subscriber System (HSS)/Home Location Register (HLR) for authentication and routing. And the Visited GMLC (V-GMLC) can get UEs location through a 3G network element such as a Serving General Packet Radio Service (GPRS) Support Node (SGSN) by Media Access Protocol (MAP).

Because nowadays many mobile devices have GPS receivers and can acquire location information, in IMS network, location sensing and notifications can also be handled in an end-to-end manner. In general, the location signaling needs to go through the CSCFs in IMS to reach LCS clients (watchers). The supplementary location related services, which reside on AS and connect to S-CSCF, may then be able to access and filter location events.

6.6.7 VoiceXML

VoiceXML [32] is a W3C standard for specifying interactive voice dialogues. VoiceXML uses different tags to instruct voice browsers to provide speech synthesis, automatic speech recognition (ASR), and audio playback. It is popularly used in Interactive Voice Response (IVR) systems for call centers and voicemail systems. For legacy phones that does not have big display, VoiceXML can also be the tool to provide voice interface to access Web 2.0 features.

6.6.8 XML-Based Event Notification and Call Control Mechanisms

As discussed in Section 6.2.1, click-to-call service can be handled either by first-party call control or third-party call control. For first party call control, people usually use XML-based protocols to control devices or telephone switches. There exists many different propriatory or open protocols for event notification and call control. For example, many popularly used IP phones, such as Avaya's SIP phones or Cisco's IP phones, use XML-based protocols to extend the capabilities of phones and allow external applications to access call information and control the phones.

The Computer-Supported Telecommunications Applications (CSTA) [23] standard maintained by ECMA International is a popularly used standard to control telecommunication devices by computers. The ECMA TR 87 of CSTA standard specifically defines a way of using CSTA for SIP User Agents. It embeds XML-based CSTA messages in the content of SIP MESSAGE requests to convey call events and control actions. For example, Microsoft Office Communicator uses TR 87 to support call control functions.

With the XML-based call eventing and controlling mechanisms, a Web 2.0 widget can easily update people's presence information (such as "in a meeting") based on their calling status. The call control functions, such as call transfer, can also be handled through a web interface.

6.7 Challenges and Existing Solutions

Previous sections presented several examples of integrating Web 2.0 technologies and communication sessions. This section will focus on three main challenges to achieve the integration, namely, crossing domain signaling; crossing domain feature interaction handling; and crossing domain security and privacy handling.

6.7.1 Crossing Domain Signaling

It is easy to access web domain information from telecommunication domain because web information is usually open and can be easily accessed by using Hypertext Transfer Protocol (HTTP). While accessing telecommunication domain information from web domain used to be difficult for the following reasons.

First, in traditional telecommunication network, telecommunication domain is usually closed for external access. Usually, a web page cannot directly access telephone switches to get communication session information. There must be an entity that can access communication events and present the events to web components. With VoIP, by using SIP dialog events notification or other XML-based call event notification protocols, such as CSTA, this problem can be well handled.

Second, telecommunication sessions are usually persistent and have session related events flowing during the sessions. Traditional web technologies (HTML over HTTP) are transaction-based, stateless, and do not support push technologies. Therefore, the traditional web technologies are not suitable for accessing and representing telecommunication sessions. With Web 2.0 enabling technologies, such as Ajax, HTML 5, Mashups, and browser plug-ins, we can better handle telecommunication sessions on web browsers.

Third, people may use different devices in different context for different communication purposes. When introducing telecommunication session information to web domain, not only should we provide basic information such as caller and callee information, we must also provide device capability information, context information, and if possible, communication purpose information. In a SIP network, there are many different ways of conveying device and context information from endpoints to a server [36, 37].

Figure 6.8 shows the architecture that uses different technologies to bring communication information to users' web browsers. In the figure, the call eventing and controlling iframe uses different means (e.g., long polling, WebSocket, or SIP dialog events) to get call information. The other widgets

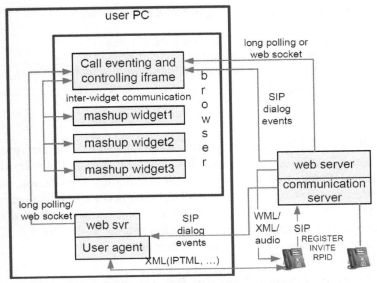

Figure 6.8 Bringing call information to web domain.

then use inter-widget communication (e.g., Web Messaging) to obtain call information from the call eventing and controlling iframe.

6.7.2 Crossing Domain Feature Interaction Handling

Feature interaction (FI) is introduced by composing modularized features. When we integrate multiple services, one service may break the assumption of another and we say that these two features interact. Feature interactions has been well studied in telecommunication domain [9, 25, 27]. Interactions between Web services have also been explored by Weiss et al. [28, 41, 42] in recent years. But only few researchers investigated crossing domain feature interaction handling. Researchers from Avaya Labs Research proposed to use the Event-Condition-Action (ECA) model to define both web features and telecommunication features [44]. The model is defined as below:

Based on the ECA model, we distinguish Communication Widgets (CW) and Communication Enabled Widgets (CEW), and introduced a centralized widget coordinator to handle feature interactions among communication related web widgets. The major difference between CWs and CEWs is that CWs can perform communication actions, such as transfering a call or hold-

feature ::= (trigger, pre-cond, action, post-cond)
where:
  pre-cond ::= (states, action parameters)
  action ::= f(trigger, action parameters)
  post-cond ::= (new triggers, new states, affected values)

Figure 6.9 Event-Condition-Action (ECA) model.

ing a call, while CEWs cannot directly change call states. We then proposed a two-step FI detection mechanism. It handles FIs among CWs first by using existing FI detection algorithms in telecommunication domain, such as Distributed Feature Composition (DFC) [24] and gets a dynamic sequence of CWs. The CEWs are then inserted into the CW sequence and we can then use graph theory to identify loops or routing branches. The loop or routing branches may introduce new feature interactions.

6.7.3 Crossing Domain Security, Privacy, and Availability

Allowing information crossing telecommunication domain and web domain may impose security problems for applications in both domains on all security aspects, including identity management, privacy and data confidentiality handling, authentication, data integrity, non-repudiation checking, access control, and availability management.

- *Identity management*: People may use different identities in different domains. Usually, people use email or user name in web domain, while telephone number in telecommunication domain. Meanwhile, the mapping between telephone number and user name or email may not be a one-to-one mapping. This many-to-many mapping may introduce security or privacy holes. For example, if a telephone number belongs to multiple users, the context information of an incoming call may show information of all the users, which may not be desirable.
- *Privacy and data confidentiality handling*: Privacy is commonly referred to the right of individuals to control what information should be collected, stored, and released. People usually use policy-based means (e.g., black list, white list, file permissions) or technical means (e.g., cryptography) to ensure their privacy. When services crossing telecommunication and web domains, both policy-based means or technical means may be compromised (as the example shown in identity management)

or unavailable (e.g., the encryption key shared in telecommunication domain may not be available in web domain).

- *Authentication and authorization (access control)*: Authentication is to prove the claimed identity of an entity is true. Authorization is to allow access to certain information. Both authentication and authorization are related to identity management. The goal of authentication and authorization management is to largely provide convenience to users (e.g., users authenticated in one domain should not be authenticated again in the other domain) while still hold the security and privacy boundaries (e.g., a team member with his manager's web domain privilege for document sharing should not have his manager's telecommunication domain's privilege to make international calls).

- *Data integrity*: Data integrity is to ensure that data have not been altered by unauthorized entities. For example, when a telecommunication widget acquires call information (e.g., using HTML 5 Web Socket or Ajax long polling) and relay the information to other widgets (e.g., using HTML 5 Web Messaging), the telecommunication widget should put a signature on the information.

- *Non-repudiation checking*: Non-repudiation is to prevent users from denying what they have performed, such as making or receiving calls, joining a web conference, and sending or receiving messages. The integration of Web 2.0 with telecommunication services can in fact facilitate non-repudiation checking. In the example we showed in Section 6.3, the PhoneMash system can perform context-based search and present related conversations to users. This can help to find the proof for non-repudiation checking.

- *Availability*: Availability means authorized access and communication actions should be performed in a timely manner. People usually have different delay expectation in web domain and telecommunication domain. For call setup delay, E.721 [40] recommends an average delay of no more than 3.0 seconds, 5.0 seconds, and 8.0 seconds, for local, toll, and international calls, respectively. For post-pickup delay, E.721 recommends an average delay of 0.75 seconds for local, 1.5 seconds for toll and 2.0 seconds for international calls. For web pages, studies [15] show that users' intention, attitude, and performance will quickly decrease as delays increased above 8 seconds. While handling crossing domain messages may increase the delay, the delay should not exceed user expected values.

There are still many other challenges we need to face for integrating communication services with Web 2.0, such as crossing domain media transmission. We will not discuss all these challenges. As evolution or revolution happens in both Web and telecommunication domain, new challenges or new solutions will keep emerging and provide better services to users.

6.8 Conclusion

In summary, this chapter discusses different aspects of integrating Web 2.0 and communication services. The Web 2.0's SLATES (Search, Links, Authoring, Tags, Extensions, and Signals) features and techniques can greatly enrich people's communication experience, while communication services can also enhance Web 2.0 sites' capabilities. We also discussed the enabling technologies and three main challenges for the integration. As new user requirements and new technologies emerge, the convergence of web domain and communication domain will provide more value to users as well as service providers. A foreseeable trend is to bring semantic information and augmented reality into the converged services. Technologies such as semantic web [7] (which is considered as the pillar for Web 3.0 by some researchers) and many online virtual worlds have already been used by people. I believe that in the near future, we will be able to discuss Web 3.0 and communication services, and hopefully our discussion in this chapter will still hold.

References

[1] The 3rd Generation Partnership Project (3GPP), http://www.3gpp.org.
[2] Facebook, http://www.facebook.com.
[3] Google wave, http://wave.google.com.
[4] LBS usage survey, http://blog.compete.com/2009/06/02/location-based-services-applications-carriers-advertisers/.
[5] Mobile location based services applications, forecasts & opportunities 2010–2014, https://www.juniperresearch.com/reports/mobile_location_based_services.
[6] Real-time conversation streams in google wave powered by ribbit, http://www.ribbit.com/wave/.
[7] Semantic web, http://semanticweb.org/.
[8] Twiliobot, click-to-call bot for google wave using the twilio API, http://code.google.com/p/twiliobot/.
[9] A. Alamri, M. Eid, and A. El Saddik. Classification of the state-of-the-art dynamic web services composition techniques. *International Journal on Web and Grid Services*, 2(2):148–166, 2006.

[10] G. Armitage, B. Carpenter, A. Casati, J. Crowcroft, J. Halpern, B. Kumar, and J. Schnizlein. A delay bound alternative revision of RFC 2598. RFC 3248 (Informational), March 2002.

[11] ComScore Whitepaper. The 2009 U.S. digital year in review.

[12] Mozilla developer center. Same origin policy for javascript, https://developer.mozilla.org/En/Same_origin_policy_for_JavaScript, November 2009.

[13] W3C Working Draft. Xmlhttprequest, http://www.w3.org/TR/XMLHttpRequest/, November 2009.

[14] Roy T. Fielding and Richard N. Taylor. Principled design of the modern web architecture. In *ACM Transactions on Internet Technology (TOIT)*, May 2002.

[15] Dennis F. Galletta, Raymond Henry, Scott Mccoy, and Peter Polak. Web site delays: How tolerant are users? *Journal of AIS*, 5:1–28, March 2003.

[16] Jesse James Garrett. Ajax: A new approach to web applications, http://www.adaptivepath.com/ideas/essays/archives/000385.php, February 2005.

[17] Ian Hickson. HTML microdata, http://dev.w3.org/html5/md/, June 2010.

[18] Ian Hickson. HTML5 web messaging, http://dev.w3.org/html5/postmsg/, June 2010.

[19] Ian Hickson. Server-sent events, http://dev.w3.org/html5/eventsource/, June 2010.

[20] Ian Hickson. The websocket API, http://dev.w3.org/html5/websockets/, June 2010.

[21] Avaya Inc. Communication-enabling social networks, http://images.tmcnet.com/expo/west-08/presentations/cd214-alperin-avaya.pdf.

[22] Avaya Inc. Echoes: A collaborative tagging system for conversations in the enterprise, http://support.avaya.com/css/P8/documents/100076129.

[23] ECMA international. Computer supported telecommunications applications (CSTA), http://www.ecma-international.org/activities/Communications/TG11/cstaIII.htm, September 2007.

[24] Michael Jackson and Pamela Zave. Distributed feature composition: A virtual architecture for telecommunications services. August 1998.

[25] N. Milanovic and M. Malek. Current solutions for web service composition. *IEEE Transaction of Internet Computing*, 8(6):51–59, December 2004.

[26] C. Mohit, X. Wu, and V. Krishnaswamy. Integrating enterprise communications into google wave. In *Proceedings of IEEE Consumer Communications and Networking Conference (CCNC'10)*, January 2010.

[27] M. Mrissa, D. Benslimane, Z. Maamar, and C. Ghedira. Towards a semantic- and context-based approach for composing web services. *International Journal of Web and Grid Services*, 1(3–4):268–286, 2005.

[28] M. Weiss, A. Oreshkin, and B. Esfandiari. Method for detecting functional feature interactions of web services. *Journal of Computer Systems Science and Engineering*, 21(4):273–284, 2006.

[29] W3C Recommendation. Document object model (DOM) level 3 core specification, http://www.w3.org/TR/2004/REC-DOM-Level-3-Core-20040407/, April 2004.

[30] W3C Recommendation. Simple object access protocol (SOAP) version 1.2, http://www.w3.org/TR/soap/, April 2007.

[31] W3C Recommendation. Xhtml 1.0 the extensible hypertext markup language (second edition), http://www.w3.org/TR/xhtml1/, August 2002.

[32] W3C Recommendation. Voice extensible markup language (voicexml) version 2.0, http://www.w3.org/TR/voicexml20/, March 2004..

[33] W3C Candidate Recommendation. Cascading style sheets level 2 revision 1 (CSS 2.1) specification, http://www.w3.org/TR/CSS2/, September 2009.

[34] A. B. Roach. Session Initiation Protocol (SIP)-Specific Event Notification. RFC 3265 (Proposed Standard), June 2002. Updated by RFCs 5367, 5727.

[35] J. Rosenberg, H. Schulzrinne, G. Camarillo, A. Johnston, J. Peterson, R. Sparks, M. Handley, and E. Schooler. SIP: Session Initiation Protocol. RFC 3261 (Proposed Standard), June 2002. Updated by RFCs 3265, 3853, 4320, 4916, 5393, 5621, 5626, 5630, 5922.

[36] J. Rosenberg, H. Schulzrinne, and P. Kyzivat. Indicating User Agent Capabilities in the Session Initiation Protocol (SIP). RFC 3840 (Proposed Standard), August 2004.

[37] H. Schulzrinne, V. Gurbani, P. Kyzivat, and J. Rosenberg. RPID: Rich Presence Extensions to the Presence Information Data Format (PIDF). RFC 4480 (Proposed Standard), July 2006.

[38] 3GPP Specification. Ip multimedia subsystem (IMS); stage 2. http://www.3gpp.org/ftp/Specs/html-info/23228.htm.

[39] OASIS Standard. Web services for remote portlets specification v2.0, http://docs.oasis-open.org/wsrp/v2/wsrp-2.0-spec-os-01.html, April 2008.

[40] International Telecommunication Union. Network grade of service parameters and target values for circuit-switched services in the evolving ISDN, recommendation E.721. In *Telecommunication Standardization Sector of ITU*, Geneva, Switzerland, May 1999.

[41] M. Weiss and B. Esfandiari. On feature interactions among web services. *International Journal on Web Services Research*, 2(4):21–45, 2005.

[42] M. Weiss, B. Esfandiari, and Y. Luo. Towards a classification of web service feature interactions. *Computer Networks*, 51(2):359–381, February 2007.

[43] Wikipedia. Web 2.0. http://en.wikipedia.org/wiki/Web_2.0.

[44] X. Wu, J. Buford, K. Dhara, M. Kolberg, and V. Krishnaswamy. Feature interactions between internet services and telecommunication services. In *Proceedings of IPTComm'09*, July 2009.

[45] X. Wu, K. Dhara, and V. Krishnaswamy. Providing content aware enterprise communication services. In *Proceedings of Principles, Systems and Applications of IP Telecommunications (IPTComm'08)*, July 2008.

[46] X. Wu and V. Krishnaswamy. Widgetizing communication services. In *Proceedings of IEEE International Communication Conference (ICC'10)*, May 2010.

7

Self-Organizing IP Multimedia Subsystem

Ashutosh Dutta[1], Christian Makaya[1], Subir Das[1], Dana Chee[1],
Fuchun Joseph Lin[1], Satoshi Komorita[2], Tsunehiko Chiba[2],
Hidetoshi Yokota[2] and Henning Schulzrinne[3]

[1]*Telcordia Technologies Inc., 1 Telcordia Drive, Piscataway, NJ 08854, USA;
e-mail: ashutosh.dutta@ieee.org*
[2]*KDDI R&D Laboratories, Japan*
[3]*Columbia University, New York, NY 10027, USA*

Abstract

While there have been tremendous efforts to develop the architecture and protocols to support advanced Internet-based services over 3G and 4G networks,
IMS is far from being deployed in wide scale. Effort to create an operator controlled signaling infrastructure using IP-based protocols has resulted in a large
number of functional components and interactions among those components.
Thus, the carriers are trying to explore alternative ways to deploy IMS that
will allow them to manage their network in a cost effective manner while offering the value-added services. One of such approaches is self-organization
of IMS. The self-organizing IMS can enable the IMS functional components
and corresponding nodes to adapt them dynamically based on the features like
network load, number of users and available system resources. This chapter
introduces such a self-organizing and adaptive IMS architecture, describes
the advanced functions and demonstrates the initial results from the prototype
test-bed. In particular, we show how all IMS functional components can be
merged and split among different nodes as the network demand and environment change without disrupting the ongoing sessions or calls. Although it is
too early to conclude the effectiveness of self-organizing IMS, initial results

*Anand R. Prasad et al. (Eds.), Advances in Next Generation Services and Service
Architectures,* 123–147.

are encouraging and it may provide additional incentives to the operators for network evolution.

Keywords: IMS, self-organization, load balancing, adaptivity, context aware.

7.1 Introduction

The current version of IMS (IP Multimedia Subsystem) has several short-comings that may act as deterrent for wider deployment. Depending upon the architecture there may be a need to have one-to-one mapping between functional component and a physical node where each node is equipped only to perform a certain function. This would require the deployment of a large number of network nodes which in turn will make the network management and operation difficult, in particular, the environment where dynamic adaptation is required based on the functionalities, processing, network loads and node failures. One way to circumvent these issues is to allocate redundant network resources, but that will not help the service providers to achieve the goal of offering services at a lower cost. Alternatively, one can define new techniques that support dynamic adaptation with the nodes merging and splitting the IMS functional components. By node merging, we mean all the IMS functional components merge and operate on a single physical node; whereas, by node splitting we mean the functional components get distributed across different physical nodes. In its current form, IMS architecture and protocols do not have the mechanisms that can easily help IMS functionalities (e.g., P-CSCF, I-CSCF, S-CSCF) to easily migrate from one node to another. Thus, it needs additional features and mechanisms to support self-organizing capability. These additional features will enable the IMS network to adapt and distribute the functionalities based on the network condition and operational environment. Currently, IEEE NGSON (Next Generation Service Overlay Network) [9], ATIS SON [2] and ITU-T SG13 [18] are working on defining the requirements that would enable the IMS to support several features such as self-organization, context awareness, and peer-to-peer services and will help the service providers to better manage their IMS network while reducing the cost of operations.

This chapter proposes and develops methodologies that are needed to support several self-organizing features of IMS. These methodologies can provide additional flexibilities to the current IMS architecture and help service providers to better manage the IMS components in a dynamic environment where nodes or links are prone to failure and the system grows or sinks with the demand. This chapter also describes the laboratory prototype test-bed and provides some initial results to demonstrate the feasibility of such a system.

The rest of the chapter is organized as follows. We introduce related work and standards efforts in Section 7.2. Section 7.3 introduces the basic IMS architecture and then describes the self-organizing features and the associated functions that are needed to support this architecture. Section 7.4 describes the operations of different function nodes in the self-organizing architecture. Section 7.5 describes two possible deployment scenarios and the call flows associated with these scenarios. Section 7.6 describes the test-bed proto-type for self-organizing test-bed and the load balancer. Finally, Section 7.7 concludes the chapter.

7.2 Related Work

The concept of self-organizing IMS networks is relatively a new topic and has not been widely studied. It is important to understand that the self-organizing IMS is different than P2P-SIP concept [12] where significant research results are available. Bessis [3] describes performance analysis and benefits of running multiple SIP servers on the same host. That paper shows how to design the IMS networks in order to maximize IMS server co-location and explains which types of SIP calls can benefit from the co-location of IMS servers. Fabini et al. [6] describe a minimal optimal IMS configuration with respect to architecture and QoS aspects. A virtual IMS test-bed (i.e., any IMS component is assigned its own virtual host), with different domains has been setup on one physical machine. It demonstrates the feasibility of an IMS system implementation within a single device (all-in-one). Matus et al. [13] propose a distributed IMS architecture by representing network functional elements in Distributed Hash Tables (DHT) overlay networks. The main focus was to distribute S-CSCF, I-CSCF and HSS functionalities by using an overlay network where these functionalities are merged in one node (called IMS DHT). Manzalini et al. [11] describe a platform to provide auto-nomic and situation-aware communication services. However, none of these papers have looked into methods of supporting self-organizing capabilities of IMS. Furthermore, these papers did not take into account reconfiguration of the functions when, for example nodes fail or for load balancing reasons. Most recently, the Next Generation Service Overlay Network (NGSON) [9], a standards activity launched by the IEEE Communications Society in March

2008, is specifying a framework for service overlay networks. This framework will include context-aware, dynamically adaptive, and self-organizing networking capabilities including advanced routing and forwarding schemes that are independent of underlying transport networks.

This chapter works around the basic concept of NGSON that is based on self-organizing IMS and has drawn some inspiration from [13]. In particular, this chapter develops the mechanisms and framework to facilitate self-organizing capabilities such as self-configuration, self-optimization and self-recovery at the service layer.

7.3 IMS Overview

At a functional level, IMS architecture primarily consists of several signaling entities such as P-CSCF (Proxy-Call Session Control Function), I-CSCF (Interrogating CSCF), S-CSCF (Serving CSCF), and HSS (Home Subscriber Server). Figure 7.1 shows one such functional architecture where P-CSCFs are distributed across the visited networks. In this specific architecture, there are four networks labeled as Home Network, Visited Network 1, Visited Network 2 and Visited Network 3. Two of the networks have CDMA2000 access while the third network supports either WiFi or WiMAX. Details of this architecture can be found in [5]. We provide a description of some of the functional components.

MN: MN is the mobile node that moves across the networks.

DHCP Servers: DHCP servers are configuration agents and help the mobiles with the configuration of network layer identifiers such as IP address, address of P-CSCF, and DNS servers. Each visited network may be equipped with a DHCP server.

HA: HA serves as the anchor point for a mobile in mobile IP environment and maintains the mapping between the mobile's home address and the new care-of-address that the mobile obtains in each network. This care-of-address is obtained either from a stateful DHCP server, or from a foreign agent or by means of stateless auto-configuration.

S-CSCF: The S-CSCF is the central node of the signaling plane. It is a SIP server and performs session control functions. It is always located in the home network. It can either query the HSS or the DNS server to locate the appropriate P-CSCF for outgoing intra-domain calls and the appropriate I-CSCF for inter-domain calls.

P-CSCF: The P-CSCF behaves as a SIP proxy and is the first outbound proxy for a mobile in the visited network. The P-CSCF routes REGISTER

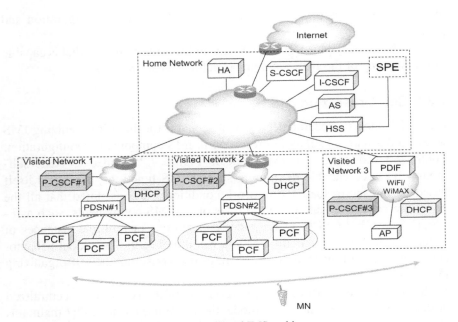

Figure 7.1 A distributed IMS architecture.

requests to the I-CSCF and caches the S-CSCF address so that it can route the rest of SIP signals directly to the S-CSCF.

I-CSCF: The I-CSCF is another SIP proxy that provides forwarding of messages to the correct S-CSCF, through HSS or DNS look-ups. I-CSCF also acts as an entity that hides information for inter-domain calls.

HSS: The HSS stores information about the subscribers, their addresses, and their services such as user account, contact URI of the user, address of P-CSCF for each mobile, E.164 number. The HSS is located in the home network and communicates with the S-CSCF.

Service Provisioning Environment (SPE): The SPE is implemented in combination with the HSS, but logically it is a separate component. It provides a mechanism to view the services that are deployed on the AS and system administrators can use it to provision services automatically for each user.

Application Server (AS): The AS sends messages to the HSS defining the applications deployed on the AS, along with the parameters needed to configure an instance of the service. This description is used by the SPE to generate the system administrators screens for service configuration and provisioning.

IMS Node: Any physical node within an IMS infrastructure that is capable of running any of the IMS functions is defined as an IMS node.

7.3.1 Self-Organizing IMS

Figure 7.2 shows a specific deployment scenario for the self-organizing IMS where nodes and links may fail resulting in IMS component reconfiguration. Figure 7.3 shows how an IMS node in Figure 7.2, termed as basic self-organizing node capable of running several IMS functions, can adapt itself based on the server load and network conditions. It is assumed that all the nodes are capable of taking on the roles of the IMS components, such as P-CSCF, S-CSCF, I-CSCF. These nodes can run one or multiple instances of different IMS functions. In this section, we introduce the functional behavior and interactions of different logical entities that constitute the self-organizing IMS architecture.

Self-organizing IMS can be based on one of the two modes: centralized or distributed. In the centralized mode, there is a master node that maintains a database with operator policy and state information for all nodes under its control. For example, this master node can be the HSS (Home Subscriber Server) as specified in [15]. The reason behind this choice could be that HSS is the master database of the cellular network and the availability and reliability of such database is much higher than ordinary nodes.

The master node database is updated when (i) a new node notifies the master node about its capabilities, (ii) a specific IMS role has been assigned to the node, and (iii) a node changes roles due to overloading or failure. An efficient policy-based mechanism should be defined in the master node to assign the functionalities or roles.

In the distributed mode of operation, the new node announces its presence through a multicast message. Existing IMS nodes reply to this message if they need to transfer some of their current functionalities. This may result in conflicts but it can be resolved by a simple rule such as, first in first out (FIFO) based on response message. If a node receives no responses, it should assume that there exists no other node and it takes over all roles. In either the centralized or distributed mode, negotiation or capabilities exchange and event notification protocols should be defined. In fact, the decision to assign a role to a new node will be based on its capabilities (e.g., CPU, load, processing power, memory). We focus on the centralized mode in this chapter.

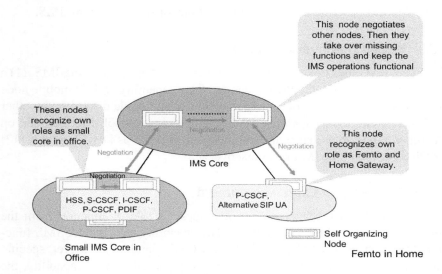

Figure 7.2 Applicability of self-organizing IMS.

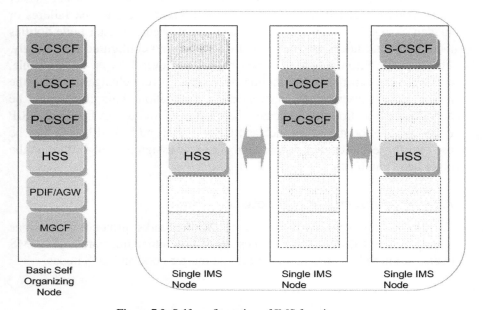

Figure 7.3 Self-configuration of IMS functions.

Some of these functionalities can be easily applied to a distributed mode as well. We next describe the important functions of self-organizing IMS.

7.3.2 Nodes Discovery Function

There are two kinds of node discovery function in Self-Organizing IMS: (1) a master node discovers the IMS nodes and their capability and (2) mobile node discovers the IMS node for registration. The master node discovers the other IMS nodes and their capability via a negotiation protocol. Mobile node on the other hand can use standardized mechanisms such as DHCP options [17] or DNS query [7] for IMS node discovery.

7.3.3 Role Request and Assignment

When an IMS node comes online, it requests a role or function from the master node by sending its capability. The capability may include IMS functionality such as P-CSCF, S-CSCF or I-CSCF, and other server specific capability such as processing power, memory. Based on this capability, the master node assigns a given role to the new node. The role assignment can happen due to load balancing, node failure and other node management purposes that can be governed by high level policies configured at the master node. Alternatively, it can be network event-based, such as link failures or sudden change in bandwidth. When such event occurs, master node notifies the appropriate nodes and the network will be auto-configured accordingly. The network auto-configuration can happen in several ways. For example, an IMS node can notify the mobile node about a role change provided the mobile node subscribes to an event after registration. This event could be defined as a function that indicates the change of role for an IMS node. SIP Event notification messages such as SUBSCRIBE and NOTIFY [16] may be used for this event change notification. The role assignment algorithm on the other hand can be implementation specific.

7.3.4 Protocol Interaction between Nodes

To allow deployment of self-organizing IMS networks, protocol interactions among different IMS nodes are necessary. These interactions are among IMS nodes and mobile nodes, and between the master node and IMS nodes.

7.3.5 Node Monitoring

To allow the master node to determine the functional behavior of all other nodes in the network, a periodic message (heartbeat) will be sent to all known nodes. If a response is not forthcoming within a specific time, the given node is marked as troubled or failed, and the master node will distribute its functionality to the remaining nodes dynamically.

7.4 Self-Organizing IMS Node Functions

In this section, we describe the functionality supported by each node involved in self-organizing IMS network.

7.4.1 Operations of the Master Node

The master node is the main component of the centralized approach. In addition to assigning roles to nodes, it has other functions depending on the type of event. As an example, in Figure 7.4, if the P-CSCF (e.g., P1) role changes, the master node notifies all S-CSCF nodes and provides them the information about the new P-CSCF. When the S-CSCF receives this notification, it establishes a list of mobile nodes assigned to this P-CSCF and notifies them by way of the new P-CSCF (e.g., P2) (Receive Role Change). Mobile node (MN) and User Equipment (UE) are used interchangeably and have the same meaning in this chapter. The notification message sent by S-CSCF to the mobile nodes contains information about the new P-CSCF. Upon receipt of notification from the S-CSCF, mobile nodes re-register to the new P-CSCF and re-subscribe to event state change for future changes.

On the other hand, if the S-CSCF role changes, the master node notifies all P-CSCF followed by notification of mobile nodes. The notification message includes information about the new S-CSCF. Upon receipt of this notification message, the mobile node re-registers and subscribes to event state change. To handle load balancing, the master node may request each of the IMS components to support a given percentage of mobile nodes previously attached to the IMS node with role change functionality.

7.4.2 Operations of Mobile Node

The operation of the mobile node is based on SUBSCRIBE and NOTIFY methods. The mobile node should be able to receive and process the notification message (NOTIFY) received either from S-CSCF when P-CSCF

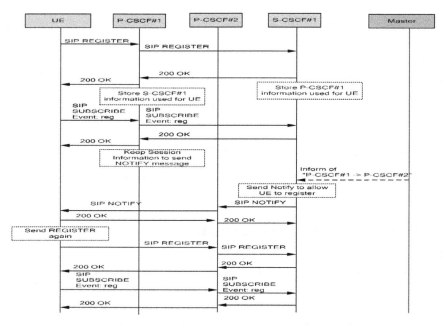

Figure 7.4 Role of master node during P-CSCF change.

role changes or from P-CSCF if S-CSCF role changes. Upon receipt of this NOTIFY message, the mobile node uses such information to register and subscribe again to event change. Mobile node's ongoing session including the QoS of the media is not affected since the access network that the mobile node is attached to is not changing. However, if the mobile node needs to establish a new session, it should include the Replace-header field in the INVITE message or use re-INVITE message.

7.4.3 Operation of the P-CSCF

To allow self-organizing IMS networks, P-CSCF components should be able to support the following additional operations:

- Receive notification of role change from the master node;
- Process (proxy) and store mobile node's subscription information during SUBSCRIBE operation;
- Send NOTIFY message with information about the new S-CSCF;

- Retrieve all mobile nodes registered to S-CSCF upon receipt of event change from the master node;
- Process mobile node's re-registration request after a role change event.

7.4.4 Operation of the S-CSCF

To allow self-organizing IMS networks, S-CSCF components should be able to support the following additional operations:

- Receive notification of role change from the master node;
- Store mobile node's subscription and profile information;
- Send NOTIFY message with information about the new P-CSCF;
- Retrieve all mobile nodes registered to the P-CSCF upon receipt of event change from the master node;
- Update HSS with the mobile node information after the role change.

7.5 Self-Organizing Deployment Scenarios

In this section, we first illustrate a few migration scenarios of functional components for self-organizing IMS network and then describe the associated call flows and proof-of-concept from a prototype test-bed. Ideally, self-organizing IMS should be able to support all kinds of reconfiguration in the network as shown in Figure 7.5. For example, Scenario 1 in Figure 7.5 shows that all

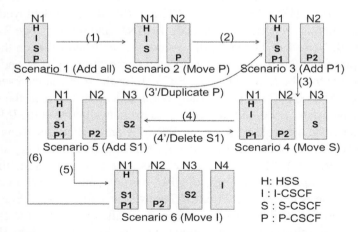

Figure 7.5 Possible re-configuration scenarios for IMS nodes.

the IMS components are running in one node, N1. Scenario 2 shows that P-CSCF functionality is migrated to node N2. Scenario 3 corresponds to a case when the P-CSCF functionality gets split into two nodes, P1 and P2. Scenario 4 depicts when the S-CSCF functionality is relocated to a different node. Scenario 5 corresponds to a case when the S-CSCF functionality gets split into two nodes S1 and S2. Scenario 6 shows I-CSCF's functionality getting migrated to a new node, and finally, the transition step (6) shows the merging functionality where all the functions merge with a single node. It is important to note that our goal is not to affect the end users (whether they have an ongoing session or initiate a new session) due to these changes in the configuration within the core of the network. In Figure 7.5, H stands for HSS, I-CSCF is termed as I, S-CSCF and P-CSCF are denoted as S and P, respectively.

We discuss two types of self-configuration scenarios: Scenario (i) where mobile node is aware of reconfiguration and participates during self-organizing process and Scenario (ii) where mobile node is not aware of self-organizing IMS and does not participate during the reconfiguration process.

7.5.1 Self-Organizing with UE Involved Case

In this section, we illustrate several self-organizing IMS scenarios that could be applicable to both UE involved and UE non-involved cases.

Figure 7.6 shows the call flows for Scenario 1, where all SIP-based IMS components (i.e., HSS, S-CSCF, I-CSCF, and P-CSCF) are deployed in the same node. In this scenario, mobile node 1 (MN1) and mobile node 2 (MN2) register with Node 1 (N1) and a call session is established between them. To allow deployment of self-organizing IMS networks, after registration, mobile nodes must subscribe to S-CSCF and P-CSCF for the role change event.

We use MESSAGE method for any communication between the mobile node and the master node. Initial capabilities exchange between the IMS node and master node and role assignment functions are carried out using MESSAGE method. We show below a sample MESSAGE method [1] showing "Role Request" and "Role Assignment". As part of role request, the IMS node offers its capabilities in terms of CPU, memory, and network load. Role Query response is not shown. Figure 7.7 shows the call flows for Scenario 2 as depicted in Figure 7.5, which corresponds to the case when the P-CSCF functionality is relocated in a new node (i.e., Node 2 (N2)) while HSS, S-CSCF, and I-CSCF remain in Node 1. In this scenario, when Node 2 comes

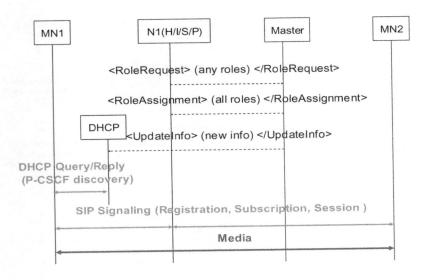

Figure 7.6 Call flow for Scenario 1.

online, it exchanges its capability information with the master node and it is found that Node 2 has the ability of serving as P-CSCF.

Hence, the master node decides to transfer or move the P-CSCF functionality to Node 2 which acts now as P-CSCF for both mobile node 1 and mobile node 2. The procedure is described as follows:

- The old IMS Node (N1) is providing the IMS functionalities for the Mobile Node.
- When a new IMS Node (N2) comes online, it sends a request to the master node for role assignment. According to the capabilities provided in the role request response message, the master node decides the role assignment to N2 governed by the operator policy. The master node provides information about the current IMS entity if the role has been already assigned to other IMS nodes.
- Upon deciding on the role assignment, the master node notifies the S-CSCF (N1/S) about role assignment changes and provides information about the new IMS Node (N2/P). The S-CSCF will notify all mobile nodes registered to the previous P-CSCF.
- The new IMS Node (N2/P) retrieves information (state or context) from the old IMS Node (N1).

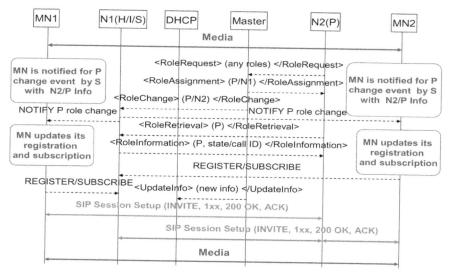

Figure 7.7 Call flow for P-CSCF relocation (Scenario 2).

- In order to allow new mobile nodes to discover the correct P-CSCF, the master node updates the DHCP server configuration.

Scenario 3 corresponds to the situation when P-CSCF functionality gets split among two nodes (P-CSCF1 in Node 1 and P-CSCF2 in Node 2). In this scenario, the master node decides to assign P-CSCF role to Node 1 while earlier P-CSCF role is located in Node 2 (from Scenario 2). Mobile node 1 is assigned to Node 1 as its P-CSCF (i.e., P-CSCF1) while mobile node 2 is assigned to Node 2 as its P-CSCF (i.e., P-CSCF2). After these steps, a call session is established between mobile node 1 and mobile node 2. Note that the P-CSCF functionality splits happen after the initial registration of mobile node 1 and mobile node 2. In other words, mobile node 1 and mobile node 2 were associated to the same P-CSCF (Node 2).

Figure 7.8 shows the call flows for Scenario 4 that corresponds to the situation when S-CSCF functionality is relocated in a new node (i.e., Node 3). In this scenario, when Node 3 comes online, it exchanges its capability information with the master node and it is found that Node 3 is suitable for serving as S-CSCF. Then, S-CSCF functionality of Node 1 is transferred to Node 3 and Node 1 (i.e., P-CSCF, HSS or I-CSCF) is notified for this change.

Transition step (6) depicts a situation when IMS main nodes (P-CSCF, S-CSCF) functionalities move back to one node due to failure of other nodes

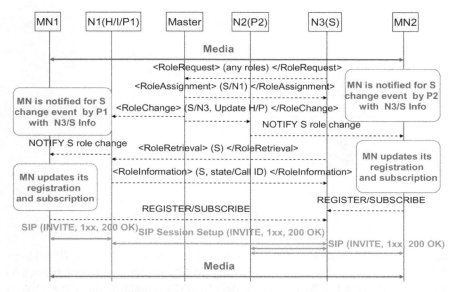

Figure 7.8 Call flow for S-CSCF relocation (Scenario 4).

or load balancing reasons. The master node detects this state change or event and restores all functionalities previously supported by each IMS node to Node 1. Mobile node 1 and mobile node 2 re-register with Node 1 again. Before making such merging decision in one node, the master node needs to consider several factors such as time to restore the functionalities and service disruption.

7.5.2 UE Non-involved Scenarios

In this section, we describe the techniques needed to support UE non-involved scenario. UE non-involved method requires usage of the Load Balancer (LB) in order to hide any change in the core IMS network. With the increasing penetration rate of real-time application such as VoIP, it is necessary to have load balancing support at SIP level since IP load balancing cannot be used for SIP-based application. In fact, with IP load balancing technology, the LB is not able to add SIP headers such as Via, Route and Record-Route in the SIP messages. This information is required in SIP message to allow adequate routing and session activation.

In the proposed solution, the LB appears as a virtual P-CSCF to UEs. In other words, from the UE's perspective, the LB is the P-CSCF and

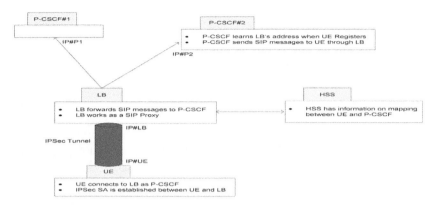

Figure 7.9 SIP load balancing in IMS architecture.

the UE sends all request or SIP messages to the LB's virtual IP address (VIP). Moreover, rather than establishing an IPSec security association (SA) between UEs and P-CSCF as specified by IMS standard, the IPSec SA is established between UEs and LB. Figure 7.9 illustrates our proposed SIP-based load balancing procedure for IMS networks. When the LB receives SIP messages from the UE, it forwards these messages to the selected P-CSCF. The selection of P-CSCF is done by a scheduling algorithm defined in LB and by using information provided by the master node or HSS. There is no direct communication between UEs and the real P-CSCF since all communications are handled by the LB. In order to correctly route the SIP packets to P-CSCFs and maintain the session persistence, the LB needs to intercept the SIP packets and modify the headers accordingly. For these reasons, the LB works as a SIP Proxy and processes SIP messages received from/to UE and the real IMS node (e.g., P-CSCF). Having a SIP-aware LB, SIP headers (i.e., Via, Record-Route and Route) in SIP messages will be set appropriately by the LB.

7.5.2.1 Registration Procedure

When UE initiates registration procedure, it sends a SIP REGISTER request to the LB. Since LB is acting as an outbound SIP proxy, it processes this message and adds itself in (the topmost) Via and Record-Route headers of SIP REGISTER message before forwarding the message to the selected P-CSCF. By adding its information to the Via header, the LB will receive the response to the request. In the opposite direction, the LB removes the Via

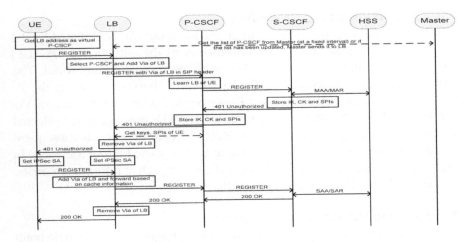

Figure 7.10 SIP registration procedure.

header which contains its information before to send the message to the UE. The selection of P-CSCF is based on different scheduling algorithms, such as hash over Call-ID, hash over From URI, or Round-Robin. In our prototype, the selection and session persistence are based on two methods: hash over Call-ID and hash over From URI since they are more consistent and provide better session persistence. Figure 7.10 shows the SIP registration procedure for UE non-involved case. For sake of simplicity, only main registration steps are illustrated. For example, interaction with I-CSCF has been ignored. The LB gets a list of P-CSCF from the master node. Since the master node has knowledge of IMS nodes, getting the list of P-CSCFs from the master node allows a reduction of signaling due to IMS node monitoring by LB. The remaining registration procedure process follows the IMS specification. The LB retrieves security association parameters (SPIs), integrity (IK) and confidentiality keys (CK) from the P-CSCF.

With all security credentials, the UE will compute the response to the challenge and establish an IPSec SA with the LB before it can send the second SIP REGISTER message. The LB then forwards this message to the previously selected P-CSCF based on the session persistence approach. Similarly, to the processing of initial registration message, the LB adds/removes its information in the Via header.

7.5.2.2 Session Setup Procedure

After a successful registration, a call or session setup can be initiated by the UE. The UE pre-loads the stored information of the outbound proxy (i.e., LB) into Route header of SIP INVITE message before sending it out. The LB uses the same procedure as for REGISTER message to select the P-CSCF, removes its own entry from Route header, adds it own entry in Via and Record-Route headers and forwards the INVITE message to the selected P-CSCF. By adding its own entry in Record-Route header, all subsequent requests within the established SIP dialog will be routed through the LB. In order to guarantee session persistence, LB can use cached information on the previous selected P-CSCF for a given UE.

7.5.2.3 Load Balancer Failure Support

Since the load balancer is a main anchor point, it might become a single point of failure for the SOIMS architecture. To avoid the whole system from being out of service due to the LB failure, a backup of LB might be deployed for redundancy. The heartbeat technique or Virtual Router Redundancy Protocol (VRRP) [8] can be used between the primary and the backup LB and they communicate periodically to inform each other that they are still alive. The primary and backup LBs are synchronized in order to share the ongoing session information (e.g., SIP dialog, list of P-CSCFs).

7.5.2.4 Session Continuity with P-CSCF Failover Support

Service provisioning or session continuity is one of the main requirements for real-time application, for example when the UE roams from one access network to another or in presence of IMS node failure. In this section, we will describe session continuity when IMS role change event (e.g., IMS node failure, role assignment) occurs during ongoing session. Let us assume that the P-CSCF's (e.g., P-CSCF#1 in Figure 7.11) role changes due to failure, the LB will be notified by the master node about network configuration change or since LB gets list of P-CSCFs from the master node, it will discover role assignment change or failure. When the master node detects failure of P-CSCF#1, it notifies the S-CSCF about this event. Upon this notification, the S-CSCF can retrieve information of the new P-CSCF (e.g., P-CSCF#2) from the HSS. At the same time, the S-CSCF updates registration status (e.g., association and mapping) of LB and UE through P-CSCF#2. Then P-CSCF#2 can restore registration information and update mapping between LB and UE for subsequent SIP messages. After restoration of registration information, the S-CSCF sends a message to P-CSCF#2 with information about media nego-

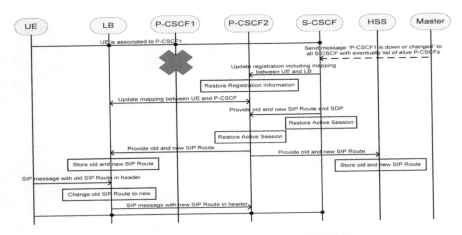

Figure 7.11   Session continuity during P-CSCF failure.

tiation (i.e., SDP), new and old SIP Route. This information exchange allows restoration of the ongoing SIP session state in the S-CSCF and P-CSCF#2, and reconfiguration of IMS core network. When P-CSCF#2 completes the update of session information, it informs the LB and HSS about the new SIP Route. The old and new SIP Routes are stored in the LB and HSS to allow mapping of previously established SIP dialog with P-CSCF#2. All of these changes are transparent to the UEs. In fact, the LB hides the change and reconfiguration of IMS core network and UEs have no direct communication with IMS components. Any subsequent SIP message sent by the UE will be sent with the old SIP Route. It is the LB's responsibility to change the old SIP Route (e.g., Service-Route) in SIP message to the new SIP Route before to forward this message to the first IMS entity (i.e., P-CSCF). Since any change in the core IMS network is transparent to UEs, with the proposed solution, there is no need for UEs to subscribe for events notification. In other words, there is no need to exchange SIP SUBSCRIBE/NOTIFY messages, leading to minimal signaling overhead and network resources usage.

7.5.2.5   Session Continuity with S-CSCF Failover Support

The S-CSCF failover support is similar to P-CSCF failover as illustrated in Figure 7.12. When the master node detects the failure of S-CSCF#1, it notifies all active P-CSCFs about this event and provides a list of available S-CSCFs. The P-CSCF updates registration information. When the new S-CSCF (e.g., S-CSCF#2 in Figure 7.4) receives the notification from the P-CSCF, it re-

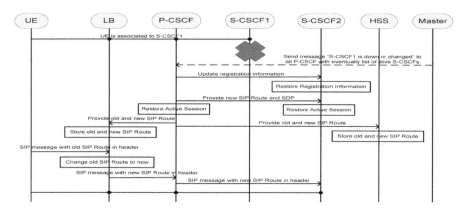

Figure 7.12 Session continuity during S-CSCF failure.

stores the registration information associated to the UEs registered with the failed S-CSCF. When S-CSCF#1 fails or changes, the P-CSCF transfers ongoing session information to S-CSCF#2 to allow session restoration and continuity. The P-CSCF notifies also the LB and HSS about this event by sending information about old and new SIP Route. The UE will continue to send any SIP message with the old SIP Route information to the LB. The LB will perform all required mapping with the new SIP Route for the ongoing SIP dialog.

7.6 Implementation and Test-Bed Prototype

We implemented both UE (mobile) involved and UE (mobile) non-involved cases to realize the self-organizing IMS. UE involved case requires that the mobile is involved in signaling exchange during reconfiguration process. UE non-involved case does not require mobile's involvement during network node reconfiguration, network entities such as load balancer and SIP proxies help re-registering the mobile when the network components change their roles.

Figure 7.13 shows the test-bed architecture used for prototyping the proposed solution. IMS node in Figure 7.13 refers to an IMS component (e.g., P-CSCF, I-CSCF, S-CSCF or HSS) running in one or different physical nodes. The home network is equipped with all SIP-based IMS components: HSS, I-CSCF, S-CSCF and P-CSCF. The master node is located in the router that also acts as the DNS and DHCP server. The edge routers act as 3GPP's

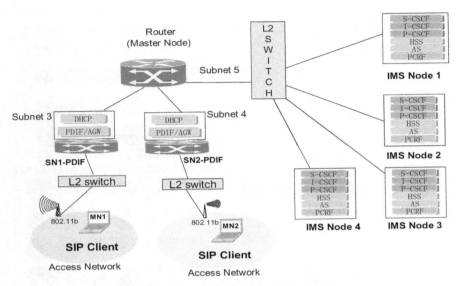

Figure 7.13 Self-organizing IMS test-bed.

PDIF (Packet Data Interworking Function) and DHCP relay agent. SIP stack used in the test-bed is based on NIST implementation [10]. Mobile nodes use SIP user agent based on SIP Communicator [4] to communicate with the IMS nodes. We use "dibbler" as a DHCP client on the mobile. P-CSCF discovery procedure is based on DHCP as specified by IMS specification. We have used XML-based query-response mechanism to obtain the required information from the master node and for role assignment. Mobile nodes communicate with the DHCP server that resides in the router via the DHCP relay agent in the edge router in order to obtain the IP address and discover SIP servers.

Figure 7.14 shows the screen shots of different operations in the test-bed. As shown in Figure 7.14, light grey illustrates that the specific functionality (e.g., P-CSCF, S-CSCF, I-CSCF) in the current node is active. Dark grey indicates that the nodes are inactive. It also shows how initially all the IMS functions are running in Node 1 but some changes in network condition migrate some of these functionalities to Node 2 (e.g., P-CSCF) and Node 3 (e.g., S-CSCF). Preliminary performance results for server reconfiguration indicate that it takes less than one second to re-assign the respective server functionalities by the master node. This reconfiguration time plays an important role in determining the extent of service interruption for the new calls.

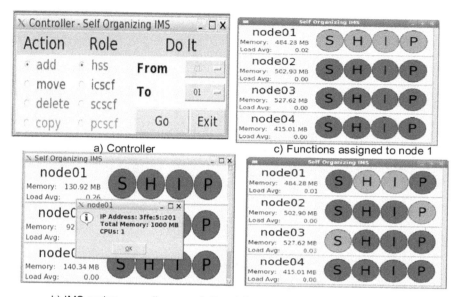

a) Controller

c) Functions assigned to node 1

b) IMS nodes are online   d) P and S roles are assigned to node 2 and node 3

Figure 7.14 Screen shots from self-organizing IMS.

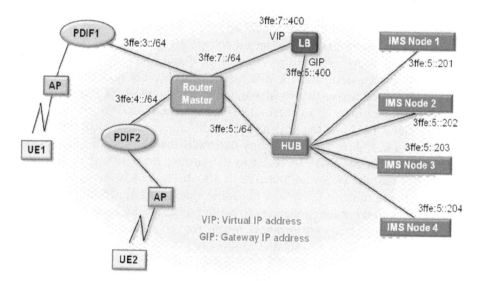

Figure 7.15 Self-organizing IMS with a load balancer.

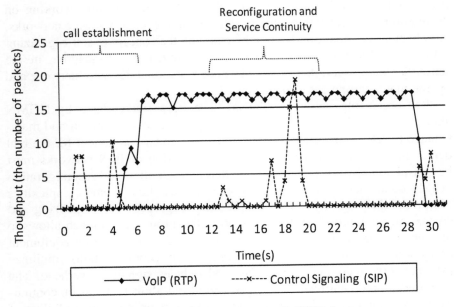

Figure 7.16 SIP control signal during P-CSCF change.

Figure 7.15 shows how the load balancer (LB) is used for UE non-involved scenario. During this reconfiguration delay, ongoing session is not interrupted. For LB, Open SIP Server (OpenSIPS) [14] has been extended with a new module in order to support the UE non-involved method. This module implements the scheduling algorithm and session persistence approaches based on hash over Call-ID and From URI. SIP Communicator, which has been extended to support 3GPP IMS requirements, is used as the IMS SIP client. We used SIPP [19] to generate SIP traffic load. Figure 7.16 shows the amount of additional control signals (e.g., SIP) generated when P-CSCF is changed although there is no interruption in media traffic (e.g., RTP).

7.7 Conclusions

We believe that self-organizing IMS is an important step towards the successful deployment of IMS networks. A policy-based configuration and reconfiguration of IMS components will help operators reduce the cost and complexities of current networks. While there are standards groups such as

IEEE NGSON, ATIS SON and ITU-SG13 that are currently working on the requirements and architecture of self-organizing, context-aware networks, this chapter goes one level beyond by realizing the concept of self-organizing IMS in a laboratory environment. Our approach is self-adaptive in the sense that we preserve the basic IMS node functionalities with network and operational environment change.

Self-organizing networks (SONs) have been standardized by NGMN alliance and 3GPP and is an important step to improve operation and maintenance (O&M). Since IMS is considered as a main component for the real deployment of 4G/LTE networks, self-organization of IMS networks must be considered also. In fact, a policy-based configuration and reconfiguration of IMS components will help operators to reduce the cost and complexities of current IMS networks. This chapter has proposed a self-organizing and context aware approach to IMS networks. The proposed approach allows to hide any change in the core IMS networks to end-users (i.e., user equipment or agent does not participate in the IMS network reconfiguration), minimize IMS networks reconfiguration delay, and reduces signaling overhead. The proposed solution has been prototyped and results show that session continuity can be guaranteed when IMS network topology changes due to failure of network entities or IMS role re-assignment with minimal control overhead. As a result, the IMS network's efficiency and scalability have been improved. Although, in this paper we have focused on P-CSCF and S-CSCF, the proposed solution can be applicable to other IMS core network elements (e.g., I-CSCF, Application Servers).

References

[1] Session initiation protocol (SIP) extension for instant messaging. RFC 3428, Internet Engineering Task Force, December 2002.

[2] ATIS SON, http://www.atis.org/son.

[3] T. Bessis. Improving performance and reliability of an IMS network by co-locating IMS servers. *Bell Labs Technical Journal*, 10(4):167–178, 2006.

[4] SIP Communicator, http://sip-communicator.org/.

[5] A. Dutta, K. Manousakis, S. Das, T. Chiba, and H. Schulzrinne. Mobilty testbed for 3GPP2-based multimedia domain networks. *IEEE Communications Magazine*, 45(7):118, July 2007.

[6] J. Fabini, P. Reichl, A. Poropatich, R. Huber, and N. Jordan. "IMS in a Bottle": Initial experiences from an OpenSER-based prototype implementation of the 3GPP IP multimedia subsystem. In *Proceedings of International Conference on Mobile Business, ICMB'06*, pages 13–13, 2006.

[7] A. Gulbrandsen, P. Vixie, and L. Esibov. A DNS RR for specifying the location of services (DNS SRV). RFC 2782, Internet Engineering Task Force, February 2000.

[8] Ed. Hinden. Virtual router redundancy protocol (VRRP). RFC 3768, Internet Engineering Task Force, April 2004.

[9] IEEE NGSON, http://grouper.ieee.org/groups/ngson.

[10] NIST implementation, https://jain.sip.dev.java/.

[11] A. Manzalini and F. Zambonelli. Towards autonomic and situation-aware communication services: The CASCADAS vision. In *Proceedings of IEEE Workshop on Distributed Intelligent Systems: Collective Intelligence and Its Applications, DIS 2006*, pages 383–388. IEEE, 2006.

[12] E. Marocco, A. Manzalini, M. Sampò, and G. Canal. Interworking between P2PSIP overlays and IMS networks – Scenarios and technical solutions.

[13] M. Matuszewski and M.A. Garcia-Martin. A distributed IP multimedia subsystem (IMS). In *Proceedings of IEEE International Symposium on a World of Wireless, Mobile and Multimedia Networks, WoWMoM 2007*, pages 1–8, 2007.

[14] OpenSIPS, http://www.opensips.org/.

[15] Third Generation Partnership Project, http://www.3gpp.org/.

[16] Adam Roach. Session initiation protocol (SIP)-Specific event notification. RFC 3265, Internet Engineering Task Force, June 2002.

[17] Henning Schulzrinne. Dynamic host configuration protocol (DHCP-for-IPv4) option for session initiation protocol (SIP) servers. RFC 3361, Internet Engineering Task Force, August 2002.

[18] ITU-T SG13, http://www.itu.int/itu-t/studygroups/com13/index.asp.

[19] SIPP, http://sipp.sourceforge.net/index.html/.

8

Blending Services over IP Multimedia Subsystems (IMS)

Madan Pande[1], Raghavendra Sunku[2], Rohan Pascal Goveas[2]
and Debabrata Das[1]

[1]*Department of Information Technology, Networking and Communication Stream,
International Institute of Information Technology, 560100 Bangalore, India;
e-mail: mpande@iiitb.ac.in*
[2]*CMS, Hewlett Packard, 560048 Bangalore, India*

Abstract

3GPP defined an access independent overlay network called the IP Multi-media Subsystem (IMS), which helps access to the services from different networks. The focus group on Next Generation Networks, envisages the use of the IMS overlay network. The main advantages which IMS based Telecom networks offer over the web world are Session control, User interaction, User Status, Device location, Identity management, Subscriber Charging, Service provisioning etc. In spite of having all these advantages the telecom networks are still not able to churn out new services at the pace of Web 2.0 world. In the current scenario, Telecom Service Providers (TSPs) offer their customers services bundled in the form of service packages. A critical success factor for next generation telecom networks would be to break the silo paradigm and provide an environment for delivering dynamic blended services which use proven Telecom and Internet services. In order to create and deliver such services there is a need for a flexible entity like a Service Broker (SB) which can bridge IMS and Web 2.0 worlds. This chapter presents a few prominent architectures for blending of services. The authors also present part of their

Anand R. Prasad et al. (Eds.), Advances in Next Generation Services and Service Architectures, 149–174.

ongoing work on architectures for dynamic blending of services and inclusion of charging for multimedia content, which they have implemented and tested.

Keywords: SCIM, service brokering, blended services, charging.

8.1 Introduction

The Third Generation Partnership Project (3GPP), defined an access agnostic overlay network called the IP Multimedia Subsystem (IMS), which could allow access to services from different networks, such as GSM, GPRS, WLAN, PSTN and Internet. In particular, it was designed to address the delivery of multimedia services to mobile users in a comprehensive manner, keeping in view that the future networks will be dominantly IP networks.

Internet today offers not only much lower cost voice services ("VoIP") but also a host of very popular services like Video and Audio Content ("You-Tube"), video conferencing via tools like SKYPE, ubiquitous e-mail, content search and browsing and location based services. The main advantages of telecom/IMS networks over the web world are session control, user interaction, user status, device location, identity management, subscriber charging, service provisioning, etc. The success of future IMS networks depends on the ability to break the silo paradigm and provide an environment for delivering dynamic blended (described in Section 8.1.1) services which use proven telecom and Internet services. In order to create and deliver such services there needs to be a flexible entity like a Service Broker (SB) which can bridge IMS and Web 2.0 worlds.

3GPP IMS standards have envisioned the need of SB for controlling and blending services which reside in various application servers (AS) [5]. In this chapter we would like to introduce the concepts of service blending and discuss the various service blending architectures available today (along with our and others contributions). Also we discuss various challenges involved in blended service creation and execution.

8.1.1 Service Blending Paradigm

With the rapid growth of the network infrastructure and technologies such as SON (Service Oriented Network), IMS (IP Multimedia Subsystem) and Web 2.0, end-users are looking for richer communication services to suite their lifestyles and needs. These services can be a blend of various service capabilities, which are hosted across different Application Servers (ASs).

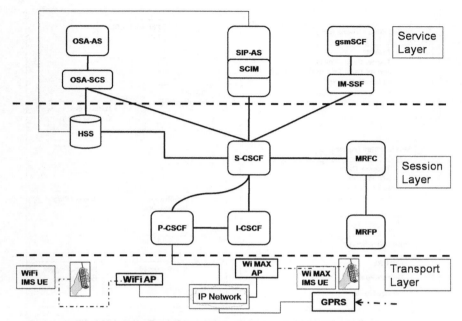

Figure 8.1 Simplified IP Multimedia Subsystem (IMS) architecture.

Services composed by coordinating two or more individual services in a single user session, are termed blended services. An example of such blended service can be pushing advertisements to subscribers in the form of either multimedia content like videos or MMS based on their location, interests and user equipment capabilities. This typical blended service involves chaining of multiple application servers with logical invocation of services, like Location Based Server, XML Document Management System (XDMS), Multimedia Content Server and MMSC.

The two paradigms for delivery of blended services to the mobile user are Web 2.0 and IMS. The Web 2.0 is a read-write web, and has a built-in orchestration mechanism, using which it is possible to combine two or more services to deliver a new service. This is termed as service Mash-Up. On the other hand, IMS blended services can be either static or dynamic in nature. The static blended service is invocation of service enablers in a predefined order. Whereas dynamic blended services represent a complex activity specified by composing multiple services enablers spanning across diversified networks based on the user contextual data and operator defined rules associated with in a single session.

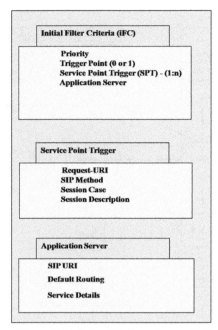
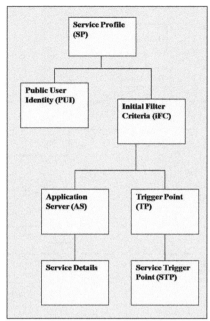

Figure 8.2 A simplified service profile in IMS and service registration components.

3GPP IMS specifications mandate that all services of the subscriber will be delivered from its home network. Further, each service would be delivered by a SIP application server, though it could use an inter-network gateway, such as IM-SSF (IMS Service Switching Function) for legacy intelligent network (IN) services from Mobile-Network (e.g., GSM), OSA-AS (Open Service Access – Service Capability Server; see Figure 8.3) or a Parlay-X gateway for services from the Internet. Such restrictions on service delivery from each application server could often result in conflict and complexity, during multiple service invocations within one session.

To enter an IMS network an IMS subscriber registers with the Home Subscriber Server (HSS) with proper steps through Proxy-call session control function (P-CSCF), Interrogating-CSCF (I-CSCF) and Serving-CSCF (S-CSCF), respectively. The S-CSCF which is the final entry point into the concerned IMS network, retrieves the service profile of the IMS subscriber from HSS and starts to evaluate this profile in order to find out which Application Servers (ASs) must be invoked. IMS service invocation mechanism is based on static service invocation rules defined by initial filter criteria. The

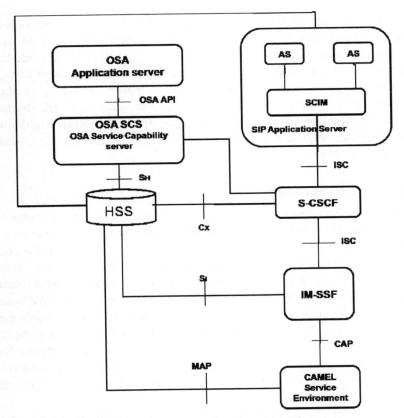

Figure 8.3 A schematic of SCIM in IMS architecture.

service profile contains the initial Filter Criteria (iFC) for each service, a trigger point (TP) as a Boolean expression and the application server details associated with this service (see Figure 8.2).

The Boolean expression for the service TP may be comprised of element from the SIP header, content or even from the SIP message body. When a Trigger Point is met, S-CSCF sends the request to the Application Server associated with the corresponding Trigger Point.

IMS uses separate control and media delivery path, which provides the flexibility inherently required for service routing. Thus, the natural way to blend services in IMS would be to provide a Service Capability Interaction Manager (SCIM) module or element in the service delivery path. The definition of SCIM in 3GPP/3GPP2 documents, give a large freedom to implementers, i.e., the SCIM could be a part of the S-CSCF, or part of a SIP application server that works with the S-CSCF. Conceptually, a SCIM would typically comprise of a SIP back-to-back user agent (B2BUA), a rule-base, a light weight real-time database, SIP and Diameter stacks as well as work-flow logic. A point inherent in SCIM implementation would be an ability to define the work-flow in some simple language, such as BPEL (business process execution language), and have a graphical tool to quickly define a blended service. Figure 8.3 shows a schematic of SCIM [4] as defined in 3GPP Release 7 stage II.

8.1.2 Blended Services

An interaction between an IPTV service and a phone call, such as automatic pause, record and resume of the IPTV service during the telephonic conver-sation without any user intervention, is one of the examples of a blended service. In the current telecom world one AS delivers one specific service. These AS may be SIP application servers in IMS. For delivering a blended service, multiple application servers which are spread across multiple bound-aries needs to be invoked. Figure 8.4 shows a generic call flow for delivering a blended service. The service blender or broker is the key component in delivering a blended service. The service logic for chaining multiple ASs is programmed as a work flow in service broker with some predefined rules. Based on the incoming user request and rules configured, an appropriate blended service is invoked.

A frequent problem in blending service is the incongruence between the protocols that are used for service invocation. For example, the session con-trol in IMS uses the SIP protocol, whereas most audio visual (AV) servers use RTSP protocol for session control. SCIM should be designed in a generic manner so that the workflow implementing the blended service is free from the protocol intricacies and the underlying network complexities. Distributed implementations of SCIM could involve multiple S-CSCF and SCIMs with suitable workflow coordination to avoid inter-service conflicts between ser-vices invoked within one session. In IMS, Service information in respect of a user, can be fetched from the Home Subscriber Server (HSS) database.

On the other hand, Web Services are located by searching the Universal Directory and Definition Interface (UDDI) repository. The service invoca-tion of web services may be done by parsing the XML document fetched from the UDDI to carry out the workflow needed for the step by step exe-cution of the single or blended service. Programming techniques like AJAX

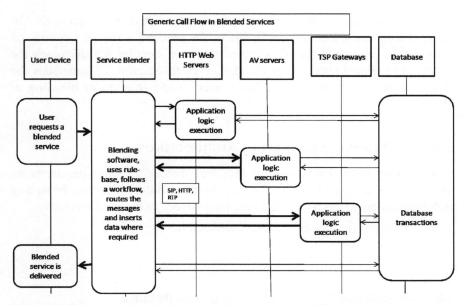

Figure 8.4 Generic call flow for delivering a blended service; TSP: telecom service provider; audio-visual (AV) servers.

(asynchronous java script and XML) and architectural paradigms like REST (representational state transfer) hide most of the lower level aspects of such interactions and make building of Web 2.0 Mash-Ups relatively easy. AJAX is a popular web development technique that makes web pages interactive using JavaScript. REST allows exchange of information with elements defined as resources, with each resource having a unique resource identifier (URI). REST also permits action on resources using established HTTP methods (like GET, PUT).

3GPP based IMS standards define a well known set of service enablers or service building blocks such as presence, location, instant messaging, conferencing etc. Popular multimedia services such as multiparty gaming, IPTV etc can be blended with the above said IMS service enablers. This blending can be done using Web 2.0 approach as well as IMS SCIM based approach. Whether one uses Web 2.0 Mash-Ups or IMS SCIM based approach, blending of services is done with extensive use of J2EE Application Servers or SIP Application Servers.

The remaining part of this chapter is organized as follows: Section 8.2 presents 3GPP proposed service brokering architectures. Section 8.3 discusses various approaches to blend of services. Section 8.4 presents implementation and execution environment for blended services. Section 8.5 presents the use cases for blending of services. Section 8.6 discusses the architecture related charging of blended services. Section 8.7 concludes the chapter with a revelation of the latest trends and evolution of blending of services.

8.2 3GPP Service Brokering Architectures

Release 8 of 3GPP IMS standard proposes the architectural requirements for service brokering [5]. Following are the three different service brokering architectures proposed by 3GPP to enable delivery of blended services:

1. Centralized Service Brokering Architecture.
2. Distributed Service Brokering Architecture.
3. Hybrid Service Brokering Architecture.

8.2.1 Centralized Service Brokering Architecture

In centralized service brokering architecture, the service broker (SB) [5] can reside in a separate application server to coordinate and control the delivery of blended services. This SB will control and streamline the interactions with the various application servers (AS) for the S-CSCF and takes the responsibility to chain the various AS based on the service logic programmed in it. In S-CSCF point of view this SB is a normal AS supporting ISC (IMS Service Control) interface. The above description is depicted in Figure 8.5.

8.2.2 Distributed Service Brokering Architecture

In distributed service brokering architecture, the SB can reside in individual AS which are involved in delivery of blended services (Figure 8.6) [5]. These SBs can interact and coordinate among themselves to deliver the blended service. S-CSCF views each SB as different AS supporting the ISC (IMS Service Control) interface. S-CSCF relays the messages among all these SBs until all the ASs finish their functions. The standards do not discuss the interaction mechanism and protocols in between the SBs. This is left to scientific community to research and propose some mechanisms required for consistent and coherent interworking of the various SBs.

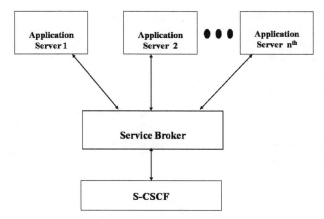

Figure 8.5 Centralized service brokering architecture.

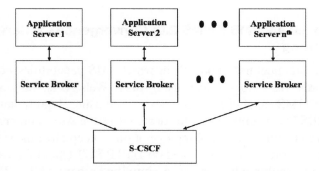

Figure 8.6 Distributed service brokering architecture.

8.2.3 Hybrid Service Brokering Architecture

This architecture is combination of both centralized and distributed service brokering architectures [5]. In this architecture the SBs need to interact with various AS to deliver the blended service and also need to interact with peer SBs.

8.3 Approaches to Service Blending

From the literature available in the open domain on blending of services, a few selected approaches are discussed below.

Simplified WIMS Architecture Schematic

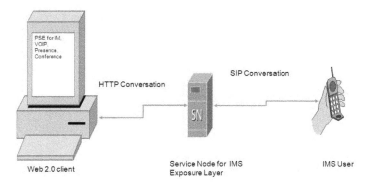

Figure 8.7   Simplified WIMS architecture schematic.

8.3.1 Web 2.0 and IMS (WIMS 2.0) Convergence for Service Blending

WIMS is an architectural approach in which IMS capabilities could be exposed towards Web 2.0 services using open Web APIs. It is argued that session based IMS capabilities could be exposed using Representational State Transfer (REST) concepts [18]. This design uses the AtomPub protocol. The proposed architecture uses a polling operation to keep alive the state information of the session and utilizes standard HTTP PUT () and GET () methods to access, say a multimedia service. A simplified schematic for WIMS given in [15] and has been presented in Figure 8.7.

In order to find out IMS capabilities, the use of "IMS-Widgets" which could be embedded into Web Services has been proposed. These widgets then may be used to exploit the communication capabilities offered by IMS, e.g., a Presence capability or a VoIP call button. Such IMS-widgets have been called Portable Service Elements (PSEs). The use of PSEs can be extended to create blended services by combining the features of a Web Service and an IMS service, e.g., Instant Messaging using Presence information to invoke web delivered video services (e.g., YouTube).

For using Web 2.0 artifacts in IMS, it is proposed to use web services multimedia content events to transfer contextual social content or even advertisement alerts towards an IMS client. Following REST guidelines, each resource is associated with a specific IMS service, with attributes like state, list of participants, media description and components and Web based thin-

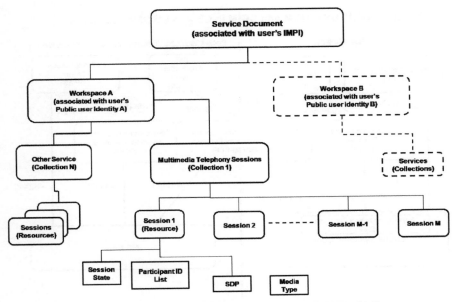

Figure 8.8   Sample AtomPub format for a service in REST guidelines.

clients can invoke such services using the REST APIs defined. The service document is organized as an AtomPub format. Generic structure suggested is of the form, "OpenAPIsRoot/IMPU/Service/SessionID". An AtomPub description of an IMS service using such a mapping is shown in Figure 8.8.

There are two predominant issues in this approach, viz., the approach does not cover session based IMS capabilities, and has to use polling to keep session information updated. Further, there is no GUI for service creation and deployment. Usage of HTML5 could be one of the possible alternative solutions for polling; thereby maintaining the active state information of the session.

8.3.2 Ericsson Composition Engine for Service Blending

The Ericsson Composition Engine (ECE) [7, 17] starts out as an interaction mechanism between intelligent networks (IN), IP Multimedia Subsystem (IMS), and Internet based services. It uses a J2EE server platform for delivering circuit-switched and IMS-triggered services. It is argued that it is possible to compose a single trigger based services of different technologies.

Concepts of IM-SSF, TRIM and SCIM in CS and PS (IMS) Domains

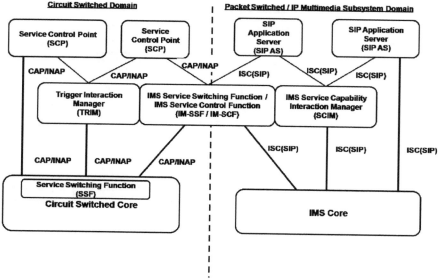

Figure 8.9 Ericsson Composition Engine schematic [17] for services integration from circuit switched and IMS networks.

The Ericsson Composition Engine follows the 3GPP architecture components like

- SIP application server (SIP-AS) for service delivery.
- SCIM for service call routing.
- IMS Service Switching Function (IM-SSF) and Service Control Function (SCF) to access Legacy IN services.

In addition, it uses a trigger interaction manager (TRIM) module (Figure 8.9). The Ericsson Composition Engine can receive indications from a multimedia telephony service (MMTEL). This information can provide information to aid further routing of the call or even add information to the IMS service control (ISC) channel. ECE supports a variety of services like SIP services, CAP/INAP-based IN services, SOAP/web services, and JSON RESTful (Java script Object Notation Representational State Transfer (REST) conformant) services.

ECE provides a GUI tool for service modeling, building and deployment. This is claimed to make the development and deployment of a new blended

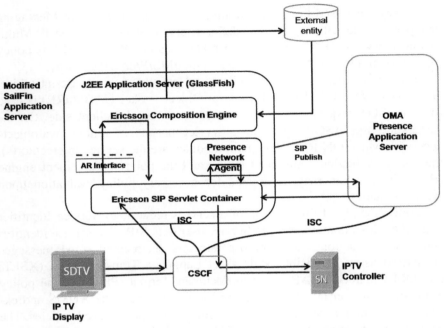

Figure 8.10 ECE blending presence service in an IPTV service request.

service considerably easier. The use of ECE in blending Web 2.0 services or Internet services has not been described for ECE in any published detail. Figure 8.10 shows the likely execution path while using ECE to blend an IPTV service with IMS Presence.

An interesting approach using SOAP-SIP gateway for mapping SOAP (Simple Object Access Protocol) messages from the Web Services on the Internet side to the SIP messages on the IMS side has also been demonstrated by Ericsson [13]. The gateway maps corresponding fields from a SOAP message to equivalent fields in a SIP message and sends it to the IMS network, and vice-versa.

8.3.3 Web 2.0/Telco Service Broker Approach for Blending of Services

FOKUS (a research group of Fraunhofer Institute for Open Communication Systems) have proposed an approach to include the existing telecom service enablers into Web 2.0 Mash-Ups by defining a policy-based service broker

[16]. They have used Presence, Group Management and Instant Messaging as Open Mobile Alliance (OMA) defined service enablers for the IP Multimedia Subsystem. FOKUS Service Broker uses an OMA XDMS as policy repository and Parlay-X Web Services as the IMS/Web 2.0 gateway.

The communication between client and gateway uses Javascript Object Notation JSON-RPC [16]. A JavaScript Object Broker (JABSORB) is used to send JSON requests to remote Java objects on the client side. On the server-side, the JASORB provides a Servlet that makes simple Java objects accessible via JSON-RPC. The main modules are the service interceptor(s), the policy evaluation component (PEC) and the policy enforcement engine (PEE). Interceptors may be applied for necessary policy evaluation upon service requests.

Policies apply at three levels of a service request, to Service Identifier (e.g. IP address, domain, etc.), to Service user (e.g. SIP URI, service identity) and to Service actions and allowed parameters. Intercepted SOAP messages are parsed using Extensible Style sheet Language Transformation (XSLT) into OMA defined SOAP interface document format (PEM-1). The policy evaluation component retrieves applicable polices from the XDMS and executes a matching process between the request and the retrieved policies. The Boolean result controls access granted or denied outcome. This results in either request getting forwarded to the corresponding service enabler or a reply with HTTP 403 forbidden code to the requestor. Policy enforcer entity uses service description files (WSDL-S) retrieved from the UDDI repository. At the time of writing of this chapter and to the best of our knowledge, the FOKUS approach does not address dynamic composition of blended services.

8.3.4 Service Oriented Networks (SON)

SON was envisioned to help the telecom operators to expose their network capabilities or service enablers such a Presence, Location, XML Document Management Server (XDMS) to Web 2.0 based service providers. The key concept behind SON is reusability and convergence. SON uses the best of both of worlds, on one side it leverages the IMS architecture, which decouples the application layer from the network layer; on the other side it borrows the concept of reusability from the SOA architecture, which is a business centric IT architectural approach.

The developer user community and applications in the Web 2.0 space is growing at a rapid rate. In such a scenario exposing the telecom based network resources to various web applications can create rich and complex

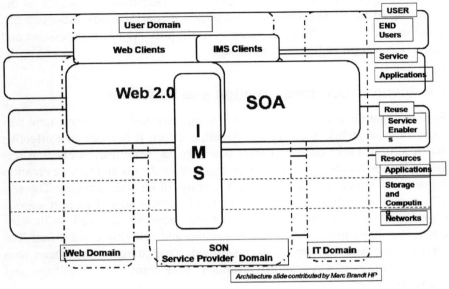

Figure 8.11 Service Oriented Network (SON) architecture.

services. SON opens a new set of opportunities by facilitating Web 2.0 based Mash-Ups or applications to reuse the various network resources and enrich the user experience. Reference architecture for Service Oriented Networks (SON) is shown in Figure 8.11 [3]. The architecture highlights the convergence of the three major domains i.e. Telecom, Internet and Information Technology (IT). The primary goal of SON is to build applications that compose service enablers across these three major domains and ensure the interoperability across application domains.

The adaption of SON based networks is claimed to address the following needs:

- User Mobility.
- Device agnostic applications.
- Reduced cost and time in introducing new services.
- Leverage the best of both the telecom and Internet based networks in a converged world.
- Leveraging on the success of web.

In this section we described different approaches made by various research communities for blending services; each with its own level of success and challenges. But none can be called as a standardized approach as the definition of SB/SCIM itself is broad in 3GPP specifications. How one can avoid use of OSA or Parlay-X gateway to simplify service composition and delivery is a point of futuristic interest.

8.4 Blended Services Creation and Execution

Service creation and deployment is not an easy task. To create new services from scratch is a costly and daunting effort. It is more cost effective and results in lower time to market new services from proven existing services. This demands powerful service delivery platforms and service creation frameworks which can facilitate rapid creation of blended services. The service execution environment for such blended services should support various capabilities such as providing multiple interfaces to interwork with various AS, policy enforcement, data federation and charging capabilities. The service creation environment should provide programmatic languages with various features like readability, usability, state transition capabilities and extensibility.

Research is still going on in finding out the best programming methodology for blended service creation. In this section we shall discuss few such programming approaches which are available today. Since the IMS networks extensively support SIP; one of the programming methodologies for creating blended services can be by using SIP Servlets [1]. SIP Servlets are very similar to HTTP Servlets and also they are extensions of HTTP Servlets. Like HTTP doGet()/doPost() methods these SIP Servlets also provide doxxx() method for each request messages [20]. For example all the SIP INVITE messages are processed by doInvite() method. One of the main advantages of using SIP servlet methodology is that it allows dealing directly with underlying network signals and can be easily leveraged with Java language to call web services. Thus blended services involving both SIP and web services can be composed. Performance wise this programming methodology is better compared to other languages like pure JAVA and BPEL (Business process execution language). SIP Servlet methodology is considered to be complex as it deals with many details of protocol layer [20]. One of the drawbacks of using SIP Servlets is that it requires the service developer to be proficient in SIP protocol. While using SIP Servlet approach, creating the blended service in graphical form is not supported.

BPEL is another paradigm in web services composition [20] and is advocated by many companies like IBM, Microsoft etc. It supports GUI based programming with drag-drop facility which allows the service developer to express his/her business logic in a pictorial form. BPEL based service composition is preferred when one wants to create new blended services using the existing service capabilities which are exposed as Parlay X or Web Services. But usage of BPEL as telecom service composition programming language is still debatable as BPEL does not support efficient state transition mechanisms. BPEL based approach is very time consuming during service development and does not deliver a good performance [20] with respect to service execution in SIP based environments.

8.4.1 Workflow Based Blending

One of the viable approaches to create blended service is usage of JBoss jBPM(Java Business Process Management) [12]. The JBoss jBPM provides a platform for graph-based execution languages. Jboss jBPM provides a native work flow language called jPDL (jBPM Process Definition Language) [2]. An advantage of jBPM is that it allows seamless integration with other services exposed as Java APIs. An example would be invoking a SIP function in the context of a process graph via a SIP Servlet Java API. JBoss jBPM execution engine supports multiple process languages like BPEL, jPDL and Pageflow [2]. JBoss jBPM builds all its process languages natively on top of a single Process Virtual Machine. Thus JBoss jBPM engine facilitates the blending of IMS telecom or Web 2.0 services.

jBPM engine can be a good choice as an orchestration engine for blending of services. Using the jPDL "programming model" service developers can easily and efficiently create a blended service. An example blended service is described using jPDL based service representation as shown in Figure 8.12 [12] using the jPDL Graphical designer. The blended service which is deployed will be packaged as a process archive file. The process definition file "processdefiniton.xml" will have the representation of the blended service. The "processdefinition.xml" also contains information about various actions and tasks involved in the blended service. This process archive file can also contain other process related files such as Java classes which implement specific service logic which is executed as part of the service graph.

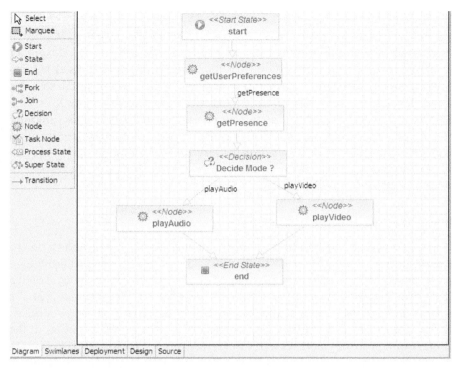

Figure 8.12 Blended service logic representation using JBPM IDE.

8.4.2 Implementation Issues in Blended Services

IMS architecture provides an easy to blend SIP based services, such as Presence, Conferencing, Multimedia Messaging and Location services to create complex blended services. But blending of SIP and NON-SIP services is also more important to create a rich user experience. The service profile stored in HSS describes how statically defined service can be individually invoked but does not give any clues on, how within a session more than one services can be invoked and blended to deliver a new service, e.g., using a Presence Service along with Conferencing Service to invoke a gaming service. To support these kinds of blended services current HSS architecture needs to be extended to include NON-SIP protocol based services, their initial Filter Criteria (iFC) and Trigger Points (TP). What such iFC and TP shall constitute, is not yet defined.

It is also necessary to define criteria for service conflict resolution when multiple services are invoked (as would be the case in a blended service)

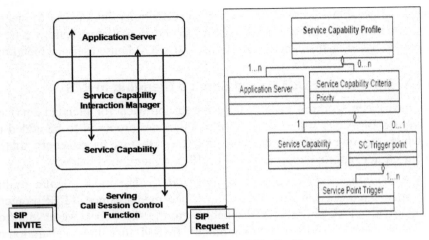

Figure 8.13 Service Capability Profile.

within one IMS session. The session service interaction and conflict management has been investigated at length in published literature [8, 9]. However, a greater clarity is needed, on how to invoke non-SIP services within the IMS network using a SCIM, though gateway based API approaches, like OSA and Parlay-X have been defined for individual non-SIP service invocation. One proposition made [9] is that a "Formal Model" could be stored in the real-time (RT) Database within SCIM or S-CSCF, which could specify the nature of required service capabilities and service interactions between service nodes and the IMS UE, at least for statically defined blended services.

This Service Capability Profile could be used to indicate under what condition, which Service Capability is allowed to be used (especially so, if a formal model is stored). If the condition is met, the related Service Capability will be considered valid for the Application Server concerned. For example, if the Presence of the subscriber at the destination is Do Not Disturb (DND), then the service capability interaction rule should skip the call forwarding error code 509 for this destination and return "Conflict".

It stands to reason that service implementation in the case of multiple service invocation, the service execution be performed on the basis of some service priority fetched from the HSS stored subscriber profile. The Rule-Base in S-CSCF could store such rules when it downloads the service profile of a user on IMS Registration. Adhering to the above clauses, it is argued that the addition of the Service Capability element in the Service Profile makes a valuable improvement to Service Capability Interaction Manager. The authors of this book chapter have proposed in the later sections a fresh approach based on their earlier experience with SCIM based service blending.

8.4.3 Enhancing SCIM Based Service Blending in IMS

We have been working on an SCIM based approach for blended services (Figure 8.14). Over time a few more modules are proposed to be added to it, while retaining its simplicity. Three interfaces that we have recently added to our earlier reported approach [10–12, 19] are mentioned below.

- *A Multimedia Request Redirection Module.* The most popular multimedia requests are for Audio and Video media. Since RTSP protocol provides PAUSE/RESUME capability to the subscriber, we have introduced a multimedia redirection module (MRM) that allows separation of the RTSP control from delivery of media over RTP and is Media Server agnostic. Details of MRM usage are given in [19].
- *A Charging Interface for Blended Services.* To create CDRs for multimedia subscribers, who may Pause/Resume the media session (e.g. when a phone call arrives on their cell phone), we use the MRM interface with SCIM to capture the media type, the start, pause, resume and stop of the media session, the network type/BW from where the UE originated the request.
- *Web 2.0 and IMS Interaction Using SCIM.* To establish two way communications between Web 2.0 and IMS, we have conducted successful experiments to invoke multimedia reception from the Internet, while making the request from an IMS UE. We are in the process of proving invocation of a VOIP call from a remote Web Browser that would call an IMS UE after due registration.

8.5 Use Cases

Blending of services enables network operators to provide rich set of services to the end users. In this section we would like to discuss some use cases where blending of services is involved.

8.5.1 A Multimedia Service Blended with Presence

Since IMS networks are targeted to cater rich multi-media applications to end users; in this section we discuss an interesting use case which involves blend-

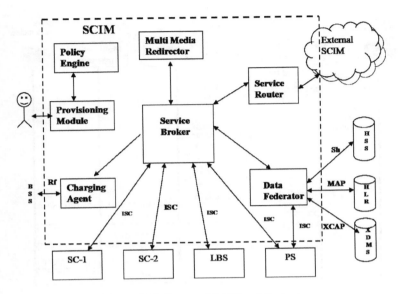

Figure 8.14 Proposed SCIM architecture.

ing of multi-media streaming services and presence service. This use case is about wishing friends on their birthday by streaming an audio/video birthday song based on their presence. It represents a dynamic blended service where chaining of multiple services is involved like presence and audio/video content servers. It has been implemented and tested in IMS Innovation Lab., at IIIT-Bangalore.

Consider two IMS users, Alice and Bob. Alice wants to wish Bob on his birthday by using an audio/video stream. The birthday wishes blended service is created by the service provider by blending the Presence Service and the Audio/Video Streaming Service for birthday songs/videos. The network operator providing the birthday wishes service would provide a web interface where users can book his/her request. Alice as an IMS user would activate this service by entering the day, when she would like to congratulate Bob. Alice also selects a particular video and audio birthday song through which she would like to congratulate Bob. These video/audio files are part of the content stored in a streaming server. The content server has access to the content directory of the streaming server. Based on Alice's request the SCIM would execute the service by chaining multiple services. The birthday blended service is triggered to execute once the timer set for Bob's birthday pops

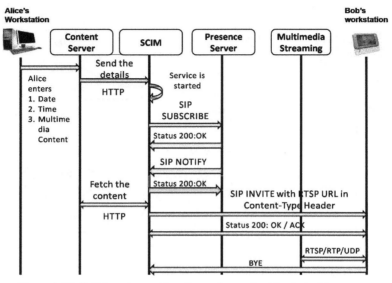

Figure 8.15 Call flow for multimedia media service blended with presence.

up. Once the birthday service is started SCIM would fetch Bob's presence status by contacting the presence application server. If Bob's presence status is "Available", then SCIM would contact the content server via HTTP interface and fetch the corresponding RTSP URL of the audio/video file which Alice wishes to stream to Bob. Next, SCIM would send a SIP INVITE request to Bob encapsulating the RTSP URL that was fetched from the content server. On the establishment of the RTSP session, Bob's client would start receiving the birthday video/audio song stream which Alice had mentioned in her request. Thus Alice would be able to congratulate Bob on his birthday with the help of this blended birthday service. In the described use case, SCIM performs the service blending by chaining the presence service and the audio/video streaming service without having to alter different protocol interfaces. Figure 8.15 represents the call flow for the above discussed use case.

8.6 Charging of Blended Services

Efficient and easy charging of blended services is essential for revenue assurance. Charging of such services can be done either offline or online. Charging of blended services involves collecting or reporting of relevant charging information that is scattered across different application servers. Gathering and processing such information for a blended service is a major challenge. This task is critical for network operator to avoid revenue leakage.

Offline charging is a mechanism where the service is charged after its usage. The blended service can be charged based on various factors like type of application servers involved in delivery of service, the order of invocation of application servers, media consumed, bandwidth used, QoS for that particular service, subscriber profile, etc. According to 3GPP standards every entity that is capable charging offline has a CTF (charging trigger function) entity which collects the relevant charging information from the incoming message and conveys this to CDF (charging data function) entity. CDF later dumps the CDR (call detailed report) and sends it to mediation device or CGF (charging gateway function). In the mediation device/CGF various CDRs are correlated, processed, filtered and final consolidated CDR is constructed which is later processed by billing system. Thus a subscriber is charged offline.

Since SCIM is entitled by 3GPP for delivery of blended service, it is the responsibility of SCIM to generate relevant charging information for a particular blended service to enable offline charging. Since SCIM interacts with various application servers for delivery of a blended service it is necessary to have a CGF entity in SCIM. This entity will pull all the CDRs from the relevant application servers which were invoked during the delivery of blended service and a final consolidated CDR will be generated. This is the major challenge and one of the important tasks for SCIM.

Online charging is a mechanism where the charging information is reported on a periodic basis. Before execution of any service the Online Charging System (OCS) needs to be contacted for knowing the authorization details. Generally OCS grants some token and usage of these token needs to be reported by the concerned AS in a periodic basis. In order to achieve this AS need to support the Diameter Ro interface where real time charging information is communicated to OCS and further OCS does the accounting operations based on the charging information received. In such a scenario the entities like SCIM should be intelligent enough to communicate to Online Charging System (OCS) for real time control of blended service execution. Also charging is not confined only to reporting of used units. There can be operator defined charging policies which can alter the service execution of a blended service. Efficient charging and charging architectures for blended services is still a grey area where many research works are in progress.

8.7 Trends and Evolutions

Service delivery for the evolving networks still remains a target for developers, and there has been no universally acceptable solution, whether based on Web 2.0 or based on IMS. The situation is evolving and will see a number of paradigms proposed and exemplified by major interested developers [6, 14].

8.7.1 Issues in Service Delivery That Need Resolution

- There is a need for a universally acceptable easy solution exists for creating and delivery of services that combine the advantages of IMS enabled Telco-like services with the advantages of Web 2.0 enabled services.
- The IMS subscriber profile needs augmentation to include elements in the schema for non-SIP services; this may be easier than enhancing the less popular UDDI repository for SIP enabled services.
- SIP is an asynchronous protocol and is the lingua-franca of 3GPP IMS networks; on the other hand, Web 2.0 services use HTTP, a stateless protocol. REST API is not IMS widget oriented, but its usage could provide a possible bridge.
- Any solution for blending services from both IMS and Web 2.0 world to be acceptable needs to provide a neat way of Service Capability and Service Profile Management, well defined Trigger Points and initial Filter Criteria, some way of defining the workflow for blending services, and a GUI for developing and deploying them.
- Invariably solutions proposed, include modifications to the software running on both Network Nodes and on the IMSUE (IMS User Equipment). This may require arriving at some detailed IMSUE specifications by 3GPP that exist today. One possible approach could be to define "common service enabler blocks", for both the SIP AS and IMSUE.
- Routing of service calls for blended services requires both workflow management and resolving issues with redirection in protocols like RTSP 1.0 (a very popular protocol used for multimedia delivery).
- While Application Servers are the dominant nodes for service delivery in both IMS and Web 2.0 architecture, most developed products have become proprietary. Hence, the contributions from the huge community of Open Source developers has been limited; and has mostly been around
- "SailFin", essentially the Open Sourced J2EE server from Sun Microsystems called "Glassfish", integrated with a SIP container from Ericsson. Recent releases of the once popular Open Source JBoss Ap-

plication Server have adopted the JSLEE (Java Service Logic Execution Environment) route instead of the J2EE, which makes it less acceptable to many developers.

8.7.2 Evolutionary Directions for Service Delivery

Recently published papers indicate two dominant directions. The first relates to enhancing the HSS schema in IMS architecture to include elements for "Service Capability", and elements to support "Non-SIP" services. The second relates to extending the notion of Presence Service, by enhancing the XDMS schema. Rule Engines will definitely play a large and definitive role in the development of blended services, to overcome the limitations of the current implementations of the S-CSCFs. Whether these would use workflow software (and its programming language) or will be developed using session beans based workflow control in J2EE servers, or Java Connector architecture, coupled with REST API is a moot question to be answered. Perhaps the answer for integrating IMS, Web 2.0 based services may lie in the implementation of a GUI programmable gateway for mapping IMS and Web 2.0 messages, a rule base or workflow tools for blended service composition and a way to enhance existing repositories (e.g. XDMS, HSS) for defining web resources for the IMS world and vice-versa.

References

[1] JSR-000289 SIP Servlet 1.1 [Online]. Available: http://jcp.org/aboutJava/communityprocess/final/jsr289/index.html, Accessed: October 11, 2010.

[2] Manage your business processes with JBoss jBPM [Online]. Available: http://www.javaworld.com/javaworld/jw-05-2006/jw-0522-jbpm.html. Accessed: October 11, 2010.

[3] *Atis Service Oriented Networks (SON) Assessment and Work Plan*, Abridged Version, 2009.

[4] 3GPP. Ip multimedia subsystem (IMS), IM call model 23.228. Release 6.

[5] 3GPP. Study on architecture impacts of service brokering, TS 23.810. Release 8.

[6] May El Barachi, Arif Kadiwal, Roch Glitho, Ferhat Khendek, and Rachida Dssouli. The design and implementation of architectural components for the integration of IP multi media subsystem and wireless sensor networks. *IEEE Communications Magazine*, 8:42–50, 2010.

[7] Torsten Dinsing et al. Service composition in IMS using Java EE SIP servlet containers. *Ericsson Review*, 3:92–96, 2007.

[8] Anahita Gouya and Noel Crespi. SCIM (service capability interaction manager) implementation issues in IMS services architecture. In *Proceedings of International Conference on Communications (ICC)*, pages 1748–1753, 2006.

[9] Anahita Gouya and Noel Crespi. Service invocation issues in IP multimedia subsystem. In *Proceedings of IEEE International Conference on Networking and Services*, 2007.

[10] Rohan Pascal Savio Goveas, Raghavendra Sunku, Pranay Arian, Madan Pande, and Debabrata Das. IMS service broker SCIM enriching rest based Web 2.0 mashup. In *Proceedings of International Conference on Internet Multimedia Systems Architecture and Applications*, 2010.

[11] Rohan Pascal Savio Goveas, Raghavendra Sunku, and Debabrata Das. Centralized service capability interaction manager (SCIM) architecture to support dynamic-blended services in IMS network. In *Proceedings of IEEE International Conference on Internet Multimedia Services Architecture and Applications*, 2008.

[12] Rohan Pascal Savio Goveas, Raghavendra Sunku, Madan Pande, and Debabrata Das. Workflow based design for blending services over IP multimedia subsystems (IMS) with service capability interaction manager. In *Proceedings of IEEE International Conference on Internet Multimedia Services Architecture and Applications*, pages 1–5, 2009.

[13] Roman Levenshteyn and Ioannis Fikouras. Mobile services interworking for IMS and XML webservices. *IEEE Communication Magazine*, 44:80–87, 2006.

[14] Salvatore Loreto, Tomas Mecklin, Miljenko Opesenica, and Heidi-Maria Rissanen. Ims service development API and testbed. *IEEE Communications Magazine*, 48:26–32, 2010.

[15] David Lorzenov, Luis A. Galindo, and Luis Garcia. Wims 2.0: Converging IMS and web 2.0. designing rest APIs for the exposure of session-based IMS capabilities. In *Proceedings of IEEE Second International Conference on Next Generations Mobile Applications, Services and Technologies*, 2008.

[16] N. Blum, T. Magadenz, and F. Schreiner. Services, enablers and architectures: Definition of a connected web 2.0/telco service broker to enable new flexible service exposure models. *Proceedings of International Conference on Intelligence in Networks (ICIN)*, 2008.

[17] Jorg Niemoller and Ioannis Fikouras et al. Ericsson composition engine – Next generation in service layer solution for IN, IMs and other service technologies. *Ericsson Review*, 2:22–27, 2009.

[18] C. Pautasso, O. Zimmermann, and F. Leymann. Restful web services vs. "big" web services: Making the right architectural decision. In *Proceedings of WWW*, 2008.

[19] Raghavendra Sunku, Rohan Pascal Savio Goveas, Madan Pande, and Debabrata Das. Offline charging for multimedia blended services using service capability interaction manager in IMS network. In *Proceedings of International Conference on Internet Multimedia Systems Architecture and Applications*, 2010.

[20] Yuan Yuan, Jia Jia Wen, Wei Li, and Bing Bing Zhang. A comparison of three programming models for telecom service composition. In *Proceedings of IEEE Third Advanced International Conference on Telecommunications*, 2007.

PART 2

IPTV AND VIDEO SERVICES

9

An Overview of IPTV and Video Services

Mischa Schmidt

*NEC Laboratories Europe, Kurfürsten-Anlage 36, 69115 Heidelberg, Germany;
e-mail: schmidt@neclab.eu*

Abstract

This chapter provides an overview of IPTV and video services in managed and unmanaged IP networks. For this, it first separates IP-based services from the traditional TV and video services. This separation is intended to draw the awareness to the commonalities and the differences of both approaches. It also introduces an understanding of the new capabilities implied by delivering TV and video services over IP-based networks. To build a common understanding of the innovative services in the IP-based environment, a collection of typical IPTV and video services discussed in the industry is presented afterwards.

Subsequently, we discern managed from unmanaged IP networks for providing IP-based TV and video services. Here the effects on IPTV and video services of both network models are discussed: their potentials, benefits and advantages as well as their shortcomings, threats and challenges. For the managed network approach a dedicated section to sketch the two main architectural approaches to IPTV services currently discussed in the industry is introduced: IMS-based IPTV and non-IMS-based IPTV. This section leads to an understanding of their main differences and the implication of using one or the other.

In conclusion of this chapter, the most prominent obstacles for the rising IPTV and video services are summarized. Here we also describe the concept of hybrid approaches – mixtures between traditional broadcasting and IP ser-

Anand R. Prasad et al. (Eds.), Advances in Next Generation Services and Service Architectures, 177–196.

vices – that are investigated at present in the industry to address some of the challenges mentioned earlier in this chapter.

Keywords: TV, video, Internet, IPTV.

9.1 Introduction

Today television is an omnipresent, mass-market, multibillion dollar, multinational service with a long history and well-established business models.

On the other hand, over the last decade services offered over the Internet have become increasingly popular. They are connecting us to one another, allowing us to access information around the globe on demand and enable doing business 24 hours a day. With the availability of high bandwidth Internet access for consumers, accessing multimedia content in the Internet has become more and more widespread. Nowadays, access to Internet video services has become a reality and first TV sets even support access to Internet-based video portals such as YouTube [12]. The Internet Protocol (IP) is the underlying protocol enabling all this and as reported in, e.g., [23] the IPTV market will grow dramatically over the next years: the number of global IPTV subscribers will grow from 41.2 million at the end of 2010 to 101.7 million in 2014, a compound annual growth rate of 25.3%.

Considering the current gradual convergence of Internet and TV services, it is important to understand the motivation behind using IP technology to deliver TV services instead of relying on the traditional, proven broadcasting technologies. To do so, we investigate in which way IP technology can stimulate an evolution of TV services. Afterwards we discuss IP-based video Services in unmanaged and managed networks and reveal some of the related challenges. Finally, we conclude this chapter by briefly discussing current trends such as hybrid TV models and highlighting potential obstacles for the realization of IP-based video and TV services in the mass market.

9.2 The Internet Protocol's Added Value: Personalization of Multimedia Services

Traditional TV transmission excellently serves users in a broadcast fashion, i.e. all users in a certain area receive the same TV signals. This makes traditional TV signal distribution highly efficient, but from a user's point of view it lacks personalization and interaction capabilities. These capabilities would allow the creation of more attractive services and more efficient forms of advertising. Besides creating revenues, personalized innovative services are a tool for TV service providers to differentiate their service offers from competitors. There are several approaches in traditional TV that already try to address these shortcomings. For example, they are tailoring the mix of advertisements sent during breaks of a certain program slot to the majority of the audience (e.g. beer and car commercials during sports shows) or providing dial-in capabilities to the audience to cast votes in TV shows. They are even offering movies "on demand" via purchasable one time PINs that enable users to decrypt the associated movie which itself is delivered in broadcast fashion during a time window known in advance. This particular approach can be seen as a non-IP variant of near Content on Demand as described in Section 9.3.2. These examples are evidence that the capabilities to personalize the TV experience and interact with the program itself exist, but they are very limited in traditional TV. At the same time they prove that business cases exist to offer personalized and/or interactive TV.

Changing the underlying technology to IP – both transmission and reception equipment – enables a *bidirectional communication channel between user and service provider*. Said channel is inherent to IP technology and can be used for personalization or interaction purposes. Chapter 10 outlines how personalization aspects can lead to an increase in user acceptance of IPTV services if the so-called "semantic gap" can be reduced.

IP transmission comes in two forms: unicast and multicast. While in unicast end-points communicate point to point, multicast is a point to multipoint communication. In the context of IPTV Service Personalization, unicast is considered the most individualized form of communication, whereas multicast is less personalized since multiple end-points receive the same data stream. On the other hand, multicast transmissions are more efficient in terms of network resource utilization than unicast transmissions. *Therefore, the service provider needs to make a trade-off between transmission efficiency and personalization.*

9.3 Video and TV Services over Internet Protocol Networks

9.3.1 Stakeholders

When defining the IPTV and video landscape, the different IPTV related fora and standardization organizations aim to reflect multiple stakeholders, roles and business models.

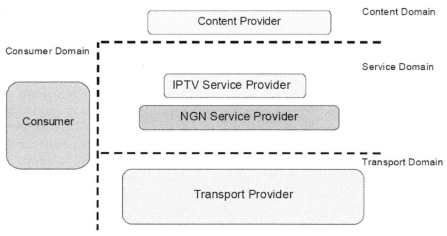

Figure 9.1 IPTV domains [14]. © European Telecommunications Standards Institute 2009, 2010. Further use, modification, copy, and/or distribution are strictly prohibited. ETSI standards are available from http://pda.etsi.org/pda.

For example, Figure 9.1 from ETSI [14] shows four different relevant domains:

- *The Content Domain* provides the actual content to be delivered via the IP network.
- *The Service Domain* provides service level access to IPTV services. Subscription management, service authentication and charging are important functionalities of this domain.
- *The Transport Domain* transports service related IP traffic between the Consumer and the Service Domain and thus connects the consumers to the desired services.
- *The Consumer Domain* models the end user using the IPTV service in his home network consuming the video content.

As mentioned before, this separation allows a provider to assume multiple business roles and it is possible that, e.g., a transport provider offers also NGN and IPTV services. Furthermore, this separation also allows models where one IPTV service provider can have different relationships with multiple content providers allowing it to offer, e.g., live broadcast content as well as relying on Internet-based video sources at the same time.

9.3.2 IPTV Service Categories

To build a common understanding for later discussions and thoughts, we present an overview of common IPTV service categories and features:

- *Content on Demand (CoD)*: Delivers content on user request. While this can refer to any kind of content, video is the most prominent type of content delivered when using this term. Therefore, *Video on Demand* (VoD) is a more common name. Typically, CoD/VoD starts content delivery immediately after the user request – a variation where content delivery is delayed is referred to as *nCoD* (see next item) in order to increase transmission efficiency. In CoD, there exists a one to one communication between user and service provider and thus this service can be considered as highly personalized. Often, CoD services come with *Trick Play* support.
- *Near Content on Demand (nCoD)*: This service starts content delivery at fixed points in time, e.g. at every full hour. Users select content and wait for content delivery until the starting time. This enables the service provider to deliver the content in a multicast fashion to save network resources. Note that this however makes this service less personalized than CoD as there is no one to one communication anymore. If Trick Play is supported, then unicast capabilities or device built-in capabilities are necessary to realize Trick Play (similar to the Linear TV with Trick Play service described below).
- *Download*: While CoD typically implies that media is delivered to the user's device fast enough to ensure play-out in real time, download services refer to the case where the content is stored on the end device. This can relax the network resource demands, i.e. the content might be delivered at a delivery rate insufficient for real-time consumption. A variant of the Download Service delays content delivery to times of low network utilization which is then referred to as *Deferred Download*.
- *Electronic Program Guide (EPG)*: Strictly, this is not a video or TV service, but it is the entry point (often in form of a portal website) for the user to select IPTV services and content: it triggers the content consumption. Generally speaking, EPGs are the virtual analogue of paper-based program guides. Often EPGs can be personalized e.g. based on past user selections of content, or the user can order it according to his/her preferences. Ideally, the EPG offers features like a search option and/or the ability to bookmark favorite shows so that the user can find desired content conveniently.

- *Gaming Services*: In the context of IP-based video/TV services we refer here to services where the rendering computations of video games are performed in a datacenter in the operator network. The rendered game graphics are then streamed as video to the user's TV. This requires that the signals to control the game, the man-machine-interface, need to be relayed to the network component hosting the gaming logic. While there are several different technical approaches to this service concept – each with drawbacks, challenges and benefits – this service is very compelling since it mitigates legal issues like software piracy. Furthermore, it enables interesting business models such as having 5 minutes of game play (with full features) for free. This kind of service, while first offered via the Internet (e.g. [8]), is now also considered in Standardization (e.g. ETSI MCD is studying this category in [2]). Also the game graphics streamed to the players as video can be streamed to other consumers as well. These might be spectators of the increasingly popular electronic sports [1] category.
- *Linear TV*: In essence this is traditional TV delivered via IP unicast or multicast. The choice of one or the other delivery method is a trade-off between personalization and transmission efficiency. Variations are also possible, e.g. to use multiple different multicast groups to cluster different categories (or segments) together. This approach allows targeting content, e.g. commercials, to particular segments.
- *Linear TV with Trick Play*: This addition enables Trick Play commands on Linear TV. This can be realized on user equipment by recording the streamed content on a PVR or nPVR (see the descriptions below) and playing it back from there.
- *Personal Video Recorder (PVR)*: A personal storage device for (video) content. This functionality can also be offered in the operator network, referred to as Network PVR (nPVR). While the name PVR refers to video media only, it often can store different types of media content as well, e.g. audio or pictures.
- *Trick Play* is not a standalone IPTV service but a feature for IPTV services allowing users to stop, rewind or fast forward the received content. Depending on the service some of the Trick Play actions might not be available. For example fast forwarding or skipping through advertisements could be prohibited by the service provider.
- *User Generated Content (UGC)*: This service allows users to share content they generated themselves with other users. To do so, they typically upload their content together with a description to a UGC portal. This

Table 9.1 Traffic characteristics of video/TV services over IP.

Service Name	Unicast	Multicast
CoD	X	
nCoD		X
Download	X	
Gaming Services	X	X (for Spectators)
Linear TV	X (rarely)	X
Linear TV with Trick Play(network Trick Play)	X	
nPVR	X	
UGC (upload)	X	

kind of service has become very popular and successful in the Internet (e.g. YouTube [12]), but may provide some legal challenges if access procedures are too lax. For example, consider the question of responsibility in case of copyright infringements. On the other hand, tight regulation of this service might reduce user friendliness and thus negatively impact user acceptance. In traditional TV, the TV show category "Home videos" can be seen as similar to this service. The uploaded content can then be accessed by other users in a CoD, nCoD or Linear TV fashion.

Table 9.1 considers the traditional distinction of IP delivery modes (unicast and multicast) for above services.

9.3.3 IPTV and Video via Unmanaged Internet Protocol Networks

9.3.3.1 Characteristics

The letters IP in IPTV stand for "Internet Protocol" and as the name suggests, the most prominent IP network today is the Internet. According to Wikipedia [4], the Internet is

> a global system of interconnected computer networks that use the standard Internet Protocol Suite (TCP/IP) to serve billions of users worldwide. It is a network of networks that consists of millions of private, public, academic, business, and government networks of local to global scope that are linked by a broad array of electronic and optical networking technologies. The Internet carries a vast array of information resources and services, most notably the interlinked hypertext documents of the World Wide Web (WWW) and the infrastructure to support electronic mail.

Services offered over the Internet evolved quickly in the past years as the Internet is based on an end-point oriented, i.e. user equipment and/or server technology oriented, approach. This approach allows an evolution of services through upgrading end-points rather than upgrading the communication network connecting the end-points. As long as the network connectivity between the end-points meets the services' requirements, the speed of evolution is very high.

On top, IP transmission equipment improved drastically and broadband access to the Internet became more and more a commodity during the past decades. This resulted in rich multimedia Internet services being reality today.

Further, the Internet is a multinational network which makes it possible, e.g., for users in Germany to access end-points in Japan, e.g. to exchange ideas or to offer new services.

These characteristics are some of the reasons for the highly innovative and competitive nature of the Internet. Traditionally, many services in the Internet have been offered for free (e.g. financed through advertisements) and users have been willing to accept glitches in service quality for this reason. This made Internet users tolerant to e.g. asynchronous video reception across different locations or waiting for video buffering to complete.

9.3.3.2 Challenges
9.3.3.2.1 Lawfulness   The open, unregulated, unstructured, multinational nature is one of the strengths of the Internet and at the same time opens doors to electronic forms of criminality. *Rights infringements* are unfortunately a reality and difficult to overcome: for more than one decade the media industry has been fighting against the sharing of illegal copies of books, music and movies distributed through Internet file sharing applications such as Napster (see e.g. [6]), BitTorrent [5] (especially in relation to the "Pirate Bay", a BitTorrent tracker) or free file-hosting services [3].

Despite violations of laws, it is also easily possible to *access violent, inappropriate, mislabeled and offending content* by accident or even deliberately. Since traditional TV and video services are subject to national regulation the consequences of video services over the Internet on the traditional TV services or vice versa are currently not predictable.

9.3.3.2.2 Quality of Service   The Internet gives *no guarantees for the Quality of Service (QoS)* available to the traffic to be transported – especially

in video services this is problematic as real-time requirements for video are relatively strict and bandwidth demand is high. This QoS issue arises partly from the fact that the network infrastructure routing the IP traffic is not aware of the service this traffic belongs to – the service is typically run from a different entity than the Operator owning the infrastructure equipment. Traffic often traverses multiple networks and the owners of the networks often do not allow business partners to, e.g., prioritize traffic within their own network.

There are various techniques and efforts to address QoS issues – e.g. buffering video at the client side to be able to cope with packet jitter or deploying sophisticated cache/CDN (Content Delivery Network) technology in networks to distribute server and traffic load. This also opens new markets: Akamai, a provider of CDN technology to ISPs and Operators reported revenues of $240 million in the first quarter of 2010 [17]. Generally speaking CDNs rely on duplication of content over multiple content caches. They assign users to caches that are located close to them. This way, media traffic paths through networks are shortened and network resources are thus saved. Despite the effectiveness of the state of the art CDN technology used on the Internet, unfortunately, users using proxies for anonymity reasons (such as e.g. onion routing [7]) typically cannot profit from that technology because the anonymization achieved by these proxies makes it difficult for the CDN to locate the user equipment and thus prevents the assignment of the most suitable (i.e. closest) cache for the requesting user. Another challenge for content caches is the so-called "long tail content" (niche content) – one of the key differentiators for the Internet video services. Content caches cache only popular (mass market) content to reduce the inter-network IP traffic load – infrequently requested content is removed from the caches and thus the benefits of caching are not applicable for "the long tail".

Other approaches of addressing QoS issues are the development of *Scalable Video Coding* (SVC) [9] and *adaptive HTTP streaming*. They allow the adaptation of the displayed video quality depending on available bandwidth during video playback by separating a traditional video stream into either multiple complementary video streams of different detail levels (in the case of SVC) or multiple equivalent video streams of different detail levels (in case of adaptive streaming). Despite differing in technical details, both approaches have in common that if enough bandwidth resources are available, higher degrees of detail are received and displayed and in case of scarce bandwidth, less detailed videos are shown to the user.

Depending on the Network QoS and impacted by video coding schemes, buffer techniques and other technical measures to mitigate potentially negative effects of network transmission on the video perception, the concept of *Quality of Experience (QoE)* resembles a measure of the user's experience when consuming the IPTV and video services. Generally speaking, a high QoE increases user acceptance of IPTV services. Chapter 10 aims to improve QoE of IPTV services through moving towards "semantic IPTV" and Chapter 12 discusses QoE optimized IPTV networks in more depth.

9.3.3.2.3 Heterogeneity   Given the open, unmanaged nature of the Internet and the fact that services are realized through logic in the end-points (typically PCs and Servers), reality shows that a very diverse set of user equipment is in use simultaneously (e.g. [16] states that 60% of enterprises still used Internet Explorer 6 in 2009). This *heterogeneity of user equipment* is a challenge to be taken into account for the creation of new services, the introduction of new standards (such as HTML 5), new codecs or new content protection schemes such as DRM. The problem is that when introducing new features or technologies to existing services, always a compromise between backwards compatibility for supporting legacy user equipment – which comes at the cost of higher service complexity – and reducing service complexity by neglecting backwards compatibility – which comes at the cost of losing users with legacy equipment – needs to be found. For IPTV in particular even more heterogeneity is to be expected as two worlds converge:

- The PC driven IP world: Here, PCs are frequently upgraded by the end users and it can be expected that new services can be accommodated relatively soon by potent hardware.
- The traditional TV world: Here Set-Top-Boxes and receivers often have a lifespan of 10 years and the market is very price sensitive. New Services cannot necessarily be expected to become widely adopted very soon unless these services are compatible with already existing user devices.

9.3.3.2.4 Personalization   Service providers may wish to distinguish from traditional TV offerings through personalized services. Adopting Identity Management (IdM) concepts can help service providers to allow personalized yet privacy protected services to their subscribers offering e.g. single-sign on functionalities as described in [24]. While IdM research has progressed considerably, the TV use case is posing an interesting challenge: Imagine a service provider would like to personalize the IPTV services to the user's preferences and, e.g., pre-filter the list of services, channels and content offered. How would this system accommodate the fact that users sometimes consume TV services alone and at other times together with other people? Certainly services advising content to users should offer different content to an adult watching TV alone than to the same adult watching TV jointly with his or her two underage children. The state of the art concepts of Internet Applications towards service personalization do not provide a satisfactory answer, yet. Current research in the IPTV field tries to overcome these kinds of problems but solutions are not yet ready for the mass-market.

9.3.3.2.5 The Business Model The business model itself is also a challenge: video data is resource demanding and – as mentioned above – has high QoS requirements. Thus, service platforms offering video services are costly. Service providers need to decide whether they charge users for accessing the video or whether they offer their services for free – e.g. financed through advertisements. Of course reality allows for mixtures of both models, but if a service provider intends to charge the users he also risks losing them to competitors that offer content for free.

Service providers also need to consider how to actually charge users: there is an *increase in fraud on the Internet and no unique billing scheme*. Credit card payments are common in the Internet today, but they are risky and also not ubiquitous. According to e.g. the Internet Crime Complaint Center (IC3) Report 2009 [22], credit card fraud is third most reported offense to law enforcement. As reported in [19], losses of online revenue due to fraud increased over the past decade and amounted to four billion US$ in Canada and the US in 2008 alone. In relation to video services, fraud can harm service and content providers by delivering video to fraudulent users as well as users whose credit cards have been maliciously charged.

Considering advertisement funded services, of course the service provider also runs the risk of losing certain users to competitors with more attractive models of advertising (e.g. inserting fewer advertisements). Generally when using advertisements to fund a service, targeting advertisements to the individual consumer is very valuable. It is attractive from the advertisers' point of view as it increases the chance of the advertisement being relevant to the consumer. Moreover it is also attractive to the service provider since he can offer this capability to advertisers and generate revenue from that. However, this presents the challenge of identifying the user (who might desire to remain anonymous) and profiling his behavior. Depending on national legislation this may be controversial or even prohibited. Certainly, this profiling of consumer behavior is problematic from the perspective of consumers concerned with their privacy.

Another angle on the consideration of the business model of (premium) video services over the Internet is the fact that carriers delivering the video traffic typically do not get premium rates for this service – often they charge flat rate tariffs to end users for Internet access. This indeed led carriers to treat particular IP traffic differently (e.g. with reduced QoS) – which depends on the broader discussion on *"Net Neutrality"* with various different opinions and rulings (see e.g. [13, 18]).

9.3.4 IPTV and Video via Managed Internet Protocol Networks

9.3.4.1 Characteristics

Looking at the benefits of IP technology compared to traditional broadcast TV and looking at the challenges for video and TV services over the Internet discussed in the previous section, telecommunication and cable TV Operators strive to leverage the fact that they own delivery infrastructure, host customer data and are already experienced in billing customers for services (e.g. on a per use or flat-rate tariff). From that perspective, operators are in a good position to offer video and TV over their own, managed IP-based networks.

The evolution of fixed telecommunication networks is often referred to as "Next Generation Network" (NGN). While traditionally one service was run over one dedicated network, NGN is an activity to consolidate existing networks, keeping and migrating the customer base to triple or even quadruple play services, i.e. phone, Internet, (IP)TV and mobile. Operators currently already possess some capabilities to address the challenges mentioned in Section 9.3.3.2:

- Operators have the capabilities to prioritize traffic to meet QoS requirements of the services they offer (e.g. in case of Voice over IP) and can use this to address the QoS issues of IP networks. If the traffic traverses other networks, the overall service performance depends also on the treatment of the traffic in those networks.
- Operators already have infrastructure to charge their customers for services in place. Thus they can leverage this infrastructure to also charge for new services through the normal billing process. This removes the risk of, e.g., credit card fraud. Also, customers trust operators more than service providers in the Internet as they often have already a long lasting relationship with them (especially in case of former incumbents).

- Operators are in control of the network ecosystem, i.e. what kind of services are accessible via the managed part of the network as well as what kind of user equipment is certified and able to access this part of the network (e.g. in terms of codec capabilities, content protection mechanisms supported, etc.). This allows for smooth but slower migration of services (this is also a challenge as described in Section 9.3.4.3). If operators offer multiple services, they can offer them in an integrated way (e.g. showing the caller ID of an incoming VoIP call on the TV screen). In a managed environment this can be controlled much more easily than in an unmanaged network.
- Operators already know their customers and how to authenticate them securely. Operators could even profile them, however regulation and legislation have to be considered here requiring explicit user consent or preventing user profiling altogether. Depending on these issues, the targeted form of advertisement may or may not be available to the IPTV operator.
- As outlined in [24], operators could establish themselves as (or co-operate with) so called identity providers, offering users secure but anonymous access to third party video services in a single sign-on fashion. On top of that, they could offer billing/charging services to third party services.
- Due to the managed nature of the network, the challenge of mis-labeled content (see Section 9.3.3.2.1) or access to inappropriate content violating local regulation is typically prevented.

9.3.4.2 Special Case: IMS and Non-IMS-Based IPTV in International Standards

International Standardization Organizations like ITU-T, ETSI or ATIS have already defined systems for delivery of IPTV services over managed networks and are evolving their specifications. Typically, they distinguish between a dedicated platform for IPTV services and an IPTV platform based on 3GPP's IP Multimedia Subsystem. From an end user perspective both approaches achieve similar results through different means.

The exemplary ETSI TISPAN architecture of the dedicated IPTV platform [15] as shown in Figure 9.2 focuses only on offering IPTV services. To consume services, a device first needs to discover the IPTV services offered by the IPTV service provider through the so called Service Discovery and Selection (SD&S) module. After the user of the device selects the content and service to consume, the device interacts with the Customer Facing IPTV

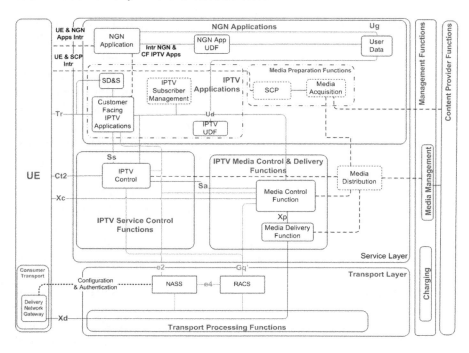

Figure 9.2 Dedicated IPTV subsystem functional architecture [15]. © European Telecommunications Standards Institute 2009, 2010. Further use, modification, copy, and/or distribution are strictly prohibited. ETSI standards are available from http://pda.etsi.org/pda.

Application and the IPTV control functions which then connects the user's device (User Equipment or UE, typically this is the Set-Top-Box or TV set) to the appropriate Media Function serving the selected content. Other components depicted in Figure 9.2 are not described here and the interested reader is referred to [15].

Additional features like integration of IPTV with other operator services (e.g. telephony or presence) require gateways to these services. Also, the interactions with components for charging support, access authentication, QoS control support and subscriber management need to be implemented in the dedicated platform approach.

The exemplary ETSI TISPAN IMS-based IPTV functional architecture [21] as shown in Figure 9.3 is realized by relying on 3GPP's IP Multimedia Subsystem (IMS, denoted "Core IMS" in Figure 9.3). Before a device can use IMS-based IPTV services, it needs to register with IMS and discover

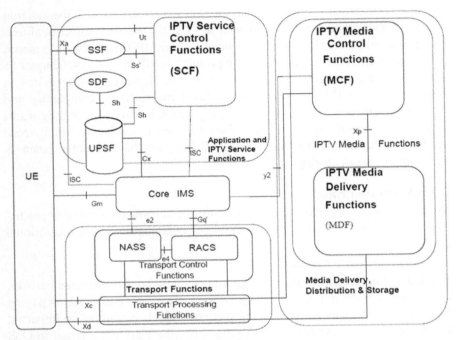

Figure 9.3 IMS-based IPTV functional architecture [21]. © European Telecommunications Standards Institute 2009, 2010. Further use, modification, copy, and/or distribution are strictly prohibited. ETSI standards are available from http://pda.etsi.org/pda.

services offered by the provider. The registration step follows standard 3GPP procedures and for service discovery the device contacts a SIP application server called Service Discovery Function (SDF). Through the discovery step the device (UE) receives a list of one or more Service Selection Functions (SSF). These SSFs serve Electronic Program Guides containing (potentially personalized) lists of IPTV services and content. After the user selects a service, the UE invokes another SIP Application Server called Service Control Function (SCF) which checks the user's subscriptions, does charging and locates content. Then, it routes the UE's SIP requests to Media Functions (MF) which deliver the media to the user's device. The other depicted entities in figure 9.3 such as the UPSF, the NASS and RACS provide ETSI TISPAN Next Generation Network specific functionality for service level platforms and are not explained in detail here.

Since IMS already supports charging, QoS control, access authentication and subscriber management IMS-based IPTV can reuse these functionalities. Integration with other IMS-based services such as SIP Telephony or Presence is less complex as the service signaling protocol is SIP in all cases. Chapter 11 depicts converged video services in more detail. However it is necessary to consider that the IMS platform itself adds some features, complexity and levels of abstraction that are not necessarily needed if an operator only wants to deploy IP-based video services. Therefore, each service provider needs to choose which platform/model for IPTV best suits his individual business needs and network strategy.

9.3.4.3 Challenges

While managed networks can present solutions to some of the challenges listed in Section 9.3.3.2, some still remain to be challenging and also additional challenges arise.

9.3.4.3.1 Service Innovation   The first challenge is the managed network itself – services are not deployed lightly: there are existing subscribers paying for certain services delivered over this network which should not be impacted by the new service. Additionally, paying subscribers are not tolerant to QoS issues or prolonged maintenance periods. Thus, *service introduction and evolution is a delicate issue for Operators, which needs careful (network) planning and execution – in turn slowing down innovation.* This motivates operators to follow and mimic successful services on the Internet, but given that typically operators are subject to regulation and legislation, not every service can be ported to the managed network easily. This might decrease user acceptance since it may delay service functionality. For example, consider a YouTube-like service which required 24 hours for checking content prior to making it available to other users.

9.3.4.3.2 Limiting User Freedom   Another challenge is *convincing the subscribers to pay for the managed network service when the Internet offers similar or even more flexible services at lower cost*: on the Internet users can get access to all kinds of content (e.g. via file-hosting services [3]), often for low cost and with flexible usage. For example, today PVR services exist on the Internet which legally record broadcast TV content for users. These services offer the user to download the content and watch it an unlimited number of times and to copy it across multiple devices. Comparing this to existing IPTV services in market today of which some limit Trick Play capabilities (e.g. to prevent skipping advertisement breaks), disallow content being moved from the recording device or limit the time of usability of the recording. These limitations appear hard to justify towards the (paying) IPTV consumers when compared to the services offered over the Internet.

Unfortunately, said flexible access to content or services in the Internet is not necessarily in consent with content rights owners who often price, restrict, cut or package movies differently according to regions. This is the afore-mentioned legal challenge in the Internet as described in 9.3.3.2.1.

9.3.4.3.3 QoS Challenge: The Home Network   While network QoS for IPTV is typically not an issue in managed networks since the operator con-trols the delivery infrastructure, the *QoS inside the home* – today complex in-home infrastructure situations are common – can be an issue impacting the overall quality of video delivery. For example, consider a subscriber paying a monthly fee for an IPTV service. While the IPTV data might be delivered with sufficient QoS to the home, quality of service within the home network might be drastically compromised. For example, the user might have star-ted several parallel file transfers inside his home (e.g. to backup the family video archive). This additional data transmission might cause congestion on in-home equipment which in turn negatively affects the IPTV traffic. Even though the poor user experience in this case is not the operator's fault, it may be enough for a non-technical user to call customer care or actually terminate his contract. Exactly for this reason, ETSI TISPAN studied the interaction between the Home Network and its Admission Control System in a technical report [20].

One solution for the IPTV service provider to address this is to use SVC or adaptive streaming technologies as described in Section 9.3.3.2.2. However, it remains uncertain whether paying customers would accept that *TV signals vary in resolution* over time depending on the traffic situation in the network.

9.3.4.3.4 Synchronicity of Delivery   A general challenge in many trans-mission systems is synchronization of the reception across different locations. When comparing satellite TV with cable TV in the same location, it is no-ticeable that the two delivery networks are not synchronized. This is not surprising, given that they are disparate technologies and transmission net-works. However, in traditional TV broadcast using the same transmission technology reception is accurately synchronized. In IP networks, this is generally not the case and special means, e.g. by forming groups of recipients and adding artificial delays to faster IP network paths are necessary to synchronize the signal reception (see [25] for a comparison of techniques). While perfect synchronicity cannot be guaranteed (e.g. because the rendering device hardware or the codec computations might differ), this minimizes at least the network's impact on the synchronicity of IPTV data packets. It remains to be proven if large scale synchronization, e.g. across a nationwide IP delivery network, is feasible this way. However, this way synchronicity can at least be achieved for smaller groups of users, e.g. a neighborhood. In managed networks with paying users, it is expected that without synchronization users will easily be upset. For example, consider the case of a football match transmission which is received earlier by a user's neighbors than by the user himself. Hearing his neighbors cheer over a goal while the user has not even seen the striker score will not lead to a satisfying TV experience.

9.3.4.3.5 Personalization   While the Operator knows the subscriber data and is technically able to track user behaviour, privacy regulations and national legislation may limit the usage of this personal data. Additionally, as mentioned also in Section 9.3.3.2.4, multi-user service personalization is not yet solved. However, this might become a compelling differentiator from traditional TV offerings.

9.4 Conclusions: Quo Vadis IPTV?

This chapter introduced TV and video services delivered over IP-based networks and explained that they conceptually differ from traditional broadcast TV in regards to personalization and interactivity – these differentiators promise to be a huge market driver for future IPTV services. Further, the previous sections also outlined the challenges implied by using managed or unmanaged IP networks for delivering IPTV services.

While up to now there is no IP-based TV service that serves as many customers at the same level of quality and reliability as the broadcasting business does, various trends in IP-based TV and video delivery are currently observable:

- Operators (especially Telecom) are offering IPTV over their IP networks while cable operators move to upgrade their networks to offer telephony and Internet access services (and IPTV-add on services probably in the long run).

- TV and STB manufacturers integrate Internet Access (e.g. the Yahoo Widgets for Yahoo Connected TV [11]) into their devices converging previously separate services and applications. Commonly this is referred to as "Hybrid TV". Often the manufacturers and/or service providers restrict the use of the Internet to their own portals in order to establish a strong position in the value chain of future TV and video services.
- Several Game Consoles (e.g. Microsoft XBOX, Sony Playstation 3) support e.g. in-home media streaming through Universal Plug and Play (UPnP, [10]) support. Given that these have also Internet capabilities it is possible that these devices evolve to multi-purpose STBs or in home Media Centers – especially since, e.g., Microsoft also offers IPTV solutions and Sony Pictures on the other hand is a prominent content provider.
- With its Android Operating System, Google offers a platform not only for mobile phones but also for STB devices and recently announced the launch of Google TV (in collaboration with Customer Electronics Powerhouse Sony). In addition Google owns YouTube and pushes the Internet-based video/TV service experience.
- For several years, Apple offers "Apple TV" as well as compelling consumer multimedia devices like the iPod, the iPhone and the new iPad devices. Innovative ideas for multimedia consumption and related services making use of these devices can be expected in the future.

The high level of activity in the field of IP-based TV and video services unfortunately is one of its major obstacles: there is a variety of deviating IPTV standards leading to a situation in which even standards-compliant pieces of equipment may not be compatible with each other. Different industry fora try to address this problem, but might further increase fragmentation. To make things worse, the field of DRM has traditionally been a landscape of competing systems. Open standards have yet to find solutions to that problem. Furthermore, as proven by the different hybrid approaches (e.g. Widget Technology) multiple companies try to position themselves in the content/service value chain. Unfortunately for end-users this often comes at the cost of incompatible services and devices also.

Nonetheless, given the prospect of interactive, personalized services in a market with strong projected growth, IP-based TV and video services are going to play a significant role in the converged triple or quadruple play service landscape of the future.

References

[1] Electronic sports, http://en.wikipedia.org/wiki/Electronic_sports.

[2] ETSI. ETSI TR 102 794: Media Content Distribution (MCD); 3D gaming graphics delivery, work in progress.

[3] File hosting service, http://en.wikipedia.org/wiki/File_hosting_service.

[4] Internet, http://en.wikipedia.org/wiki/Internet.

[5] Legal issues with BitTorrent, http://en.wikipedia.org/wiki/Legal_issues_with_BitTorrent.

[6] Napster, http://en.wikipedia.org/wiki/Napster.

[7] Onion routing, http://www.onion-router.net.

[8] Onlive, http://www.onlive.com.

[9] Overview of scalable video coding, http://mpeg.chiariglione.org/technologies/mpeg-4/mp04-svc/index.htm.

[10] UPNP forum, http://www.upnp.org.

[11] Yahoo connected TV, http://developer.yahoo.com/connectedtv.

[12] YouTube, http://www.youtube.com.

[13] FCC formally rules Comcast's throttling of BitTorrent was illegal, http://news.cnet.com/8301-13578_3-10004508-38.html, 2008.

[14] ETSI, ETSI TS 181 016: Service layer requirements to integrate NGN services and IPTV, version 3.3.1, July 2009.

[15] ETSI, ETSI TS 182 028: NGN integrated IPTV subsystem Architecture, version 3.3.1, October 2009.

[16] What browser wars? The enterprise still loves IE 6, http://news.cnet.com/8301-17939_109-10231713-2.html, May 2009.

[17] Akamai reports first quarter 2010 financial results, http://www.akamai.com/html/investor/quarterly_releases/2010/press_042810.html, 2010.

[18] Court: FCC has no power to regulate net neutrality, http://news.cnet.com/8301-13578_3-20001825-38.html, 2010.

[19] Credit card fraud trends, http://www.merchant911.org/fraud-trends.html, 2010.

[20] ETSI, ETSI TR 182 031: Remote CPN QoS control; Study on CPN-RACS Interaction, version 3.1.1, September 2010.

[21] ETSI, ETSI TS 182 027: IPTV Architecture; IPTV functions supported by the IMS subsystem, version 3.4.1, June 2010.

[22] Internet Crime Complaint Center Report 2009, http://www.ic3.gov/media/annualreport/2009_IC3Report.pdf, 2010.

[23] IPTV global forecast – 2010 to 2014 – semiannual IPTV global forecast report, http://www.researchandmarkets.com/reportinfo.asp?report_id=1246317, June 2010.

[24] Winkler et al. Enriching IPTV services and infrastructure with identity management. In *Global Telecommunications Conference, IEEE Globecom*, 2008.

[25] M. Garcia F. Boronat, J. Lloret. Multimedia group and inter-stream synchronization techniques: A comparative study. *Elsevier Information Systems* 34:108–131, 2009.

10

Towards "Semantic IPTV"

Sahin Albayrak, Juri Glass, Baris Karatas, Stefan Marx,
Torsten Schmidt and Fikret Sivrikaya

DAI-Labor, Technische Universität Berlin, 10587 Berlin, Germany;
e-mail: sahin.albayrak@dai-labor.de

Abstract

IPTV promises transferring the innovation strength of the Internet to the television platform, with its tremendous potential to enhance the experience and benefits of multimedia-based entertainment, communication, and advertisement services from the perspective of both users and providers. Today's upcoming IPTV services provide an increasing amount of video content and TV broadcasts to users. Moreover, all media resources on the Internet may also be accessible to future IPTV users. Therefore there is a need to assist people in discovering and accessing the desired content, and in filtering out irrelevant or annoying ones in the vast ocean of multimedia content. This chapter presents the semantic IPTV vision and provides a comprehensive overview of the main aspects of an IPTV system with special focus on the involvement of semantic elements. The main enabler for semantic IPTV will be the creation of a semantic space that describes the content, context, and user preferences. Unlike passive audience of standard television broadcast, semantic IPTV users will be able to interact with, search for and easily access any content of interest in the way they can currently access data on the Web. Moreover, personalization services will enable automatic transformation of user benefits and interests into recommendations or even personalized channels that are generated on-the-fly. After a comprehensive overview of all these elements comprising the semantic IPTV vision, a possible reference architecture for realizing this concept is also presented in this chapter.

Anand R. Prasad et al. (Eds.), Advances in Next Generation Services and Service Architectures, 197–230.

Keywords: IPTV, semantic media, IMS, personalization, semantic recommendation, multimedia metadata, semantic gap.

10.1 Introduction

IPTV has tremendous potential to enhance the experience and benefits of multimedia-based entertainment, communication, and advertisement services from the perspective of both users and providers, as presented in Chapter 9. Through the use of IPTV, content providers can introduce new value added services and therefore new revenue channels along with media broadcast, whereas users can enjoy the larger amount of content and services with added interactivity. As this large amount of content or broadcast channels reaches overwhelming levels, sequential channel skipping or manually searching for programs within EPGs become quite challenging and discouraging, if not infeasible, for the user. Therefore with the rise of IPTV, there will be a growing need for making the content easily discoverable and assisting the users in content selection and filtering.

IPTV enables providers to identify and address individual users (or households), which is one of the fundamental differences from traditional TV broadcast. This in turn creates an opportunity for personalization and recommendation services as well as more effective targeted advertisements to be provided to users. Exploiting this potential to its full extent demands a *semantic space* to be formed, which should collectively describe content, advertisements, user interests and context.

This chapter presents the *Semantic IPTV* vision and provides an overview of the main enablers to achieve this vision. Unlike passive audience of standard television broadcast, semantic IPTV users will be able to interact with, search for and easily access any content of interest in the way they can currently access data on the Web. Moreover, personalization services will enable automatic transformation of user benefits and interests into recommendations or even personalized channels that are generated on-the-fly.

While there exist semantic approaches developed and used in the area of IPTV, they generally lead to fragmented solutions trying to address some specific problems. In order to take full advantage of semantic technologies in IPTV and to exploit all potential benefits that IPTV can deliver, we believe in the need for a holistic approach that permeates all IPTV-related areas by interconnecting and mutually boosting them. We identify three major areas in which semantic approaches are most relevant in the context of IPTV;

multimedia information extraction and retrieval, personalization, and as a derivative problem field targeted advertisement.

The chapter starts with a survey of multimedia information extraction and retrieval approaches and the description of semantic gap, which are at the heart of a semantic space for IPTV. This is followed by an overview of user modeling, user profiling, and content selection mechanisms in order to enable personalization and recommendation services for IPTV. A discussion on targeted advertisements and how semantics can improve IPTV-based targeted advertising is provided next. An architectural view of how those semantic enablers can be potentially integrated in a standard IPTV architecture precedes the concluding section that features a summary and future research directions.

10.2 Semantic Space for IPTV

In today's IPTV landscape users get in touch with many different types of media. Movies, TV shows, program guide, photos and music are just a few examples of content available on IP television and other IPTV-based services. But these services can only be accepted by consumers if their expectations in terms of functionality and usability are met. In order to achieve the expected Quality of Experience (QoE), IPTV infrastructures depend on a solid and machine oriented understanding of the provided content. Only with such an understanding, IPTV can provide those advanced functionalities like personalized recommendations or targeted advertising.

Metadata, i.e. data about data, form the key enabling information structure that could provide this understanding. Metadata can be utilized to archive, organize and search many different information objects. The importance of metadata is widely considered in the research community and also by standardization bodies. The most prominent standards related to our vision of semantic IPTV are promoted by the Motion Picture Expert Group (MPEG), the Dublin Core Metadata Initiative (DCMI), the International Press Telecommunications Council (IPTC) and the Society of Motion Picture and Television Engineers (SMPTE).

MPEG-7 for example is a well-known standard, which provides guidance for the annotation of multimedia content. It defines elements for metadata consisting of *Descriptors* (defining syntax and semantic) and *Description Schemes* (providing a structure for semantic relations between descriptors and description schemes). The semantic description scheme [20] allows a representation consisting of semantic objects (e.g. locations or people) and semantic relations. The Description Definition Language (DDL) [45] allows describing and modifying these schemes and descriptors. These descriptors can be connected via the Web Ontology Language (OWL) and the Resource Description Framework (RDF) [37] to the emerging semantic web. For instance, the information elements of the Electronic Program Guide (EPG), which provides information such as title, channel, genre and others to the program, could be linked to existing resources on the web, thereby enriching the available knowledge base.

Multimedia content is currently described by humans to the most extent, providing an excellent quality; however, the required effort will become more and more unbearable with increasing amounts of media content. On the other hand, machine processing is cheap but at the moment it lacks the desired quality in automated metadata extraction. Furthermore, humans can describe media data on a high level (topic, mood, etc.) while machines have to use advanced pattern recognition techniques to extract such metadata from raw data. This problem, the gap between human and machine interpretation, is known as the *Semantic Gap*. We further elaborate this issue in the next subsection, after which we cover the current state of the art on metadata extraction and multimedia retrieval techniques.

10.2.1 The Semantic Gap

Machine-based automatic annotation of large amounts of multimedia content, such as text, images, speech and video, is the foundation for semantic IPTV. This requires effective methods for indexing, organizing and searching information. While machines can easily extract low-level metadata, e.g. the gamma value of a picture; high-level metadata, i.e. the human accessible interpretation of content needed for most IPTV augmenting services, are quite different to obtain. *Semantic gap* refers to the difficulty of describing audiovisual content automatically, as a result of "the lack of coincidence between the information that one can extract from the visual data and the interpretation that the same data have for a user in a given situation" [64].

The main challenge is not only the transfer of machine recognized low-level features to high-level concepts, but also the translation from human specified concepts to machine recognizable features. For instance, in the search scenario depicted in Figure 10.1, user defined queries need to be translated and matched onto a vocabulary set representing the general knowledge of the system. From this information, concepts (high-level semantics) can be selected using prior knowledge regarding concept detection techniques.

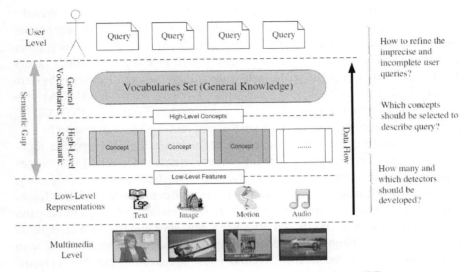

Figure 10.1 Semantic gap in multimedia content access [75].

Thereafter, objects can be extracted and transferred into specific detectors (low-level features) using object detection techniques.

Ontologies and machine learning techniques are approaches to bridge the semantic gap. Semantic data can be retrieved through manual annotation, Hidden Markov Models and Support Vector Machines (SVM). While Hidden Markov Models enable pattern recognition allowing one to describe patterns of time-based events, Support Vector Machines are used to describe problems which can be solved through machine learning techniques. These techniques require an administrator to train a system working with test data, only after which these techniques can be applied on real data.

10.2.2 Multimedia Metadata Extraction

We differentiate here between metadata extraction applied to images, to audio signals and finally to video content, noting also that video metadata extraction is mostly based on audio and image extraction techniques.

An image can be described by colors, texture and shape. A simple approach might be the quantification of the number of pixels for each color in the image, i.e. building a histogram, which then leads to an image-to-image comparability which might be used for topic recognition. Also, other transformations like discrete Fourier transforms (DFT) or discrete Wavelet transforms (DWT) enable image comparison. While statistical approaches use measures, cover contrast and directionality for variation of intensity, other approaches extract connectivity and density in image regions. IBM's Query By Image Content (QBIC) [21], Multimedia Analysis and Retrieval System (MARS) [35], Photobook [56], the Movie Content Analysis (MoCA) project [42] and Content-based Image and Video Access System (CONIVAS) [2] are projects covering visual feature extraction and spatial region search.

Audio analysis is divided into audio characterization [49], speaker identification [25] and keyword spotting. Audio characterization differentiates silence, speech, music and utilizes features like bandwidth or frequency. Speaker identification distinguishes between different people and different parts in audio streams. Speech recognition can then be applied on those parts recognized as spoken words and afterwards keyword spotting can be done as in textual natural language processing. Speech recognition works well with few users (speaker-dependent systems) while speaker-independent systems need to be trained and mostly have a limited vocabulary.

Video analysis applies content-based video analysis approaches split into video content analysis, video structure parsing, video summarization and video indexing [62]. While image analysis techniques fit well for low-level retrieval, they are insufficient to infer semantic concepts. For that reason additional information like subtitle, speech and visual characters are taken into account in order to derive metadata using various techniques. For example, subtitles can be summarized grabbing keywords, speech can be mapped into text using speech recognition software, and characters can be identified with Optical Character Recognition (OCR) solutions. Research in [33] shows advance of OCR against speech recognition. The result is a set of keywords representing the video content. Video structure parsing is the subsequent step in the video analysis chain subdividing the video content into scenes and each scene into shots. Thereby shot boundary detection techniques make use of content analysis approaches for optimization. Collecting key frames and using prior knowledge the content is summarized for representation [18]. Based on those techniques an index or a database could be generated for semantic IPTV to provide search or other data related functionalities on the extracted metadata.

10.2.3 Multimedia Retrieval

Multimedia retrieval is a well studied research area, and we focus here on multimedia retrieval improved with semantic technologies, as we envision this as a key principle for the realization of semantic IPTV. As a starting point for investigating semantic multimedia retrieval, a widely adopted standard TV-Anytime (TVA) could be exploited. TVA is a set of specifications for storing multimedia contents on digital storage in consumer end devices. It should ease the access of content by standardized formats describing content and was developed in two phases. While first phase enabled searching and accessing content on local or remote storage, the second phase allows sharing and distribution of content whereas even segment search is applicable. A step towards utilizing personalization was made in [36] by developing an architecture allowing content-based matching and retrieving information on the description of broadcast content. Tsinaraki et al. [72] enriched TV-Anytime with domain-specific knowledge applying the Web Ontology Language and ensuring interoperability among applications. Similarly, Weibel et al. adopted this idea, as one part in the semantic IPTV chain, and covered mentioned semantics applying state of the art techniques like the BBC Ontology [28]. Other metadata standards are the Dublin Core (DC) [76] and Descriptive Metadata Scheme-1 (DMS-1).

Multimedia Information Retrieval (MIR) approaches [5, 31, 41] address feature-based similarity search and started already in 1980s. Later the results of projects like Query By Image and video Content (QBIC) [21] and Virage [4] were adapted into image search services on the Internet. The Informedia project by the Carnegie Mellon University enabled content-based video retrieval including speech indexing, motion extraction and video summarization [74]. Informedia-II implemented dynamic features for extraction, summarization, visualization, and presentation of distributed video. Begeja and Van Vleck describe in [8] automatic metadata generation in contextual advertising scenarios covering features such as closed caption, face detection and speaker boundary detection. Researches identified the need to focus more on semantic concepts since earlier systems are excessively complex and not that user-friendly as desired.

Since 2001 Text Retrieval Conference Video Retrieval Evaluation (TRECVID) [63], organized by the National Institute of Standards and Technology (NIST), evaluated video retrieval supported by semantic concepts. With focus on automatic segmentation, indexing and content-based retrieval heterogeneous content, e.g. news or documentaries, were offered to the research community in order to benchmark their development by evaluating the performance of concept detection [51]. From this point on researchers have built detectors for high-level semantics. The number of semantic concepts increased from 10 in year 2002 to 39 concepts in 2005. This status was termed as Large Scale Concept Ontology for Multimedia (LSCOM) [50], in which different representatives worked collaboratively towards defining a set of 1000 concepts where each concept should cover realistic video retrieval problems by detecting objects in video data sets. With requirements such as utility, coverage, feasibility and observability it should cover physical objects including animated objects (people or animals) and static objects ranging from large-scale (streets or buildings) to small-scale (devices or basic commodities), actions (explosion), events (soccer game), locations (outdoor or nighttime), settings and graphics. The MediaMill set [65], an intermediate step, included 101 concepts, while the full LSCOM set contains over 2600 concepts and the complete LSCOM ontology.

In consideration of the semantic gap, Snoek et al. [66] propose automatic video retrieval based on high-level concept detectors. A multimedia thesaurus, a set of machine-learned concepts enriched with semantic descriptions, establishes a connection between user's vocabulary and available concepts. Therein three detector selection strategies, i.e. text matching, ontology querying and semantic visual querying, select the best detector from the thesaurus, showcasing the dependency of the quality and the size of the thesaurus. As expected, one strategy is superior to another depending on the query. On the other hand, [64] evaluates the correlation between high-level concepts and low-level features. According to [32] the quality of results is necessarily depending on the number of used concepts. Driven by the question of how many concepts are sufficient for a good retrieval system, they come to the conclusion that even if the hit rate is low, good results can be expected with enough concepts. The simulation was realized by applying TRECVID and using LSCOM concepts. By contrast the approach in [75] proposes an Ontology-enriched Semantic Space (OSS) for modeling and arguing concepts in a linear space to bridge this gap by constructing and reasoning of a multimedia-based ontology. Only modern approaches enable the semantic search by pooling a set of concepts that can bridge the semantic gap. They figure out that a minimal set of concepts is sufficient in combination with the OSS. Meanwhile IBM developed MARVEL, a multimedia analysis and retrieval system [52], enabling machine learning techniques and a multimedia search engine. In contrast, MIRACLE by AT&T [44] creates automated algorithms for media processing. An integration of semantic technologies providing personalized recommendations based on well-known ontologies is realized in [28].

In summary, the described approaches work better for domain-specific use cases, e.g. "sports", since objects, events and movements are easier to retrieve. Conversely, news work well too since additional information is utilized to retrieve metadata. In a normal user scenario when the user watches heterogeneous types of programs, there is a lack of ontologies being able to describe the whole spectrum of multimedia content. Furthermore, discussed research projects have shown that more semantic coverage and classifiers are required, more than concepts. Even if research projects achieve good results on each level of the semantic gap chain, it requires computer-driven semantic technologies and mappings to appropriate ontologies for a complete bottom-top or top-bottom solution in order to bridge the gap. Besides these aspects most approaches work on video content and not on live broadcast. In the near future this might arise as an important challenge to overcome, demanding efficient real-time semantic retrieval solutions. In order to achieve the QoE and the functionalities we envision for the viewers of semantic IPTV, unified strategies for information retrieval with simple interfaces, and a user-centric approach are required.

10.3 Personalization and Recommendation

Today's upcoming IPTV services provide an increasing amount of video content and accessible TV broadcasts with the ability to address individual users. Potentially every media resource will be accessible to IPTV users delivered over IP networks managed to support the required level of quality of experience (QoE). To optimize QoE in an environment with an ever-increasing amount of available multimedia resources, there is a need to assist people in selecting the appropriate content and to filter out all the irrelevant or even annoying ones. This can be achieved through personalization.

Personalization is the approach of providing an overall customized, individualized user experience on the basis of profile and context information associated with a user or a group of users. Therefore it requires the ability to select content satisfying the preferences of users and to decide if a given user should be provided with a particular piece of information [59]. Applied to IPTV services like electronic program guide (EPG), video-on-demand (VoD), personal video recorder (PVR) or targeted advertisement, personalization is assumed to become the key factor of differentiation and success [14, 26]. It

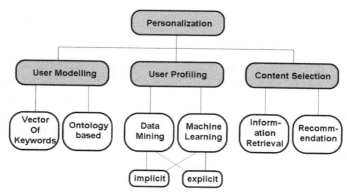

Figure 10.2 Personalization taxonomy.

adds value to IPTV services not only for the end-user, but for all players in the IPTV value chain, i.e. network, service and content providers.

In the rest of this section we briefly introduce a personalization taxonomy, based on which we discuss the problem space of IPTV service personalization. We make no claim to be complete since personalization is a multifaceted topic, but we do point out some domain specific characteristics, demonstrate how semantic technologies might help improve the QoE and motivate some future directions.

10.3.1 Personalization Taxonomy

One important goal to improve QoE in IPTV services is to provide users with what they need without them asking for it explicitly. The central technology used to achieve this goal is personalization [48]. It implies the selection and delivery of content that is tailored to interests of a particular user or a segment of users. The ability to provide appropriate content or items in a user-related fashion requires that personalized systems must be able to infer what a user needs based on previous or current interactions with that user, and possibly with other users. As stated by Pertersen et al. [27], "behind every instance of personalization is a profile that stores the user preferences, context of use and other information that can be used to deliver a user experience tailored to their individual needs and preferences".

Figure 10.2 depicts a high level view of personalization taxonomy. We distinguish three different logical building blocks: user-modeling, user-profiling and content selection. *User modeling* subsumes the representation of individual users of the system, *user profiling* denominates the process of generating specific user model instances, i.e. user profiles, by gathering and processing user-centric information, and *content selection* refers to choices made by the system on behalf of the user for delivering the content of interest. These terms are further elaborated next.

10.3.1.1 User-Modeling

A key part of personalization is the nature of user models. The purpose of user modeling is to separate user modeling functionality from user-adaptive application systems [38]. User models incorporate all the knowledge of the system about its users. User modeling can be rather simplistic, e.g. using attribute-value pairs or a vector of weighted keywords, or more sophisticated based on semantic knowledge and ontologies.

The use of semantic technologies increases the *expressiveness* of user models, which is the ability to express as many types of facts and rules about the user as possible. Personalized systems and services benefit from ontology-based user models, since ontologies represent explicit specifications of concepts and relationships that exist between them [29]. Such specifications form a semantically rich knowledge base. Furthermore semantic technologies bring strong *inferential capabilities* to perform reasoning on top of ontology-based user models resolving conflicts when contradictory facts or rules are detected [10, 38]. This improves the accuracy of expressed facts and the reliability of the modeled information.

Heckmann et al. introduce the general user model ontology GUMO for the uniform interpretation of distributed user models in intelligent semantic web enriched environments [34]. Besides gained expressiveness they point out the demand for semantic user modeling in order to enable *knowledge sharing*. User modeling commonly serves more than one application instance at a time. One major problem and challenge here is the lack of standardization corrupting interoperability and harmonization of user profiles across services [27]. As stated in [10] , "to facilitate reuse of user modeling data by multiple personalization applications, there is a need for semantically-enriched and ontology-based representation of the user models". Hackmann et al. propose the use of OWL and modern semantic web languages. They represent a proven solution for large-scale knowledge sharing and integration of external data [11].

Applied to user modeling, semantic web technologies and languages enable standardized, interoperable user models with explicit semantics, a formal representation and capabilities for formal reasoning [19]. Since user models

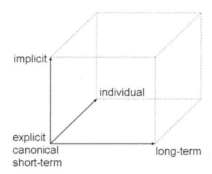

Figure 10.3 Three-dimensional space of user profiles.

are a key input of personalized systems, choosing the semantic representation has a profound effect on all personalization techniques.

10.3.1.2 User-Profiling

User profiling is the process of generating specific user model instances, i.e. user profiles, by collecting and processing user-centric data that represents the interests and activities of users. Data mining and machine learning techniques are applied to this process in order to discover, generate and aggregate user profiles based on collected profile information, e.g. user identity, preferences, main activity or state of the user [67]. Figure 10.3 visualizes the characterization space of user profiles introduced by Rich [58] defined by the three dimensions: *explicit vs. implicit, individual vs. canonical* and *short-term vs. long-term.*

Canonical user profiling refers to one generic model of a canonical user. It does not keep track of individual users but points to homogeneous groups and segments of users. Since personalized services generally deal with heterogeneous user needs, individual profiling is crucial. It is all about identification of single users and the association of their interactions with their individual profile. This introduces the need for identity management and data integration functionalities. Furthermore individual profiling has to scale with the amount of users and the size of the profiles, which grows with every stored individual facet.

While explicit information collection demands user interaction and feedback i.e. rating, voting or queries, implicit profile information is derived by the system automatically, for example through log file analysis. The choice between implicit and explicit profiling directly influences the *degree of user effort*, an important aspect considering the quality of personalization. A tradeoff exists between decreasing the degree of user effort and the reliability of user profile information, since most of the information contained in implicitly gathered profiles are essentially educated guesses rather than facts. Thus "the system must have some way of representing how certain it is for each fact, in addition to a way of resolving conflicts and updating the model as new information becomes available" [58]. We can conclude that semantic technologies with inferential capabilities are needed to handle implicit profiling well.

In short-term profiling user information is gathered solely from interactions during the current session. Such user profiles are typically not persistent as opposed to long-term profiles, which have to be stored and maintained. Short-term profiling typically deals with short-term interests of the users, whereas long-term profiling processes stable, long-term interests. This raises the need for evolutionary enhancements of user profiles, which requires advanced data integration capabilities (e.g. following linked data principles [11]).

Along all those dimensions there is an increasing need for semantic techniques in user profiling, due to the individual expression of facets, vagueness of implicit profiling and required data integration capabilities for long-term profile maintenance.

10.3.1.3 Content Selection

All the information mentioned for user profiling is used for personalized content selection. Based on the user profiles the personalization system computes a prediction of the user's level of interest in, or utility of, specific content categories or items, and ranks these according to calculated weights. Therefore personalization can be seen as a prediction task discovering what the user might be most interested in. Since personalization has this characteristic in common with recommendation systems, both are often used synonymously. In the following we will focus on recommendation systems as a variant of content selection besides information filtering or retrieval approaches mentioned earlier in this chapter (Section 10.2.3).

As mentioned by Burke et al. "it is the criteria of *individualized* and *interesting and useful* that separate a recommender system from information retrieval systems or search engines" [13]. Recommendation systems are differentiated by the basic techniques they rely on. In the following we briefly describe *content-based*, *knowledge-based*, *demographic* and *collaborative* re-

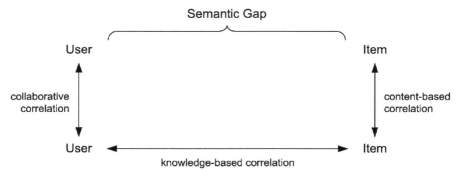

Figure 10.4 Correlations in different recommendation scenarios.

commender systems, as well as *hybrid* approaches that combine techniques for improved performance.

Content-based Recommendation – In a *content-based* system, items are recommended based upon a description of the items and a profile of items the user liked or disliked. This kind of recommendation is also called "item-to-item correlation" [60] as long as new, unknown items are correlated to already known and possibly rated items. The calculation of the correlation is heavily dependent on the description of items and how comparable these are.

Knowledge-based Recommendation – Knowledge-based recommendation attempts to suggest items based on functional knowledge about the items and inferences about preferences of users as well as knowledge about how a particular item meets a particular user need [13]. This can be seen as "item-to-user correlation" where both content models and user models are taken into account. Therefore this kind of recommendation system relies on knowledge engineering approaches, modeling the features of items that can be mapped to user interests. They often rely on semantic web technologies and languages like OWL, RDFS and RDF to describe concepts and their relations between features of items and users interests.

Demographic Recommendation – Demographic recommender systems are using statistical information of specific demographic classes to compute recommendations. Users are associated to demographic classes by categorization based on personal attributes like age, gender, social status, etc. This kind of recommender is suitable for advertising scenarios where a demographic class of users represents a specific targeted audience. Since demographic recommenders rely on background knowledge they are a subclass of knowledge-based systems.

Collaborative Recommendation – Collaborative recommender systems generate new recommendations based on comparisons of users to recognize commonalities between them. This so called "person-to-person correlation" technique is completely independent of any machine-readable representation of the items that are recommended. The assumption made by collaborative systems is that users who have similar interests would like the same items, and also that some of those users know some items that others do not yet know about. Therefore those items that are yet unknown to a specific user are recommended. Burke reports that collaborative recommendation works well for complex objects, such as music and movies where variations in taste are responsible for much of the variation in preferences [13].

Hybrid Recommendation – The class of hybrid recommenders is characterized through combination of several techniques for recommendation generation. All different kinds of recommender systems have their own problems and drawbacks, so that today's generation of recommender systems is almost always hybrid, combining different approaches to overcome weaknesses associated with individual techniques.

10.3.2 Recommendation for IPTV Services

While the management of user profiles is uncommon in today's TV networks, IPTV providers with experience from detailed subscription handling are well positioned to address user modeling and profile management challenges [40]. Friedrich et al. are describing "a so called IPTV service personalization function, that is used for gathering information on user behavior, the processing of user profile data and manual user ratings and for generating content recommendations" [26]. IPTV-related services that benefit from personalization are, e.g., EPG, PVR and VoD functionalities, parental control and targeted advertisements. Personalized EPG focuses on recommendation of upcoming live broadcasts, tailored to the users interests, while a personalized PVR might support automated recordings of relevant broadcasts identified based on calculated recommendations. Advising VoD assets helps guiding the user as well as featuring specific assets for business reasons. Parental Control performs content filtering to protect users, especially children, from improper content and intelligent recommendations applied to targeted advertisement helps realizing a win-win situation for users and advertisers. In the following we discuss recommendation in regard to IPTV services based on our taxonomy and the dimensions introduced in [57]. The discussion points out where semantic approaches improve performance and identifies challenges for further research.

IPTV service providers usually aim for a long-term relationship with their users. Therefore they will want to maintain persistent, *long-term user profiles*, which accumulate information over time. Due to such length of relationship, IPTV services revert to rich knowledge about their users. Furthermore IPTV providers hold user accounts associated with transactions and demographic information. This characterizes a *deep user relationship* resulting in knowledge which is valuable for clustering techniques. IPTV services based on next generation networks (NGN), e.g. an IMS infrastructure, adopts strong identity management capabilities [1] enabling *individual profiling* and together with NGNs extensive session control capabilities bringing great opportunities for *implicit profiling*. As we have seen, in order to exploit the value of long-term, individual, implicitly gathered profiles, the use of semantic web technologies and languages in user modeling and profiling has to be considered. The challenge is to integrate them without violating the performance of IPTV systems and services. Today's IPTV systems generally have to scale with a large subscriber base, for which reason standardization activities focus on performance optimized solutions. Another challenge IPTV service providers have to deal with in regard to the potentially rich knowledge about the user is privacy. They must provide privacy-supporting policy tools to their users and have to follow national and international privacy legislations and conventions.

The problem structure of personalization in the IPTV domain from a recommendation point of view is commonly characterized by *selection-* and *exploration problems*. Solving selection problems, e.g., for EPG and VoD services takes the burden from the user, since the comparison of concurrently accessible multimedia assets is done on behalf of the user. One major challenge here is support for multiple users. Since TV typically addresses a social multiuser viewing experience (e.g. family, friends in front of the screen), personalization has to deal with groups of users concurrently using the same TV device. In [73] van Brandenburg et al. have shown that current IPTV architectures have all the ingredients to implement concurrent, multiuser TV use cases, but that major unsolved challenges exist in the area of usability (e.g. unobstrusive identification of persons in front of the TV). In regard to exploration problems, solving them means to support users and groups in seeking something interesting, where there is no specific goal (e.g. zapping behavior). This raises the requirements for recommendations designed for serendipity. Content-based approaches lack serendipitious effects and tend to cause over-specialization, so other techniques must be considered to increase QoE. Collaborative filtering provides serendipity and is a common solution for exploration problems. Candidate-critique conversational systems are also suitable. They are enabled by IPTV's integral return channel designed for interactivity. While the mentioned techniques have a big momentum of coincidence, semantic-based approaches are able to uncover serendipity in a controlled and meaningful way. For example, Zanker et al. [78] have shown that serendipity controlled by knowledge-based recommendations outperform other techniques due to catalog coverage. We can consider an IPTV provider who wants to increase VoD catalog coverage, due to business reasons, while providing the user a serendipitious experience. Exploiting controlled serendipity in content selection and exploration is considered as a key factor to improve the QoE of IPTV services, but also as a big challenge [79].

Life-span is an important dimension of the IPTV domain describing the time period in which a recommendation is valid. Live TV broadcasts, news content, etc. are ephemeral media with a short life-span, for which content-based recommendation techniques are preferred. Collaborative-filtering involves the new-item problem [3], where frequent updates of a semantic knowledge base are costly. Collaborative and knowledge-based methods are more suitable for, e.g., VoD services, which provide a quite static solution space for recommendations. Nevertheless this marks the challenge of how to profit from advantages of semantic techniques (e.g. expressiveness, knowledge sharing) when dealing with ephemeral content.

Another dimension is *criticality* which denotes the cost of wrong recommendations. IPTV services have low criticality compared to, e.g., medical domain where understandable recommendations are demanded with a high *degree of explanation*. In critical domains the application of knowledge-based systems is crucial to achieve a high degree of explanation establishing trust in recommendation. In less critical domains knowledge-based, semantic approaches are often omitted due to maintenance costs (e.g. of ontologies). However, since we assume that trust in recommendation is a key factor of influence to the QoE of personalized IPTV services, the use of knowledge-based approaches should be considered. If we think of VoD items a user is to purchase, trust in recommendation is crucial for acceptance. A key research challenge here is denoted by the question of how to visualize an explanation of a recommendation's quality to the user.

Some further future research challenges are exposed with the application of semantic-based personalization to the domain of IPTV services, which we point to by the following questions before we conclude this section:

- How can semantic web technologies be integrated in standardized IPTV infrastructures without violating performance requirements?
- Which tools, applications and architectures are required to exploit controlled serendipity in content selection and exploration?
- How can an IPTV support concurrent, multi user personalization in an unobstrusive and usable manner?
- How can the advantages of semantic-techniques like expressiveness, accuracy and explainability be applied to ephemeral content?
- How can explanations of a recommendations quality be computed and visualized, to increase the users trust in personalized services?

10.4 Targeted Advertisement

The Internet has grown enormously with the World Wide Web, whose main content is text. Google has successfully capitalized on this main ingredient of the Web by allowing people to efficiently reach the content they are looking for. On the other hand, Google's continued growth and success stemmed from ingeniously monetizing the search for and the publication of content, which again is mainly text. The context-sensitive advertising scheme through their AdSense[1] and AdWords[2] products composed one of their main revenue streams, by letting publishers effortlessly match the advertisements to the search results or the contents of a web page. Most major players in search and online advertisement businesses follow the same trend.

Though the Internet and the Web have witnessed significant innovations and transformations both technically and economically in as little as ten years, the much older publishing platform, the Television, has shown rather slow progress. The picture quality and user experience continued to improve but there have been few, if any, innovations in terms of the interaction of users with the TV, the advertising schemes, and related business models. TV broadcast and advertisement started in an analog world with the same advertisement model still in place in the era of digital TV. IPTV promises to transfer the innovation strength of the Internet to the television platform and

---

[1] http://www.google.com/adsense
[2] http://adwords.google.com

transform the way users watch, interact with, and benefit from one of their most traditional household items.

One important characteristic of the Internet becomes available to television with the shift to IP as the distribution network; the addressable network endpoint. In other words there is an addressable set-top-box and finally the notion of a single user. This allows content provider to serve consumers with customized content and the advertisers to target their ads to demographics or even to the individual users. Research shows that targeted advertising increases the total number of purchases [9], and therefore is the choice of many IPTV infrastructures.

Targeted advertising on the Internet is a widely studied research area and many research results can be adapted for targeted advertising over IPTV. There are two main kinds of targeted advertising, contextual-targeted (CT) advertising [53] and behavioral-targeted (BT) advertising [77]. Context-targeting exploits the content in which the advertising should be inserted, for instance an advertising for sports shoes is inserted into a web page about running. This could be compared to the advertising scheme known from plain television, in which the program that the ads should be inserted, the time and channel are almost the only clues an advertiser have in its disposal to select the advertisement to show. BT advertising in contrast utilizes the knowledge about the single user, such as click history or visited sites. BT advertising can not be done for plain television without additional equipment, but BT, CT or a combination of both can be used for IPTV.

As pointed out earlier the two-way network channel enables besides targeted advertising also interactive TV, which allows the viewers to interact with the content and the ads displayed on the TV screen [7]. The input device is usually the remote control, but other devices are possible. With interactive TV users can click on advertising elements, requesting more information or buying the marketed product directly. With the feedback channel given by IPTV, the success of advertising could be measured much more precisely than with plain television. This could also lead to similar payment structures as seen on the Internet, for which the pay-per-view model (comparable with the payment model of plain television) becomes decreasingly important, as other models, like pay-per-click and pay-per-transaction, gain more weight. In the remainder of this section we state some challenges for advertising over IPTV which can be tackled with semantics and in a second step we discuss further challenges and research areas.

10.4.1 Improving Targeted Advertisement with Semantics

Let us start by raising some basic questions that characterize the effectiveness of advertisements on IPTV, which also set the ground for explaining how semantics may help improve targeted advertisements on IPTV.

How to select the best advertisement? The advertising network needs to select the advertisement which (i) maximizes the clicks or other interactions by the user on the advertisement (in online advertisement measured as click-through-rate) by matching the advertisement to the preferences of the user, to the content, or to the context respectively, (ii) maximizes the profit of the advertising network, and (iii) does not interfere too much with the Quality of Experience (QoE) for the user. For this problem many research results achieved for Internet advertising can be adopted for semantic IPTV.

Which form of advertisement should be used? The predominant form of advertisements on the TV set mostly are 30 second full-screen spots or some kind of animated overlays. But there might not be such spots for every special interest to reach the consumers in the long tail. And since the TV screen offers other display characteristics than an Internet browsing device, the primary advertising form of the Internet, the banner, can hardly play the same role in semantic IPTV.

Where and when should the advertisement be inserted? If we only use full-screen spots for IPTV advertising the answer for where the advertisement should be placed is already answered, however the question of when it should be inserted still remains. But if we assume that other forms of advertising besides the TV spot should be utilized for semantic IPTV, then we have to answer both questions. The objectives here are to (i) minimize the disruptions in the QoE for the user and (ii) identify and exploit ad placement opportunities as much as possible.

We now try to find some answers to these three questions, starting with the first one on how to select the advertisement. The major difference between online advertising and advertising in IPTV scenarios is that most of the content has no primary textual representation. Keywords or terms for matching the ad to the content have to be extracted from different sources like metadata provided by the content provider, for instance EPG, or directly from the video with image recognition and other processing methods, as discussed in Section 10.2. Once extracted, those keywords can be used quite similar to online advertising for contextual matching. However, if the content would be semantically classified and annotated as proposed, advanced contextual matching algorithms could be exploited. Broder et al. show that a topical (semantic) match introduced to contextual targeting could significantly improve the quality of the matching results, though the overall performance depends on the classification quality [12]. A further semantic approach for contextual matching, which shows advantages over syntax-based matching, is discussed in [15]. A further ontology-based approach is discussed in [71].

It was shown for online advertising by Yan et. al that behavioral targeting, thus exploiting the knowledge about consumers for ad placements, boosts the effectiveness of advertising, in their study measured with an increased click-through-rate of 670% [77] on average. Similar results can be expected for IPTV and therefore a combination of contextual targeted and behavioral targeted advertisement is the most promising approach for IPTV advertising. Konow et al., for example, propose a behavioral targeting approach comparable with collaborative recommendation techniques based on the assumption that if users share preferences for distinct movies, they could also be targeted with the same ads [39]. So it is the knowledge about the content available to the user, the user himself and the placeable advertisements, that is the foundation for targeted advertising for IPTV, and the quality of this knowledge will make or break advertisements in semantic IPTV.

The second question we try to tackle is how semantics can be used to decide which kind of advertising should be used. For this, we first present a few different advertising formats [6]:

Banner-Style Advertisements. Following their widespread use on the Web, the first and most common form of advertisements used with online video has been the use of ad banners around the video. This model is mostly tailored to web-based content and is not suitable for professional and high-quality IPTV deployments, where the user would like to maximize the viewing area for the video.

Ad Video Inserts. Not surprisingly, the other most common form of advertisement follows the traditional ad format in regular television: ad videos are inserted before, after, or between media chunks of the original video or channel stream. The main advantage of this scheme is clearly the ease of use – there is nothing but simply inserting the ad video into the main video content. It also makes sense as long as the user is already engaged in a relatively long viewing session (compared to the length of ad video). Therefore this seemed to work fine for traditional TV, where channel streams have continuous, mostly long chunks of programming. On the other hand, IP-based video systems promise also a large number of very short clips and perhaps a collection of these as personalized channels, mainly due to user created content. In such a case, the users may quickly lose interest in the main stream or channel when they have to wait, for instance, 30 seconds to watch another 60-second clip.

Contextual Overlay Advertisements. Overlay advertisements started to emerge as a wise alternative to those previous methods that are stealing explicit screen space or time from the main content. The idea is to bring up the ad content (text, image, or even video) as an overlay to the main stream, with which the user can usually interact to get further information or even buy a product. Moreover, the ad overlay is usually tied to the semantic elements contained within the main video and is placed at the right moment and place to maximize user interest.

On-demand Advertisements. We envision that a revolutionary service offered to the future users of semantic IPTV will be what we term as "on-demand advertisements", which takes the contextual advertisement scheme one step further. In the current model of business, TV program production almost exclusively precedes the advertisements for the produced video. The advertisements are simply inserted between chunks of media content or added as overlays where suitable. In future IP-based media delivery systems, we foresee a much stronger collaboration between content producers and advertisers. For instance, a content producer may agree to use an advertised mobile phone in a TV episode and semantically embed the phone advertisement info in the show, allowing the users to interact with the TV to ask questions such as "what is the model of this cell phone?" In this scheme, the user is not annoyed with "forced" advertisements, but instead takes the initiative to ask about something of interest, which then brings up more detailed information about the product and possibly a link to buy it. Note that this on-demand video advertisement scheme hides the actual advertisement and expects a natural interest of user in the showcased product, after which advertisement information is revealed.

The last question is about where and when the advertising should be inserted. With the advent of IPTV additional ad spaces appear on the TV screen. Depending on the granularity of annotations to video and TV content ads could be inserted based on elements (e.g. products like a watch or a car) in the video stream, extracted topics (e.g. sports) other top-level information about the show. Besides the ads in the live TV stream, ads can be placed in VoD (Video on Demand) streams, in program guides, in video search and in other IPTV augmenting services. López-Nores et al. for example exploit semantic technologies for their product placement architecture MiSPOT [54]. Also Mei et al. deal with the question of where to place the advertisement in the picture [47]. VideoSense [46] considers detecting in a video stream those positions where advertisements can be placed less intrusively. At the same time it extracts the current topics to aim for contextual targeting.

Semantic IPTV could also profit from the fact that semantic web technologies can help with the integration and harmonization of different data representations among stakeholders of IPTV, as the number of the stakeholders grows from plain TV to IPTV. In plain television the TV provider is in most cases the service provider, the content provider and the ad network in one. As service provider the TV provider delivers the content to the viewer, often generates the content itself and also serves the ads. On the Internet these roles are more distributed, where there is the Internet provider, the ad network and the content provider that are mostly different parties. In interactive IPTV these roles are distributed even more, as there is the Internet provider that is often also the IPTV provider, and there are content providers, ad network provider, and other service providers that are providing their services over the IPTV infrastructure. If the roles are distributed as described, the need for data integration and harmonization rises again, even more in order to achieve an effective communication between the stakeholders and focused on the user.

Besides the questions we discussed in this section, some other issues are still not considered to a great extent. First of all, with contextual targeting the ads are selected based on the context in which the ads should be inserted. One way to extend the context is a *sentiment analysis*, which could detect if the content is positive or negative about a topic [43]. For instance, in a report that discusses a product group and gives a bad rating, the ad for this kind of a product might be the top-ranked ad calculated by any contextual-targeting algorithm, but should not be placed there, since it is less effective than the same ad in a positive context. The results of a sentiment analysis could be easily integrated into a semantic annotation.

In plain television advertisements are pushed to the viewers and there is no way to let them choose their ads by themselves. With IPTV this is a valid option and should be considered by semantic IPTV. With an *ad search* the IPTV provider could showcase products and could learn more about the viewer. On the contrary, one major challenge is to deal with *ad skipping* in VOD or PVR deployments [17]. If the advertisement is inserted in the video stream and not marked as non-skippable advertising most of the users skip over ads and therefore ad placements may not be sellable by the service provider.

Displaying targeted advertising to a viewer is only the first step in the right direction, and further steps must include interactivity with the shown ad. But then arises the question what should happen when viewers, for instance, click on advertisements; should they see an extended version of the ad or get directed to the Web for further information and shopping opportunities [16]. This question is also a major one, which could have a huge impact on the evolution of *interactive IPTV advertising.*

With all those data collected by the IPTV provider, the ad network and other stakeholders, it is critical for the acceptance of targeted advertising that the privacy of the viewer is respected. Guha et al. propose an architecture that tries to solve the tradeoff between personalization (i.e. targeted advertising) and privacy [30].

10.5 Semantic IPTV – A Reference Architecture

Telecommunication operators recognized IPTV as key a service differentiator and as a service to generate new revenues from. The Next Generation Network (NGN) architecture, originally specified for telecommunication services, presents a well-suited infrastructure for IPTV service for industry and standardization bodies because of its provision of high quality and high performance services as well as the possibility to re-use infrastructures for both IPTV and telecommunication services. Several standardization bodies such as International Telecommunication Union's Telecommunication Standardization Sector (ITU-T), Telecoms and Internet-converged Services and Protocols for Advanced Networks (TISPAN) of the European Telecommunications Standards Institute (ETSI), Alliances for Telecommunications Industry Solutions (ATIS), Digital Video Broadcasting (DVB) project and Open IPTV Forum (OIPF) are currently working on the standardization of IPTV delivery, especially on the integration of IPTV and NGN. Foreseeing the growing importance and evolution of an integrated NGN IPTV infrastructure, in this section we propose a solution for a reference architecture that integrates semantic functionalities with NGN-based IPTV services.

As mentioned earlier in Section 10.3.2, managed NGN-based IPTV networks are ideal for the provision of personalized IPTV services due to their long-term subscriber relationship characteristics. For managed networks, two different NGN-based approaches are specified by ITU-T IPTV Focus Group and ETSI TISPAN: IP Multimedia Subsystem (IMS)-based IPTV and a dedicated IPTV system without IMS. Both differ in the IMS-based or non-IMS-based implementation of the Service Control Function of the IPTV architecture [55]. In the following we concentrate on the IPTV standardizations of the OIPF, which follows TISPAN and ITU-T approaches by considering the IMS as a base platform for IPTV and integrates existing standards of the other mentioned standardization bodies.

The aim of this section is to show a way to integrate the OIPF IPTV architecture with a semantically enhanced IPTV service personalization function that collects information on user profile, preferences and behavior and provide content recommendations to other IPTV applications and services which in turn use these recommendations to provide personalized content to IPTV service subscribers. It should be noted that we refrain from laying down a concrete implementation of such an integration, but instead try to describe possible extension and interaction points for an enhanced IPTV service personalization function with current IPTV functionalities specified by the OIPF. Before describing a potential integration we first want to give an overview of current IPTV services specified by the OIPF.

Specifications of the OIPF build an IPTV solution for enriched and personalized IPTV services in managed or un-managed networks. In Release 1 of the OIPF specifications these IPTV services comprise linear TV, private video recording, electronic program guide, content-on-demand, information, notification and communication services. Release 2 specifications extend this by services that realize more personalized aspects [22, 23]:

- *Content guides* provides a searchable list of Linear TV and Content On Demand items to the user according to her preferences.
- *Personalized channel* (PCh) services provides scheduled content according to the user's preferences, viewing habits or service provider recommendations.
- *User notifications* for scheduled content program are provided to the user and can be filtered based on user preferences or targeted to specific users or groups of users.
- *Information services* present tailored information to the user (related to content or not).
- *Interactive applications* such as content voting and rating are interacting with IPTV services using standardized APIs.
- *Advertising services* deliver advertisements as part of the content, in the content guide or in a separate window to the users. Interactive advertisements provide feedback to the service provider and/or advertiser.
- An *Audience research* function is a user monitoring and tracking system that collects several kinds of user activities and provides metrics such as the number of users accessing a particular content item.

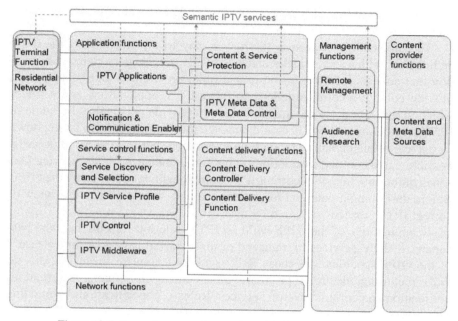

Figure 10.5 Semantic system integration with OIPF IPTV architecture.

We believe that these services could potentially be enhanced by using semantic approaches (PCh, Notifications, Information service, Advertising) or alternatively could provide input to a semantic recommendation system such as user profiling data (interactive applications, audience research).

Taking a look at the OIPF IPTV architecture, these services are mainly provided by functionalities located in the *Applications functions* layer. Figure 10.5 depicts a high level view of the OIPF IPTV architecture that is derived from the OIPF functional architecture [24], whereby the functional components are classified in functional blocks with orientation to TISPAN and ITU-T publications. In the following we point out how this architecture could be extended by semantic services by discussing the integration with several relevant functionalities in detail.

The user's contact point and view to the IPTV system is the *IPTV Terminal Function* (ITF) that is built up of several applications running on the user's end device. After switching on its IPTV end device, the user has to be informed about the services that her service provider is offering; for instance in the form of a web-based linear TV and VoD content guide. In current IPTV standards the retrieval of such information is specified by a common bootstrap procedure between user end-device IPTV functionalities and service provider's IPTV network components, called *IPTV Service Discovery and Selection*. The OIPF components involved in this procedure are the ITF and the network sided *IPTV Service Discovery* functionalities. Because of the richness and diversity of IPTV content and services, personalization in the *IPTV Service Discovery and Selection* process is needed to provide an initial filtered view and attractive entry point to the services offered by the IPTV service and content provider. This can be achieved by the combination of the *IPTV Service Discovery* functionalities with semantics approaches to enhance the filtering and personalization of provided content and services.

The OIPF functional architecture [24] defines two approaches for the realization of a PCh; either the *ITF* or an *IPTV application* interacts with the *IPTV Metadata Control* to retrieve content guide data and then generates a personalized content guide. Instead of generating the PhC on its own, we observe the possibility that *ITF* and *IPTV Applications* communicate with semantic services to retrieve recommendations as input for the personalized content guide. Scheduled content-related user notifications, such as program reminders, are sent by IPTV applications via Instant Message, email or SMS using notification enabling functionalities like Instant Messaging Application Server. As for the PhC service these IPTV applications could communicate with semantic services to provide recommendations and semantically enhanced content to the user.

An audience research subsystem can be implemented as network-application-based, signalling-based or as a combination of both. An *Audience Research Collector* receives user action data from IPTV functionalities (*IPTV Applications, IPTV Control*) or network functionalities to provide it to an *Audience Research Agency* for advertising, personalization and content recommendation solutions. The extensible set of collected data comprises service access data (type, user and content identifier, access time), trick play data (type, user and content identifier, operation time) and service interaction data (rate, vote, comment, click-to-call). The *Audience Research Collector* together with the *IPTV service profile*, where user preferences are stored, has the potential to provide perfect input for semantic recommendation services. The collection of monitored user data and actions achieves the benefits of an implicit and long-term user profiling as mentioned in Section 10.3.1.2.

While advertising functionalities are not yet specified in the OIPF Release 2 functional architecture, TISPAN [69] describes an integration of its

IPTV architecture with the Society of Cable Telecommunications Engineers (SCTE) advanced advertising solution SCTE-130 [61]. An advertising triggering function, the Advertising Management Service (ADM), is positioned in different components: the user equipment *ITF*, the *IPTV Application* and *IPTV Media Control and Delivery* function. An external Ad Decision Service (ADS) component of the advertisement system is requested by the ADM to determine which advertisement is to be inserted when considering aspects like current content delivered, user data and preferences. We see the ADS as a component interacting with semantic services to improve the selection process of best fitting advertisements to be delivered to the user. According to the proposed SCTE-130 – TISPAN integration [69], SCTE-130 *Content Information System* (CIS) and *Subscriber Information System* (SIS) could be integrated with the *IPTV Meta Data* store and *IPTV Service Profile* store of the OIPF architecture to provide personalized advertisement selection and insertion.

It can be observed that current IPTV standardization bodies recognized the growing need to enhance their IPTV service architectures with personalization and recommendation services. This has some impact on current IPTV standardizations, such as OIPF Release 2 [22, 23] and TISPAN Release 3 [68, 70], which introduce personalized channels, notifications and content guides based on recommendations generated from user preferences and collected user data. We think such IPTV services can be enhanced by the integration with semantic services that provide IPTV services and applications with a semantic knowledge sharing system by utilizing semantic approaches discussed in Section 10.3. This results in a *semantic* IPTV service architecture that is exploited by IPTV applications to deliver content to the user in a more personalized and targeted manner to assist her in consuming the growing mass of media that IPTV services will be providing.

10.6 Conclusion and Outlook

Semantic IPTV can be summarized as a vision towards more relevant, more accessible, more profitable, more dynamic content and advertising for the benefit of all players in the media broadcast value chain. Its main enabler will be the creation of a semantic space for IPTV that collectively describes the content, context, and user preferences in a coherent manner. Although significant work has been carried out in the literature in this direction, specifically on the analysis and information extraction for image, audio, and video content, there is still a long way to go. On the one hand, efficient real-time multimedia processing approaches are necessary for fully exploiting live broadcast or time-critical content for semantic services. On the other hand, improved approaches for automated metadata extraction and representation are needed in order to minimize the semantic gap. Based on the presented survey of ongoing efforts in this direction we believe that the semantic gap can be reduced to a negligible level for practical purposes, if not completely removed, forming the main next step towards semantic IPTV.

With the semantic enablers in place, the benefits of IPTV can be truly exploited through personalization and recommendation services, delivering users the most relevant content. The quality of recommendations will be one of the key determinant factors for the acceptance and success of future IPTV systems. As argued earlier, providing a true measure of the recommender's quality to the user and convincing the user in trusting the recommender output is crucial, since it would determine how the user reacts to the offers of the service/content provider, such as buying a recommended VoD content or not. Hence it is a further research challenge to not only improve the recommender systems, but also develop unified mechanisms to assess and visualize the quality/dependability of their results.

Semantic IPTV also paves the way for context sensitive, personalized, targeted advertisement over TV. Understanding the meaning of multimedia content and being aware of user context will allow designing much more efficient advertising schemes over television. Moreover, the two way communication and interactivity opportunities in IPTV provides means to quantify the actual value of advertisements served by the service or content providers, just as the click-based advertisement schemes do on the web.

All these conceptual semantic components and their benefits can only be realized if they are implemented within an IPTV architecture that will be adapted by major providers and hence reach a large user base. Therefore it becomes crucial as a further research area to also investigate how the envisioned functionality would be integrated into the architectures that are being standardized by different organizations as reviewed in this chapter. Already existing hints within these architectures for supporting knowledge bases for personalization and recommendations signal the move towards personalized semantic IPTV services.

References

[1] 3GPP. 3GPP TS 33.220 V9.3.0 (2010-06); 3rd Generation Partnership Project; Technical Specification Group Services and System Aspects; Generic Authentication Architecture (GAA); Generic bootstrapping architecture (Release 9), 2010.

[2] Mohamed Abdel-Mottaleb, Nevenka Dimitrova, Ranjit Desai, and Jacquelyn Martino. CONIVAS: Content-based image and video access system. In *MULTIMEDIA'96: Proceedings of the Fourth ACM International Conference on Multimedia*, pages 427–428, ACM, New York, 1996.

[3] Gediminas Adomavicius and Alexander Tuzhilin. Toward the next generation of recommender systems: A survey of the state-of-the-art and possible extensions. *IEEE Transactions on Knowledge and Data Engineering*, 17(6):734–749, 2005.

[4] Jeffrey R. Bach, Charles Fuller, Amarnath Gupta, Arun Hampapur, Bradley Horowitz, Rich Humphrey, Ramesh Jain, and Chiao-Fe Shu. Virage image search engine: An open framework for image management. In *Storage and Retrieval for Image and Video Databases (SPIE)*, pages 76–87, 1996.

[5] D.H. Ballard and C.M. Brown. *Computer Vision*. Prentice Hall, New Jersey, USA, 1982.

[6] Susan B. Barnes and Neil Frederick Hair. From banners to YouTube: Using the rearview mirror to look at the future of internet advertising. *International Journal of Internet Marketing and Advertising*, 5(17):223–239, June 2009.

[7] L. Begeja and P. Van Vleck. Contextual advertising for IPTV using automated metadata generation. *Proceedings of the 6th IEEE Conference on Consumer Communications and Networking Conference*, pages 437–441, 2009.

[8] Lee Begeja and Paul Van Vleck. Contextual advertising for IPTV using automated metadata generation. In *CCNC'09: Proceedings of the 6th IEEE Conference on Consumer Communications and Networking Conference*, pages 437–441, IEEE Press, Piscataway, NJ, 2009.

[9] D. Bergemann and A. Bonatti. Targeting in advertising markets: Implications for offline vs. online media. Cowles Foundation Discussion Paper No. 1758, 2010.

[10] Shlomo Berkovsky, Tsvi Kuflik, and Francesco Ricci. Mediation of user models for enhanced personalization in recommender systems. *User Modeling and User-Adapted Interaction*, 18(3):245–286, 2008.

[11] Christian Bizer, Tom Heath, and Tim Berners-Lee. Linked data – The story so far. *International Journal on Semantic Web and Information Systems*, 5(3):1–22, 2009.

[12] A. Broder, M. Fontoura, V. Josifovski, and L. Riedel. A semantic approach to contextual advertising. In *Proceedings of the 30th Annual International ACM SIGIR Conference on Research and Development in Information Retrieval*, page 566, 2007.

[13] Robin Burke. Hybrid recommender systems: Survey and experiments. *User Modeling and User-Adapted Interaction*, 12(4):331–370, 2002.

[14] Jun Kyun Choi, Gyu Myoung Lee, and Hyo-Jin Park. Web-based personalized IPTV services over NGN. In *Proceedings of 17th International Conference of Computer Communications and Networks, ICCCN'08*, pages 720–725, IEEE, 2008.

[15] M. Ciaramita, V. Murdock, and V. Plachouras. Semantic associations for contextual advertising. *Journal of Electronic Commerce Research*, 9(1):1–15, 2008.

[16] Kathrin Damian, Christian Bopp, and Lars-Eric Mann. User experience test: Interactive advertising on N-TV plus. *Comput. Entertain.*, 7(3):1–1, 2009.

[17] Michael J. Darnell. An experimental comparison of methods of skipping TV advertisements in DVR recordings. In *EuroITV'10: Proceedings of the 8th International Interactive Conference on Interactive TV&Video*, pages 67–70, ACM, New York, 2010.

[18] Nevenka Dimitrova, Thomas McGee, and Herman Elenbaas. Video keyframe extraction and filtering: A keyframe is not a keyframe to everyone. In *CIKM'97: Proceedings of the Sixth International Conference on Information and Knowledge Management*, pages 113–120, ACM, New York, 1997.

[19] Peter Dolog and Wolfgang Nejdl. Semantic web technologies for the adaptive web. In *The Adaptive Web: Methods and Strategies of Web Personalization*, chapter 23, pages 697–719, 2007.

[20] Nastaran Fatemi. MPEG-7 in practice: Analysis of a television news retrieval application. *J. Amer. Soc. Inf. Sci. Technol.*, 58(9):1364–1366, 2007.

[21] M. Flickner, H. Sawhney, W. Niblack, J. Ashley, Qian Huang, B. Dom, M. Gorkani, J. Hafner, D. Lee, D. Petkovic, D. Steele, and P. Yanker. Query by image and video content: The QBIC system. *Computer*, 28(9):23–32, 1995.

[22] Open IPTV Forum. Service and Platform Requirements v2.0, 2008.

[23] Open IPTV Forum. Services and Functions for Release 2 v1.0, 2008.

[24] Open IPTV Forum. Functional Architecture v2.0, 2009.

[25] Gerald Friedland and Oriol Vinyals. Live speaker identification in conversations. In *MM'08: Proceeding of the 16th ACM International Conference on Multimedia*, pages 1017–1018, ACM, New York, 2008.

[26] O. Friedrich, R. Seeliger, and S. Arbanowski. Interactive and personalized services for an open ims-based iptv infrastructure. In *Proceedings of Seventh International Conference on Networking ICN*, 13–18 April, pages 302–307, 2008.

[27] Mike Pluke, Giovanni Bartolomeo, and Francoise Pertersen. Personalization and profile management, *iJIM*, 2(4):25–29, 2008

[28] Juri Glass, Stefan Marx, Torsten Schmidt, and Fikret Sivrikaya. Semantic TV engine: An IPTV enabler for personalized recommendations. In *Proceedings of LREC'10, 1st Workshop on Semantic Personalized Information Management (SPIM 2010)*, 2010.

[29] Thomas R. Gruber. Toward principles for the design of ontologies used for knowledge sharing. *Int. J. Hum.-Comput. Stud.*, 43(5–6):907–928, 1995.

[30] S. Guha, B. Cheng, A. Reznichenko, H. Haddadi, and P. Francis. Privad: Rearchitecting online advertising for privacy. Technical Report, Max Planck Institute for Software Systems, 2009.

[31] R.M. Haralick and L.G. Shapiro. *Computer and Robot Vision*. Addison-Wesley, New York, 1993.

[32] Alexander Hauptmann. How many highlevel concepts will fill the semantic gap in news video retrieval. In *International Conference on Image and Video Retrieval (CIVR)*, 2007.

[33] Alexander G. Hauptmann, Rong Jin, and Tobun Dorbin Ng. Multi-modal information retrieval from broadcast video using ocr and speech recognition. In *JCDL'02: Proceedings of the 2nd ACM/IEEE-CS Joint Conference on Digital Libraries*, pages 160–161, ACM, New York, 2002.

[34] Dominik Heckmann, Tim Schwartz, Boris Brandherm, Michael Schmitz, and Margeritta von Wilamowitz-Moellendorff. GUMO – The general user model ontology. In *User Modeling*, pages 428–432, 2005.

[35] Tom Huang, Sharad Mehrotra, and Kannan Ramchandran. Multimedia analysis and retrieval system (MARS) project. In *Proceedings of 33rd Annual Clinic on Library Application of Data Processing – Digital Image Access and Retrieval*, 1996.

[36] Fotis G. Kazasis, Nektarios Moumoutzis, Nikos Pappas, Anastasia Karanastasi, and Stavros Christodoulakis. Designing ubiquitous personalized TV-anytime services. In *Proceedings of Advanced Information Systems Engineering Workshops*, University of Maribor, pages 136–149, 2003.

[37] Graham Klyne and Jeremy J. Carroll. Resource description framework (RDF): Concepts and abstract syntax. Technical Report, W3C, 2004.

[38] Alfred Kobsa. Generic user modeling systems. *User Modeling and User-Adapted Interaction*, 11(1-2):49–63, 2001.

[39] Roberto Konow, Wayman Tan, Luis Loyola, Javier Pereira, and Nelson Baloian. Recommender system for contextual advertising in IPTV scenarios. In *Proceedings of 14th International Confernce on Computer Supported Cooperative Work in Design (CSCWD)*, pages 617–622, 2010.

[40] Martin Körling. Evolution of open IPTV standards and services. In *Innovations in NGN: Future Network and Services (K-INGN2008), First ITU-T Kaleidoscope Academic Conference*, pages 11–14, May 2008.

[41] Martin D. Levine. *Vision in Man and Machine*. McGraw-Hill, New York, 1985.

[42] R. Lienhart, S. Pfeiffer, and W. Effelsberg. The MOCA workbench: Support for creativity in movie content analysis. In *Proceedings of the Third IEEE International Conference on Multimedia Computing and Systems*, pages 314–321, June 1996.

[43] Kangmiao Liu, Quang Qiu, Can Wang, Jiajun Bu, Feng Zhang, and Chun Chen. Incorporate sentiment analysis in contextual advertising. In *TROA2008 (in conjunction with WWW2008)*, pages 1–8, 2008.

[44] Zhu Liu, D.C. Gibbon, and B. Shahraray. Multimedia content acquisition and processing in the miracle system. In *Proceedings of 3rd IEEE Consumer Communications and Networking Conference CCNC 2006*, volume 1, pages 272–276, 2006.

[45] Jose M. Martinez. MPEG-7 overview (version 9). Technical Report, International Organisation for Standardisation, October 2004.

[46] T. Mei, L. Yang, X.S. Hua, H. Wei, and S. Li. Videosense: A contextual video advertising system. In *Proceedings of the 15th International Conference on Multimedia*, page 464, 2007.

[47] Tao Mei, Jinlian Guo, Xian-Sheng Hua, and Falin Liu. ADON: Toward contextual overlay in-video advertising. In *Multimedia Systems*, pages 1–10, 2010.

[48] Maurice D. Mulvenna, Sarabjot S. Anand, and Alex G. Büchner. Personalization on the net using web mining: Introduction. *Communications of the ACM*, 43(8):122–125, 2000.

[49] Y. Nakajima, Yang Lu, M. Sugano, A. Yoneyama, H. Yamagihara, and A. Kurematsu. A fast audio classification from MPEG coded data. In *Proceedings of IEEE International Conference on Acoustics, Speech, and Signal Processing, ICASSP'99*, Vol. 6, pages 3005–3008, March 1999.

[50] M. Naphade, J.R. Smith, J. Tesic, S.F. Chang, W. Hsu, L. Kennedy, A. Hauptmann, and J. Curtis. Large-scale concept ontology for multimedia. *IEEE Multimedia*, 13(3):86–91, 2006.

[51] Milind R. Naphade and John R. Smith. On the detection of semantic concepts at trecvid. In *Multimedia'04: Proceedings of the 12th Annual ACM International Conference on Multimedia*, pages 660–667, ACM Press, 2004.

[52] Apostol Natsev, Jelena Tešić, Lexing Xie, Rong Yan, and John R. Smith. Ibm multimedia search and retrieval system. In *CIVR'07: Proceedings of the 6th ACM International Conference on Image and Video Retrieval*, pages 645–645, ACM, New York, 2007.

[53] Xiaofei Niu, Jun Ma, and Dongmei Zhang. A survey of contextual advertising, In *Proceedings of the 2009 Sixth International Conference on Fuzzy Systems and Knowledge Discovery (FSKD'09)*, volume 07, pages 505–509, IEEE Computer Society, Washington, DC, 2009.

[54] Martín López Nores, José J. Pazos Arias, Jorge García Duque, Yolanda Blanco-Fernández, Manuela I. Martín-Vicente, Ana Fernández Vilas, Manuel Ramos Cabrer, and Alberto Gil-Solla. Mispot: dynamic product placement for digital TV through MPEG-4 processing and semantic reasoning. *Knowl. Inf. Syst.*, 22(1):101–128, 2010.

[55] International Telecommunication Union. Focus Group on IPTV Telecommuniction Standardization Sector. Recommendation ITU-T Y.1910 (09/2008): IPTV functional architecture, 2008.

[56] A. Pentland, R. Picard, and S. Sclaroff. Photobook: Content-based manipulation of image databases, 1994.

[57] Maryam Ramezani, Lawrence Bergman, Rich Thompson, Robin Burke, and Bamshad Mobasher. Selecting and applying recommendation technology. In *International Workshop on Recommendation and Collaboration in Conjunction with 2008 International ACM Conference on Intelligent User Interfaces (IUI 2008)*, 2008.

[58] Elaine Rich. Users are individuals: Individualizing user models. *International Journal of Man-Machine Studies*, 18:199–214, 1983.

[59] Teppo Kurki, Sami Jokela, Reijo Sulonen, and Marko Turpeinen. Agents in delivering personalized content based on semantic metadata. In *Proceedings of 1999 AAAI Spring Symposium Workshop on Intelligent Agents in Cyberspace*, pages 84–93, 1999.

[60] J. Ben Schafer, Joseph A. Konstan, and John Riedl. E-commerce recommendation applications. *Data Mining and Knowledge Discovery*, 5(1/2):115–153, 2001.

[61] SCTE. The Society of Cable Telecommunications Engineers: SCTE-130 (all parts): "Digital Program Insertion – Advertising Systems Interfaces", 2008.

[62] Nicu Sebe, Michael S. Lew, and Arnold W. M. Smeulders. Video retrieval and summarization. *Computer Vision and Image Understanding*, 92(2–3):141–146, 2003.

[63] Alan F. Smeaton, Paul Over, and Wessel Kraaij. Evaluation campaigns and trecvid. In *MIR'06: Proceedings of the 8th ACM International Workshop on Multimedia Information Retrieval*, pages 321–330, ACM, New York, 2006.

[64] Arnold W.M. Smeulders, Marcel Worring, Simone Santini, Amarnath Gupta, and Ramesh Jain. Content-based image retrieval at the end of the early years. *IEEE Trans. Pattern Anal. Mach. Intell.*, 22(12):1349–1380, 2000.

[65] C.G.M. Snoek, I. Everts, J.C. Van Gemert, J.M. Geusebroek, B. Huurnink, D.C. Koelma, M. Van Liempt, O. De Rooij, A.W.M. Smeulders, J.R.R. Uijlings, and M. Worring. The MediaMill TRECVID 2007 semantic video search engine. In *Proceedings of IEEE International Conference on Acoustics, Speech and Signal Processing (ICASSP 2007)*, volume 4, pages IV-1213–IV-1216, 2007.

[66] C.G.M. Snoek, B. Huurnink, L. Hollink, M. de Rijke, G. Schreiber, and M. Worring. Adding semantics to detectors for video retrieval. *IEEE Trans. on Multimedia*, 9(5):975–986, August 2007.

[67] M.R. Tazari, M. Grimm, and M. Finke. Modelling user context. In *Proceedings of 10th International Conference on Human-Computer Interaction*, Crete (Greece), 2003.

[68] ETSI TISPAN. Draft ETSI TS 183 063 V3.3.0 (2009-12); Telecommunications and Internet Converged Services and Protocols for Advanced Networking (TISPAN); IPTV Architecture; IMS-based IPTV stage 3 specification, 2009.

[69] ETSI TISPAN. ETSI TS 182 028 V3.3.1 (2009-10); Telecommunications and Internet Converged Services and Protocols for Advanced Networking (TISPAN); IPTV Architecture; NGN integrated IPTV subsystem architecture, 2009.

[70] ETSI TISPAN. Draft ETSI TS 183 064 V3.2.0 (2010-01); Telecommunications and Internet Converged Services and Protocols for Advanced Networking (TISPAN); IPTV Architecture; NGN integrated IPTV subsystem stage 3 specification, 2010.

[71] D. Tsatsou, F. Menemenis, I. Kompatsiaris, and P.C. Davis. A semantic framework for personalized ad recommendation based on advanced textual analysis. In *Proceedings of the Third ACM Conference on Recommender Systems*, pages 217–220, ACM, 2009.

[72] Chrisa Tsinaraki, Panagiotis Polydoros, and Stavros Christodoulakis. Integration of owl ontologies in MPEG-7 and TV-anytime compliant semantic indexing. In *Advanced Information Systems Engineering*, pages 143–161. 2004.

[73] Ray van Brandenburg, M. Oskar van Deventer, Georgios Karagiannis, and Mike Schenk. Towards personalized TV for concurrent use; unlocking the potential of IMS-based IPTV. In *Proceedings of IEEE GLOBECOM Workshops*, IEEE Communications Society, USA, December 2009.

[74] Howard D. Wactlar, Takeo Kanade, Michael A. Smith, and Scott M. Stevens. Intelligent access to digital video: Informedia project. *Computer*, 29(5):46–52, 1996.

[75] Xiao-Yong Wei and Chong-Wah Ngo. Ontology-enriched semantic space for video search. In *Multimedia'07: Proceedings of the 15th International Conference on Multimedia*, pages 981–990, ACM, New York, 2007.

[76] S. Weibel, J. Kunze, C. Lagoze, and M. Wolf. Dublin core metadata for resource discovery. Technical Report, United States, 1998.

[77] J Yan, N Liu, G Wang, W Zhang, Y Jiang, and Z Chen. How much can behavioral targeting help online advertising? In *Proceedings of the 18th International Conference on World Wide Web*, pages 261–270, 2009.

[78] Markus Zanker, Markus Jessenitschnig, Dietmar Jannach, and Sergiu Gordea. Comparing recommendation strategies in a commercial context. *IEEE Intelligent Systems*, 22(3):69–73, 2007.

[79] Ethan Zuckerman. Serendipity, echo chambers, and the front page. *Nieman Reports*, 2008.

11

A Framework for Converged Video Services in the IMS

Devdutt Patnaik[1], Aamir Poonawalla[1], Ashish Deshpande[1],
Russ Clark[1], Michael Hunter[1], Naoki Takaya[2] and Arata Koike[2]

[1]College of Computing, Georgia Institute of Technology, Atlanta, GA 30332, USA;
e-mail: devdutt.patnaik@gatech.edu
[2]NTT Labs, Tokyo 180-8585, Japan

Abstract

The IP Multimedia Subsystem (IMS) is being adopted as the standard service delivery architecture for Next Generation Networks. The IMS provides scalable, secure and high-quality multimedia services to mobile and fixed IP terminals. The World Wide Web (WWW) on the other hand has now become ubiquitous and has found its way from fixed terminals to 3G and 4G enabled mobile terminals. The availability of the WWW on mobile terminals, combined with its ease of use, makes it a popular platform for communication. Converged services and applications composed of Web 2.0 related technologies and IMS-based call-control and multimedia capabilities can provide very powerful and rich services to end-users. We focus on video service creation in the IMS with an emphasis on user generated video content. We discuss the composition of richer video-enabled services by converging technologies such as the IMS, Web 2.0 related technologies, presence, location and social networking. We first develop a framework for video streaming over the IMS using a Media Resource Function (MRF) and an Application Server (AS). We then demonstrate the flexibility and capabilities of this framework by implementing rich services around video delivery suitable for both fixed and mobile terminals.

Anand R. Prasad et al. (Eds.), Advances in Next Generation Services and Service Architectures, 231–249.

Keywords: IP multimedia subsystem, video streaming, media resource function, Web 2.0, mashup, presence, social networking.

11.1 Introduction

With the growth in the deployment of the IMS for next generation telecom services, operators are looking for services that would provide good user-experience while providing them opportunities for revenue generation. Operators have been offering basic video streaming capabilities for some time now [1]. However, there is a need for more compelling use cases and a stronger revenue generation model around these video services [2]. The basic IMS video service when combined with features and technologies like Web 2.0 mashups, presence and social networks can provide compelling services and applications to end-users. In this chapter we propose a framework for converged video service creation in the IMS using a Media Resource Function (MRF) and an Application Server (AS). The goal of the framework is to not only provide for multi-user video streaming capabilities, but to also provide mechanisms for composing richer services around the video service. This framework enables the composition of richer video-enabled services by converging technologies such as the IMS, Web 2.0, presence, location services and social networking. The flexibility and capabilities of this framework are demonstrated by implementing rich services around video delivery. These services are suitable for both fixed and mobile terminals. By combining the IMS video service with other technologies, interesting applications around the basic video service can be developed. These applications and services provide value to customers as well as revenue streams for telecommunication operators.

Section 11.2 discusses related work done to combine Internet technologies with the IMS, and recent work in the area of video streaming in the IMS. Section 11.3 gives details of our framework, its architecture and its capabilities. Section 11.4 discusses the implementation of video enabled services using the composition of Internet technologies and IMS features, such as presence and social networking, with an emphasis on user-generated content. Section 11.5 concludes with results from our work and discusses the possibility of extending this video delivery framework to support a full REST interface.

11.2 Background and Motivation

While the IMS has achieved some level of adoption, there is arguably still a need for more compelling services that would help to make it a commercial success. The WIMS 2.0 initiative [24] aims at converging the Web and the IMS, by exposing the session-based call-control capabilities of the IMS to commercially successful web technologies and vice versa. This means the integration of IMS features into Web 2.0 mashups which makes them usable from any Web 2.0 service. In this chapter we discuss a similar approach with video streaming capabilities of the IMS and user-generated content with the goal of exposing these capabilities to web-based applications. Falchuk et al. [21] extend the ideas expressed in [24] and present applications and mashups that use 'contextual information', to develop some interesting mashups. Their focus is however not specific to video-based services. Julien et al. [23] investigate methods for the delivery of personalized multimedia from an IMS network. They emphasize the delivery of advertisements along with profile and preference-based delivery of video content. Shet et al. [25] present use cases for interactive mobile-cast over the IMS. The approach they take briefly mentions some demo scenarios of the use cases. Fernandez et al. [22] present a framework for point-to-multipoint streaming of video over 3G networks. Shiroor [26] presents the design of a framework for the delivery of IPTV and VoD services over the IMS.

The above efforts are useful contributions to increase the adoption of the IMS. However, they do not focus on using the video-streaming capabilities of the IMS. They do not lay emphasis on combining the IMS video service with Web 2.0, presence and social networking, to deliver compelling services and applications. In this chapter we extend and take forward the Web 2.0/IMS convergence philosophy [21, 23, 24], and compose services and applications around the basic video delivery capabilities of the IMS. The main contribution of this work is the design and implementation of a framework for video delivery in the IMS, using an MRF and an AS. Using this framework (mainly the interfaces implemented in the AS), rich services around video delivery are implemented, that are suitable for both fixed and mobile users. We integrate the video service with Web 2.0 technologies, presence and social networking and demonstrate a number of use cases that are of interest to service providers.

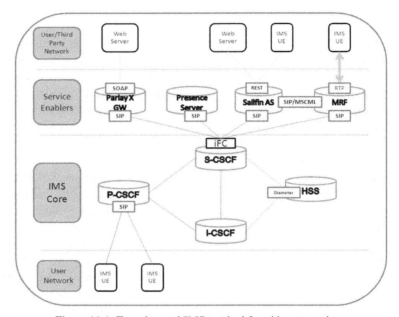

Figure 11.1 Experimental IMS test bed for video streaming.

11.3 Video Delivery Platform Architecture

Based on these goals, we designed and implemented an IMS compliant architecture for video-streaming. The reference architecture includes a Media Resource Function and an Application Server. Figure 11.1 depicts the components of our experimental IMS Testbed.

11.3.1 User Network

The User Network Layer includes IMS capable devices (User Enpoints or UEs) such as mobile handsets, laptops or a PC. It also includes third-party web and application servers, e.g. Facebook REST Server [5]. The IMS UEs are capable of connecting to the IMS network using standard authentication mechanisms and registration procedures. These devices are web and SIP/IMS capable. They have browsing capabilities. They can also stream live video captured from a built-in or attached camera, and also capable of receiving and playing media formats such as H.263/H.264 variants.

11.3.2 IMS Core

The IMS Core components include the standard IMS elements such as the CSCFs and the HSS. A CSCF (Call/Session Control Function) is a SIP server that handles the SIP signaling in the IMS core. Depending on the functionality provided, the CSCFs are divided into three types. There can be multiple CSCFs of each type in one IMS core. The Proxy CSCF (P-CSCF) is the gateway from the packet-switched network to and from the IMS core. The P-CSCF generates charging information toward a charging collection node. When a user makes a call, it is routed from the P-CSCF to the Interrogating CSCF or I-CSCF. The I-CSCF queries the HSS to find out if a Serving CSCF or an S-CSCF is allocated for this particular UE. If an S-CSCF is previously allocated, then the ICSCF simply forwards the incoming request to that S-CSCF. The Home Subscriber Server (HSS) acts as a registrar in the IMS. In our experiments we used the OpenIMS IMS core from the FOKUS group [20]. This is a standards compliant implementation that is interoperable with Presence and Application Servers from other vendors. We configure the Initial Filter Criteria (iFC) at the S-CSCF to provide rules based on the service profiles of users. Depending on the services that a user subscribes to, the iFC determines the logic for forwarding requests to the appropriate AS.

11.3.3 Service Enablement Layer

The Service Enablement Layer is a key layer in our architecture. This layer consists of servers that provide intelligent service capabilities, when combined with the basic call-control capabilities of the IMS. Our framework includes a Presence Server, an Application Server that exposes both HTTP [15] and SIP [16] interfaces, and an MRF that is capable of streaming video (point-to-multipoint and multipoint-to-point). The AS and the MRF are important components in our framework and require more detailed discussion.

1. *Application Server*

   The AS plays an important role in providing the basis for developing converged applications in the IMS, in this case for SIP and HTTP. We used the open source Sailfin Application Server [11], which is developed over the Glassfish Communications Server [6]. Sailfin implements the JSR 289 SIP Servlet 1.1 Specification [8]. This allows us to leverage Java EE services in SIP-based applications, and initiate calls from them. It provides for features such as security, replication, overload control

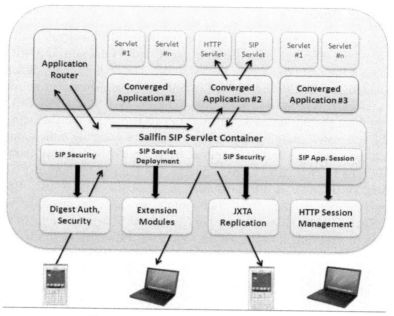

Figure 11.2 Application server architecture.

and session management, using the underlying services of the Glassfish framework as shown in Figure 11.2. For our purposes we use these capabilities of Sailfin to implement the following key functions:

(a) The IMS Exposure Layer:

In line with the goals of Web 2.0/IMS convergence, the key IMS Exposure Layer in the AS was implemented. The Sailfin AS provides basic capabilities to develop converged applications by exposing HTTP and SIP interfaces. This layer exposes the session-based call-control capabilities of the IMS to the web world using an HTTP/REST interface. Using the servlet interface, our implementation handled sessions and combined HTTP/REST and SIP interaction, thus exposing IMS capabilities to Web 2.0 applications. We used URI-based hierarchical organization and representation of resources. This allowed for the identification of applications/services based on the user's IMS Public URIs [27]. Figure 11.2 shows the architecture and components of the AS. As shown in Figure 11.2, web-based applications can access the call control features of the IMS. This is possible since the SIP Servlet handles the signaling requirements in the IMS layer. Each application/service is comprised of one or more servlets. In a typical converged application a combination of SIP and HTTP servlets are used to expose their respective interfaces.

(b) Third Party and Session-Based Call Control:

In order to support multi-party video streaming sessions based on requests from IMS UEs, the AS must provide powerful session-based call control capabilities. For this purpose features such as Third Party Call Control (3PCC) and session-based call control were implemented. These allow UEs to 'Create', 'Join' and 'Leave' existing multimedia sessions. These signaling capabilities were implemented using the SIP Servlet architecture in Sailfin, that gives control of SIP Sessions using the SIP Servlet API. The 3PCC approach and the conferencing approach was combined as depicted in figures 18 and 24 from the Media Control Draft [7], using an MRF. In our implementation, participants can dynamically join and leave the multimedia session, i.e. Create, Join and Leave operations are supported in the context of multiple participating UEs. Using the SDP [12] attributes such as sendrecv, sendonly and recvonly we can define the role that a particular UE plays in the streaming session.

(c) Media Server Control:

The AS plays an important role in the control and mediation of media sessions, as shown in figure 24 of the Media Control Draft [7]. In IMS terminology, it behaves as the Media Resource Function Controller (MRFC), and is capable of sending control commands to the MRF (which is essentially a Media Resource Function Processor or an MRFP). The AS implements the business logic to compose the video streams based on user preferences, and the required application. The AS uses a combination of SIP and MSCML [10] to send commands to the MRF. MSCML is used for advanced scenarios such as recording the live streaming session at the MRF, customized 'video mixing and composition' for individual participants based on user preferences, and playing pre-recorded content from the MRF video database.

(d) Application Routing:

The Application Router in the AS provides support for identifying the specific service/application to which the incoming requests should be routed. This allows a modular and compositional approach, where one or more independent applications can be invoked for a given request. The Application Router implementation can consult arbitrary information or data stores, such as subscriber information, business policies, etc., and thus allow flexible and programmable routing of requests. This is particularly useful for hosting multiple applications/services on the AS.

2. *Media Resource Function*

The Media Resource Function (MRF) is the most important component of our framework. The role of the MRF is to support transcoding, scalable multicast media-streaming, stream composition, and programmable control. For the implementation, we used the Dialogic IP Media Server [4] as the MRF component. It supports basic SIP and SIP + MSCML control messages to create and manage multimedia sessions. The MRF provides the following capabilities and use-cases in the Video Streaming Framework:

(a) P2MP and MP2P Live Video Streaming:

The MRF's basic responsibility is to enable multi-party video streaming capabilities. The MRF performs replication of the source stream and sends it out to the other recipients in the multimedia session. This optimizes the bandwidth utilization, and also relieves the UEs from being concerned about generating multiple streams. This is referred to as Point-to-Multipoint (P2MP) streaming. The MRF also has the capability to send streams from multiple UEs to one client UE. This is the Multipoint-to-point capability (MP2P) which uses the video composition feature of the MRF, where a single stream is generated out of multiple incoming streams. These are the 2 models of video streaming that are most common. Multiple UEs can receive the same media stream, or a single UE can view multiple video streams at once. The MRF efficiently supports both these functions. The IMS provides device independent access to this functionality, i.e. for fixed and mobile terminals, as well as between heterogeneous terminals.

(b) Recording of Live Video Streams:

The MRF supports the capability of receiving a user-generated video stream, which can be stored and viewed depending on user preference. This feature allows use-cases such as recording of important personal or professional events, video sharing services such as those available on YouTube [14] and other services. This capability when exposed though appropriate APIs, from the AS, can allow IMS UEs to push live video streams to the MRF for storage.

(c) Video on Demand:
The MRF is very suitable for Video on Demand and IPTV services. The MRF is a repository of video/audio content, which when combined with a good user front-end allows interactive VoD and IPTV services to be supported. A very interesting use-case is that of a group of friends watching the same channel and being able to simultaneously send instant messages discussing the program.

(d) Contextual Advertising:
It is possible to insert advertising content during a multimedia session in the form of overlays or on a separate portion of the screen using the video-composition features of the MRF. The content is provided using intelligent means, such as gathering profile information and preferences of users from Web 2.0 features such as Facebook [5] REST APIs. For example, users could receive targeted discounts as an incentive for participating. Another interesting use-case is to further refine the advertising content-based on the current user location.

11.4 Video Enabled Services

In this section we present different implementation experiences with use-cases. They combine video-streaming with other technologies such as HTTP, presence, location-based services and social networking, using the framework. The implemented services can be deployed on any standards compliant IMS UE. We present three different services, that were implemented to demonstrate the capability of the framework and the possible composition of richer services around the video-streaming service. In each case we present the overall use-case followed by implementation details.

11.4.1 Map and Video Mashup Service

1. *Applications*

(a) Video Surveillance and Security Monitoring: In recent years, security monitoring and surveillance has become increasingly important. Leveraging the capabilities of our video framework, we have designed a security and surveillance system.It uses multiple

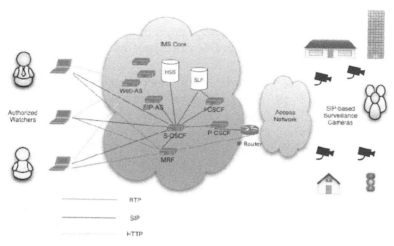

Figure 11.3 Video surveillance with Google Maps.

SIP/IP enabled cameras [9] that can send media streams to the MRF. The cameras can then be selected and controlled by authorized watchers (e.g. police) using a map interface. Figure 11.3 depicts this use-case.

(b) Traffic monitoring using GPS Devices: Currently car infotainment and GPS devices are equipped with large displays. Given a 3G enabled GPS device, a very useful service would be to provide live traffic feeds, wherein the driver can decide between a set of possible routes. The user can use the touch screen display to view the live traffic situation on different routes and make an informed decision. The IMS 3G network can be used to send live streams to the GPS device.

2. *Implementation*

We implemented the video surveillance use-case by combining the Google Maps interface with the IMS call-control capabilities. The IMS Exposure Layer exposes the call-control capabilities to the Web world as an HTTP/REST interface. We used the Google Maps API in the HTTP servlet to serve out the map and depict the location of the SIP/IP-based cameras. The SIP/IP-based cameras send registration in the SIP REGISTER message. With some degree of programmability on the camera device we can use a location service such as WhereAmI [13] to dynamically update location information. The AS uses this location information

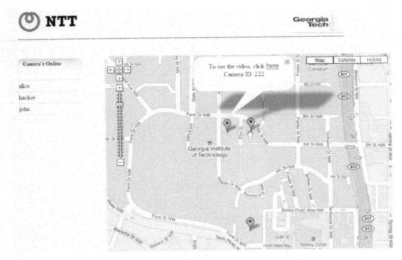

Figure 11.4 Google Map with camera location.

to display the Google Map with the cameras shown as markers in any browser. A snapshot of this application, with online cameras is shown in Figure 11.4. When the user clicks on any marker, he/she can view the content from the corresponding camera. We implemented scenarios wherein multiple users can view the content from the same camera, as well as that were a single user views content from multiple cameras. The resulting call-flow when the user clicks on a marker is shown in Figure 11.5.

11.4.2 IMS Video and Social Networking Service

Recently, there has been an explosive growth in the use of social networking sites and applications over 3G data networks. Websites such as Facebook [5] allow users to create and share applications using the Facebook API [5]. The concept of exposing IMS capabilities in social networking sites has not been explored to our knowledge. This form of convergence opens opportunities for developing converged applications that could enable users of social networking sites to make calls, send and receive instant messages, and set up video-conferences with their friends. We believe that offering several such features of great utility under a single umbrella will result in a superior user-experience.

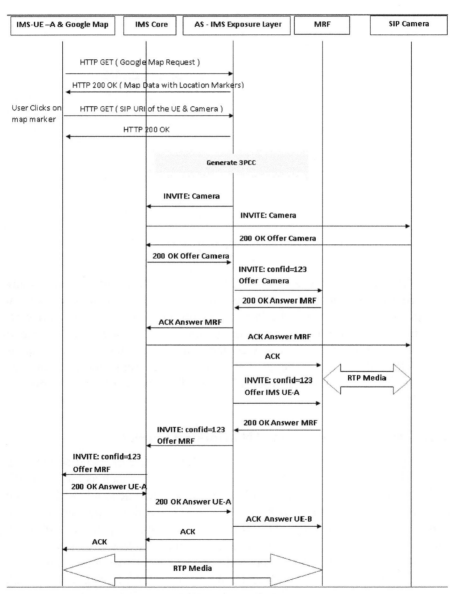

Figure 11.5 Call flow showing HTTP and SIP interaction.

1. *Applications*

   (a) Interactive Social-Networking: The use of social-networking sites for making calls and setting up video-conferences, facilitates a more interactive conversation. Currently, people use separate applications for socializing and for actual communication. The integration of social-networking with the IMS offers a one-stop solution to all these needs. Moreover, since the social networking services are now offered on 3G data-networks, such integration enables users on heterogeneous platforms to seamlessly interact with each other thereby making it possible to have video conferences between people logged on to social networking sites and those logged on via their IMS enabled smart-phones.

   (b) Professional networking: The integration of social-networking with the IMS makes professional activities like live events, classrooms or webinars possible [3]. With most active people using social networking sites, companies can hold live sessions with a large group of people thus making these sites a quick and efficient medium for both effective and real-time communication.

2. *Implementation*

   In order to demonstrate the integration of social networking with the IMS, we developed a converged application for Facebook that involves the use of both HTTP-REST and SIP. This application augments Facebook's features by providing additional video calling capabilities with online contacts. When a Facebook user logs in he/she can see the presence status of his friends and make video calls to them (even Facebook users signed-in on their 3G phones). Similarly any user who logs in to our Facebook service using a 3G IMS UE, can see the presence status of his friends currently logged in to Facebook. Thus Facebook users can use direct video communication with friends in their social network using the IMS features made available through our application. This offers a seamless unified communication platform.

   At application startup, the user logs into our Facebook application and his/her preferred IMS softphone/client application (a web-based IMS UE would be ideal for this implementation). The application retrieves the user's list of friends from the Facebook REST server. Subsequently, he is able to see the presence status of all his friends. If any of his friends have registered with the Facebook IMS application using any 3G compliant IMS UE, he can directly call them or exchange instant messages

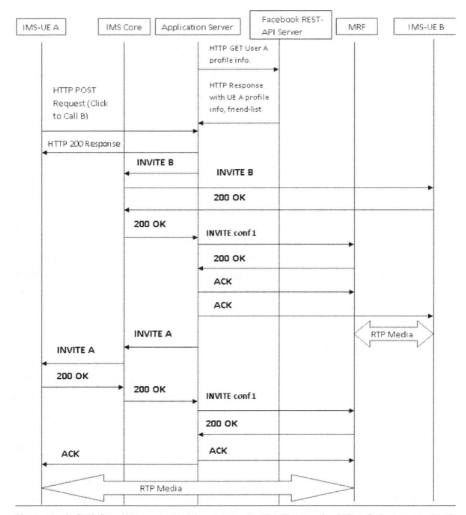

Figure 11.6 Call-flow between a Facebook user (IMS-UE A) and a SIP softphone user (IMS-UE B).

with them. Figure 11.6 explains the interaction between a Facebook user and an IMS UE. The Facebook user initiates a call by clicking on a 'Call' hyperlink in the application, for any friend who is registered with the IMS application. This triggers an HTTP request to the application-server, which then initiates a 3PCC and the user receives a call on his/her

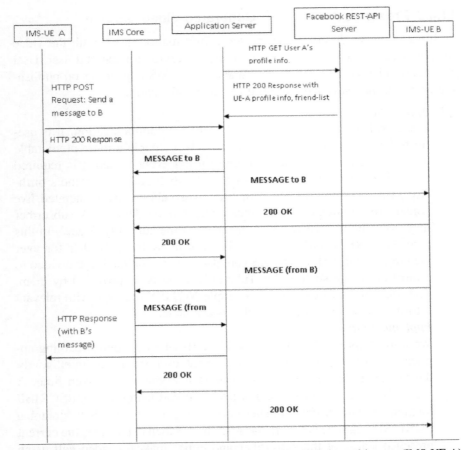

Figure 11.7 Exchange of instant messages between a Facebook logged-in user (IMS-UE A) and a SIP phone user (IMS-UE B).

IMS UE or softphone. More users can be invited to join the session. They can also simultaneously exchange instant messages using the WebUI and the IMS UE as in Figure 11.7. This can be further integrated with the regular IMS presence features to provide for seamless calling between the Facebook users and users who have registered with the IMS using standard IMS registration mechanisms.

11.4.3 Presence Aware User-Generated Video Service

In this section we describe how presence information along with the video delivery can be used to achieve an enriching service for the end user. User generated video content can be streamed to the IMS depending on publish-subscribe mechanisms and current user status information.

1. *Applications*

   Presence is an important aspect of real-time communication. It is useful to know the presence status of friends to enable a more predictable user experience. This can be useful during events when it is required to receive videos in real time, such as football games, a friend's birthday party, a safari trip, etc. Our framework allows user-generated live content to be streamed to friends using mobile devices. A subscriber may not want to receive video, as he could be currently 'Busy'. In this case the MRF records the video being sent by the publisher for later viewing by subscriber. The AS can also mediate uploading the video to a third-party site such as YouTube [14] using APIs provided by them. The subscribers would receive a simple NOTIFY message with relevant information whenever a new video is available.

2. *Implementation*

   We implemented a PUBLISH-SUBSCRIBE-based video streaming application [17–19] and also used 'current user status' to intelligently determine if the subscriber can receive live video at a given time. A user subscribes to video streams from a friend using the SUBSCRIBE method. Whenever the publisher of the live content is ready to stream, a PUBLISH message is sent to the AS. The AS then determines the current presence status of the subscriber and either makes a video call using 3PCC and INVITE if the user is available, or sends a NOTIFY message if the recipient is 'Busy'. The case for 3PCC is shown in Figure 11.8. The NOTIFY message contains SIP URI of the live stream. The subscriber can then join the session if he/she is interested.

11.5 Conclusion and Outlook

In this chapter we discussed the design and implementation of a scalable and flexible IMS compliant video-streaming framework using an MRF and an AS. The main contribution of our work is the development of a framework that enables the convergence of the video-streaming service with other applications by using the AS. The MRF enables user-generated video con-

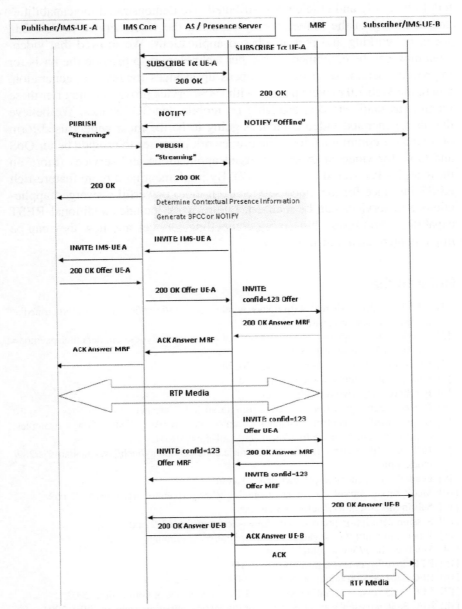

Figure 11.8 Message exchange between a Facebook logged-in user (IMS-UE A) and a SIP phone user (IMS-UE B).

tent to be easily and efficiently distributed. We demonstrated the capabilities and flexibility of the framework by exposing the video-service to Web 2.0, social networking and presence-based applications. We showed that video-streaming can be combined with other technologies to provide the basis for improved operator services. This opens up avenues for revenue generation, where the Web 2.0 players partner with IMS Service providers and offer these features as paid services, available on mobile and 3G devices. We believe that user-generated video content is going to be the most widely used form of media for communication. The framework promises to provide better QoS and QoE for video in these converged applications and services (more on these topics is covered in Chapter 12). By implementing a more feature-rich REST interface for the video service even more powerful converged applications and services can be realized. Future work includes additional REST capabilities and researching more compelling services and how they can be implemented using our framework.

References

[1] AT&T video share, http://www.wireless.att.com/learn/messaging-internet/media-entertainment/video-share-faqs.jsp.
[2] AT&T video share review, http://www.strategyanalytics.com/default.aspx?mod=reportabstractviewer&a0=3758.
[3] Classroom 2.0, http://www.classroom20.com.
[4] Dialogic MRF, http://www.dialogic.com.
[5] Facebook, http://www.facebook.com.
[6] Glassfish communications server, https://glassfish.dev.java.net.
[7] IETF media control channel framework (CFW) call flow examples, http://www.ietf.org/id/draftietf-mediactrl-call-flows-01.txt.
[8] JSR 289 SIP servlet 1.1, http://www.oracle.com/technology/tech/java/standards/jsr289/index.html.
[9] Leadtek SIP camera, http://www.leadtek.com.
[10] Media server control markup language, http://www.rfceditor.org/rfc/rfc5022.txt.
[11] Sailfin SIP AS, https://sailfin.dev.java.net.
[12] Session description protocol, http://www.ietf.org/rfc/rfc2327.txt.
[13] Where am I, http://rnoc.gatech.edu/whereami/programming.
[14] YouTube, http://www.youtube.com.
[15] RFC 2616 hyper text transfer protocol, 1999.
[16] RFC 3261 session initiation protocol, 2002.
[17] RFC 3265 session initiation protocol (SIP) – specific event notification, 2002.
[18] RFC 3856 a presence event package for the session initiation protocol (SIP), 2004.
[19] RFC 3903 session initiation protocol (SIP) extension for event state publication, 2004.
[20] The open IMS project, http://www.openimscore.org, 2007.

[21] B. Falchuk, K. Sinkar, S. Loeb, and A. Dutta. Mobile contextual mashup service for IMS. In *Proceedings of International Conference on IP Multimedia Subsytems Architecture and Applications*, Bangalore, 2008.

[22] L.L. Fernandez. Palpatine: A P2MP IMS video share architecture and implementation. In *Proceedings of International Conference on Next Generation Mobile Applications, Services and Technologies*, Cardiff, Wales, 2008.

[23] E. Julien, F. Huve, and S. Manvi. Using video mashup to deliver personalized video services in IMS networks. In *Proceedings of International Conference on IP Multimedia Subsytems Architecture and Applications*, Bangalore, 2008.

[24] D. Lozano. WIMS 2.0: Converging IMS and Web 2.0. designing rest APIs for the exposure of session-based IMS capabilities. In *Proceedings of International Conference on Next Generation Mobile Applications, Services and Technologies*, Cardiff, Wales, 2008.

[25] R.B. Shet and S. Mahalakshmi. IMS interactive mobile cast. In *Proceedings of International Conference on IP Multimedia Subsytems Architecture and Applications*, Bangalore, 2007.

[26] R.G. Shiroor. IPTV and VOD services in the context of IMS. In *Proceedings of International Conference on IP Multimedia Subsytems Architecture and Applications*, Bangalore, 2007.

[27] Naoki Takaya, Kazuhiro Tokunaga, and Akira Kurokawa. Presence with Avatar for Web 2.0-IMS services using rest interface. In *Proceedings of ICIN 2008*, Bordeaux, France, 2008.

12

Current and Optimal Cost Allocation for QoE Optimized IPTV Networks

Maximilian D. Schlesinger[1], Tobias Heger[1], Thomas Monath[2] and Mario Kind[2]

[1]*European Center for Information and Communication Technologies – EICT GmbH, Ernst-Reuter-Platz 7, 10587 Berlin, Germany; e-mail: maximilian.schlesinger@eict.de*
[2]*Deutsche Telekom AG, Laboratories, Ernst-Reuter-Platz 7, 10587 Berlin, Germany*

Abstract

With upcoming IPTV services telecommunication networks will face new challenges in terms of traffic congestion and quality of transmission. The Quality of Experience (QoE) concept offers ways to overcome congestion problems and to increase perceived quality by end users. In this chapter we used an adapted target costing analysis to evaluate the current and optimal cost structure for QoE optimized IPTV networks. Total current costs, especially access and aggregation network component costs, are found to be significantly higher than the users' willingness to pay for IPTV. The analysis shows that (1) an IP-based network solely built for IPTV delivery will not be profitable, (2) additional services using the same infrastructure might lead to profitability, (3) end users' primary interest is in availability of content and ease of use of the IPTV system, whereas network technology is of secondary interest only. Therefore, operators' future investments should reflect the customer preferences and closely investigate the possibilities that will emerge with enabling functions for quality.

Keywords: IPTV, target costing, cost allocation, quality of experience.

Anand R. Prasad et al. (Eds.), Advances in Next Generation Services and Service Architectures, 251–277.

12.1 Introduction

Major changes in customers' usage patterns of television (TV) services are taking place. In the Internet strong growth of rich media and especially video consumption can be observed. Additionally, users increasingly seize opportunities that service personalization offers along with possibilities to "watch any video content at any time on any device" [3, 15]. Due to these new usage patterns users become more sensitive to quality issues. This applies especially to paid video services such as IPTV and Video-on-Demand (VoD) [34].

IPTV combines conventional TV with new, multifaceted and interactive services. This way IPTV providers seek to win users for their offerings instead of Web services running on PCs.

The changes in user behavior and the introduction of IPTV lead to changing network traffic characteristics. Network traffic in aggregation and access networks grows heavily and increasing usage of personalized services requires unicasts instead of common broadcast connections [3]. As a result of these expected changes, network congestion can be expected to rise [5].

The combined effect of measurable Quality of Service (QoS) [36] and the more subjective perception by end-users is reflected in the Quality of Experience (QoE) concept [16]. In the EUREKA Celtic project RUBENS[1] technological possibilities related to QoE were explored and combined to exploit the entire available solution space. The aim is that all user requests should be answered in traffic peak times while considering the imminent challenges for the network.

This chapter centers on the current and optimal cost allocation for the components of an QoE-optimized IPTV reference system. We concentrate on an IPTV system for transmission of conventional TV streams without integration of further web services. The analysis builds upon data drawn from five European network operators and one network equipment vendor that were involved in RUBENS. An adapted target costing approach is used to analyze aggregation and access network, customer premises equipment and the service platform that is necessary to successfully operate an IPTV system.

---

[1] For details about RUBENS, see http://wiki-rubens.celtic-initiative.org/.

A short literature review for Quality of Experience and target costing is given in Section 12.2. IPTV is not further defined in this chapter as you can find the foundation of IPTV in Chapter 9. In Section 12.3 our research methodology is presented. Sections 12.4 and 12.5 present intermediate results of our analysis: customer needs for IPTV and the underlying network model. These provide the basis for the main result – the current and optimal cost allocation – which is presented in Section 12.6. The derived implications are outlined and discussed in Section 12.7. Finally, in Section 12.8 our results are summarized, limitations are pointed out and an outlook for further research is presented.

12.2 Theoretical Background

12.2.1 Quality of Experience

Quality sensible IP-based services require broadband Internet connections with real-time, interactivity, security and reliability capabilities. The term Quality of Service (QoS) is not used consistently in the literature. However, it usually describes the possibility to differentiate services and the possibility to allocate different quality parameters to services. Mostly four technical parameters are used to determine service classes: (1) the available bandwidth, (2) delay time, (3) jitter, and (4) packet loss [37]. Service classes range from level 0, called *best-effort*, to level 7, called *layer 2 network control reserved traffic* [36].

Quality-of-Experience (QoE) is more user-centric than QoS. It seeks to link technical parameters and the users' perception of quality. Several definitions of QoE exist; three selected ones are as follows:

- Widely used is the definition of ITU-T SG12 that describes QoE as "overall acceptability of an application or service, as perceived subjectively by the end-user" that "may be influenced by user expectations and context" [16].
- Soldani et al. define QoE as "how a user perceives the usability of a service when in use – how satisfied he/she is with a service in terms of, e.g., usability, accessibility, retainability and integrity" [31].
- Recently, Fiedler et al. defined QoE as a concept that describes "the degree of delight of the user of a service, influenced by content, network, device, application, user expectations and goals, and context of use" [10].

All these definitions have in common that service classes are defined by the user's *perception* in addition to measurable network parameters. The user's perception may be influenced by the *network*, the *context* (i.e. the kind of service used, prices and content), *usability* of services and applications and his/her *expectations*.

Technologies addressing QoE can be categorized in the following three technology fields:

1. *Video Quality Management.* Technologies for video quality management can be used in at least three cases. First, the variety of end user devices can be served with the correct resolution, minimizing CPU load on the devices. Second, downscaling of video in case of traffic peaks allows continuation of streaming instead of complete failures. Third, studies (e.g. [35]) found that the Mean Opinion Score (MOS) fluctuates depending on the kind of the movie despite of the same bit rate, resolution, etc. By implication this means that the perceived quality on a certain level can be achieved with different video parameters, potentially allowing either improving or economizing video streaming services. Technologies of this category include Equal Quality Video Streaming (EQVS) and Scalable Video Control (SVC) algorithms [18].

2. *Monitoring and traffic estimation mechanisms* allow forecasts of congestion situations and triggering adequate reactions to congestion problems at occurrence. Technologies in this field are, for example, monitoring mechanisms and Pre-Congestion Notification (PCN) [19].

3. *Transport, routing and admission control mechanisms.* These mechanisms increase the network efficiency by optimizing link usage. Micro-caching within the access network allows answering similar service requests fast and without causing traffic in higher network aggregation levels. Routing mechanisms [20] and admission control mechanisms [9, 17, 27, 28] are further examples for technologies in this field.

Summarized, research shows that QoE improvements are technically viable. However, in contrast to over-the-top-services, network support is often a prerequisite. Network operators need to adapt their networks accordingly to make them work. Whereas the different approaches promise to increase customer satisfaction and increase network efficiency it remains to be analyzed whether these effects cover capital expenditures and operating costs.

12.2.2 Target Costing

Target costing is a holistic set of management tools for product cost planning with a strong customer orientation. It is "primarily a technique to strategically manage a company's future profits. It achieves this objective by determining the life-cycle cost at which a company must produce a proposed product with specified functionality and quality of the product if the product is to be profitable at its anticipated selling price" [8].

Target costing was first applied in Japan by Toyota in 1965 [2]. Within the 1970s more and more Japanese companies implemented target costing methods and first scientific publications appeared in Japanese management literature [12, 26, 33]. Rather recently, globalization and cost pressures induced further interest in target costing in the US [6,8] and German [13,14,30] industry as well as scientific literature.

Instead of traditional cost plus methods (production costs + overhead + target profit = product price), inverse accounting is leveraged. Here, the price that the market is willing to pay is taken as upper limit (upper price set by the market − target profit − overhead = allowed production costs) [1]. The main objective of target costing is to align the company with the market, i.e. it aims to control the overall cost structure in an early stage in a market-oriented way [26]. In addition, it assures long-term financial goals and seeks to motivate employees with the use of easy "operationalizable" indicators instead of rather abstract financial ratios.

Target costing should be considered as planning method, when product complexity increases, resources are limited and a strong market-orientation is desired [32]. It is especially appropriate for markets with strong competitive pressure and high customer expectations. In this case, cost reductions are necessary to remain competitive while the quality standard needs to remain constant. Target costing is typically implemented in industries such as automotive and electronics [4]. However, it can also be found in service enterprises and companies from process industries [7,24].

12.3 Methodology

For our analysis we apply a modified target costing approach to meet the specific needs of the telecommunication industry. It is used to evaluate current costs, plan future cost allocation for IPTV components and identify needed changes in the infrastructure, all based on customer needs.

The analysis is split into the following five steps:

1. *Identification of customer needs.* The best way to determine customer needs is a Conjoint analysis [13, 30]. With this analysis a differentiated picture of the user needs, the associate degree of importance of the needs and the users' willingness to pay can be derived. Due to time and budget constraints within the project, secondary sources such as studies, reports and scientific papers were used instead. The obtained customer needs were grouped in different categories on three levels of detail. Nineteen experts were asked to add further customer needs and check the list for completeness. After including the comments by rearranging and merging the customer needs, all resulting customer needs were rated by the experts on a scale from 1 (unimportant) to 5 (very important). To accentuate the relative importance between the different customer needs the ratings were weighted with points (a 5 is worth 125 points, 4: 50 points, 3: 15 points, 2: 5 points, 1: 0 points). The differences between the point results were processed into relative importances of all customer needs. Eventually, only customer needs of the detail levels 1 and 2 and their importance were used to avoid impeding accuracy.

2. *Description of product functions and product components.* The primary aim of the target costing process is to ensure market-orientation and reduce production costs. As most of the manufacturing costs are determined in the design phase of a product, it is important to clearly define functions and components in the beginning of the product design process. Therefore, the description of product functions and components should be exhaustive. A product function usually describes separable functionality of the product, e.g. recording possibility of TV program. In this chapter, we will not describe these product functions in detail, as they are only used in an intermediate step when translating customer requirements into degrees of importance for the product components. However, product components, e.g. remote control or application server, need to be specified. For our analysis we defined a network reference model to ensure the integrity of the IPTV system to be analyzed. The IPTV system definition was completed with intermediate results from the ongoing RUBENS project. The reference model and the specifications of product functions and product components were validated by the afore-mentioned 19 experts in terms of completeness.

3. *Identification of standard costs.* The standard costs reflect costs that will occur if the product is built with the current manufacturing possibilities.

In our case, these were based on the reference model and the distribution of product components within. The costs were calculated with an Excel Tool from a previous telecommunication project [22, 23].

4. *Identification of optimal product costs.* Total optimal product costs are determined based on the market price, target profit and overhead costs. In the "target cost breakdown" the optimal costs on the component level are obtained by translating user needs into degrees of importance for each product component. The optimal cost allocation corresponds to the relative importance of each product component. Due to the notion of strong market-orientation the resulting optimal product component costs should be seen as target costs. Again, the estimation of importance of user needs was based on experts estimations in our case.

5. *Development of implications based on the analysis.* The last step of the target costing process is the analysis and interpretation of the obtained results. In this step, actual costs and optimal costs of product components are compared. Additionally, the results of the target cost breakdown are leveraged to evaluate the satisfaction of user needs in existing products, in our case existing IPTV systems. We identified the main cost drivers within an IPTV system. The implications derived at this point were commented on, validated and completed by our expert panel.

The next three sections cover our results in terms of customer needs (step 1), network reference model (step 2) and current and optimal cost allocation (steps 3 and 4). Results from step 5 are given in the discussion in Section 12.7.

12.4 Customer Needs for IPTV

The customer needs for IPTV can be categorized in eight groups.[2]

Content, quality of transmission and ease of use are revealed as the three most important customer need groups.

Content should be of high quality and there should be a wide variety of channels on the one hand and many integrated services which add value on the other hand. The demand for high quality content is a potential problem for network operators because they control the network only. Thus, in case content delivery and content provision is separated between network operat-

---

[2] A detailed list of the customer needs in sub levels 1 and 2 as well as their degree of importance can be found in Table 12.6 in the Appendix.

ors and service providers, the former cannot control or adapt the content's source in terms of quality. In this case the network operators have to rely on the service providers.

An IPTV system needs to be easy to use. Customers often prefer TV consumption as a passive "lean back" service instead of participating actively in any way apart from channel switching. In addition, the installation needs to be smooth and the system should integrate well into the living environment. During usage the handling needs to be simple, response times should be short, channel switching needs to be fast, etc.

Ease of use, customer support and data security add up to nearly 40% of customer needs. However, these "soft-facts" are often not considered in an adequate way by the actors in the market. To meet market needs adequately, they need to be considered and incorporated in future services and products.

Time sovereignty, different access types and interactive services can excite people; they are potential unique selling propositions. However, they are not as important as the basic features content, quality and ease of use. Systems that provide exciting features but do not fulfill the basic ones will usually fail to gain customer adoption.

12.5 Network Reference Model

The network reference model shows one possibility to design a network for an IPTV system. Several other possibilities may exist, but our reference model was defined to create the basis for the following analysis and our calculations. Its completeness was confirmed by the expert panel.

12.5.1 Architecture

The network reference model is organized in a hierarchical structure as depicted in the schematic overview in Table 12.1.

The network nodes group the product components and represent functionality on an abstract level. However, it does not provide information about location of the nodes and components. Therefore, two or more nodes might have the same geographical location, e.g. a Central Exchange and a Local Exchange. The nodes are connected through links that are used to model the necessary connection between nodes of the different hierarchical levels.

To obtain network costs we used a uniform hierarchical distribution model that neglects regional differences and uses average values for node-node relationships on different levels instead [21]. For example, one Local

Table 12.1 Hierarchical structure of network reference model.

Level	Link	Node		Product component	
Core-network	—	WAN	Wide Area Network	CR	Core Router
				LER	Label Edge Router
Service-platform	—	SP	Service Platform	C&A S	Control & Application Server
				Caches	Caches
				SR	Service Router
				HE	Head End
				MS	Management Server
Aggregation-network	1:4	CE	Central Exchange	BRAS	Broadband Remote Access Server
				AGS2	Aggregation Switch 2
	1:64	LE	Local Exchange	AGS1	Aggregation Switch 1
Access-network	1:256	CAB	Cabinet	DSLAM	Digital Subscriber Line Access Multiplexer
	$1:10^a$	NT	Network Termination	RGW	Residential Gateway
User	—	CPE	Customer Premises Equipment	CPE	Customer Premises Equipment

^{a} One paying customer is assumed per RGW with the possibility to use up to ten different devices.

Exchange aggregates 64 Cabinets indicated by a 1:64 relationship. Based on the distribution numbers as shown in Table 12.1, 65,536 customers can be served per Central Exchange. However, on average roughly 24% of all customers are business customers. This analysis focuses on private users, thus the maximum number of relevant residential users per Central Exchange adds up to 49,807. Ultimately, the necessary network equipment elements can be determined based on the number of customers and our distribution model.

12.5.2 Product Components

In total, 13 different components are included in the reference model.

The Customer Premises Equipment (CPE) includes Personal Computers, mobile or smart phones, all kind of screens including TV sets. Of most interest for our analysis are IPTV set-top-boxes that belong to this category as well.

The Residential Gateway (RGW) connects all devices from the lower level to the network at the Network Termination (NT). In our case, the RGW is assumed to be a WLAN router including the Digital Subscriber Line (DSL) network termination function. We assume that it is part of the access network. Therefore, the network provider has the responsibility for the device.

The Digital Subscriber Line Access Multiplexer (DSLAM) provides the termination of DSL lines, switching and multiplexing functionalities. It is located within the access network at the Cabinet. We assume a Fiber To The

Cabinet (FTTCab) deployment for higher access speed DSL due to the physical reach constraint of ADSL2+ and VDSL. If customers are served directly from the Local Exchange (LE) with xDSL, they are physically connected to a DSLAM located at the LE. This scenario is called FTTCab/Copper near LE.[3]

The aggregation network is further concentrated via Aggregation Switches (AGS1, AGS2) towards the Broadband Remote Access Server (BRAS). The Aggregation Switches are located at Local and Central Exchange, respectively. The BRAS at Central Exchange serves as a gateway to the Wide Area Network (WAN) and to the Internet. This server is responsible for Authentication, Authorization and Accounting (known as AAA) within the network.

Within the Service Platform (SP) multiple network elements are aggregated for IPTV provisioning, namely Control & Application Server, Service Router, Management Server, Head End and Caches. All these elements are used by all customers collectively. This also applies for the Core Network at WAN level which consists of the Core Router (CR) and Label Edge Router (LER).

The network reference model and the product components provide the basis for our cost analysis.

12.6 Cost Allocation

12.6.1 Current Allocation of Component Costs

The actual costs of an IPTV system are composed of three main cost segments: product component costs, overhead costs and target profit.

To approximate the latter, we chose to use average return on sales (ROS, revenues divided by EBIT) of different telecommunication companies of the last three years. To reflect greater importance of recent years for future profits these are weighted higher. In our analysis the weighted ROS, the target profit, was found to be approximately 15.9% for telecommunication operators.[4]

The overhead costs are composed of costs for general administration (7.8% of total revenues), marketing, logistics and distribution (28.3%), R&D (2.3%), service & support (2.6%) and infrastructure (2.3%). The values were taken from past annual reports and former project results.

---

[3] For more information about FTTCab, FTTCab/Copper near LE, ADSL2+, VDSL and their constraints, see [23].

[4] See Table 12.9 in the Appendix for details.

The product component costs were calculated on the basis of the network reference model with different fiber roll-out scenarios and several assumptions.

Within the calculation of product component costs QoE technologies were neglected in the first step. Their cost influence was analyzed separately.

The first fiber roll-out case is a "double play scenario" where only a limited number of cabinets are connected by fiber. As a result fast Internet, and consequently IPTV, is only accessible for few people. In the second relevant case, a "triple play scenario", the fiber network is rolled out for nearly every cabinet, making fast Internet and IPTV services possible for all customers.

The costs for telephony and common Internet connections are assumed to be identical in both scenarios independently of customer numbers. We further assume that the investments for the complete fiber roll out are only undertaken for the IPTV system. The third case, an "IPTV scenario", is an IPTV system in an existing double play network.

The costs for this scenario were simplified to be the costs for the triple play scenario minus the costs for the double play scenario.

We further limit our analysis to a seven year period for depreciation and customer numbers. The discount rate was set to 10% per year.[5] All cost and customer estimations are based on calculations conducted with a tool for approximations that is commonly in use by telecommunication operators.[6] Here, costs are assumed to be directly proportional to the number of potential customers, i.e. twice as much potential customers mean doubled costs.

The "double play scenario" starts with about 16,000 connected customers and ends with about 24,500 customers in year 7. In the "triple play scenario with normal penetration rates"[7] the number of double play customers is roughly the same. But additionally there are 355 triple play customers in year 1 and 6,285 in year 7. Thus, on average 2,661 customers are triple play customers, i.e. about 5% of all customers. This estimation is in line with recent forecasts [23].

As shown in Table 12.2, product component costs for an IPTV system add up to 12.01m €. Not included are costs for infrastructure which are considered as overhead. Over 80% of these costs, i.e. 10.04m €, occur within the access network. The costs for CPE and Service Platform are 0.4m €

---

[5] This does not have a major impact as most of the capital expenditures (CAPEX) have to be invested within the first years.

[6] The underlying model is described in detail in [22, 23].

[7] For a sensitivity analysis we calculated a "triple play scenario with high penetration rates". The detailed customer estimations can be found in Table 12.7 in the Appendix.

respectively 1.5m €. The core network costs are zero by assumption as no additional costs occur within the core network for an IPTV system. Within the access network the DSLAM causes nearly half of the costs. The other half is caused by AGS1, AGS2, BRAS and RGW.

The sensitivity analysis with higher penetration rates, i.e. more IPTV customers, revealed a low cost impact of additional customers. With a doubled penetration rate the costs increase approximately 2.5% only. Returns, on the other side, double within the seven year period with twice as much customers.

As stated above, the overhead costs add up to 43.3% of the total costs, i.e. 12.76m € for a seven year period. The net profit is 15.9% by assumptions, i.e. 4.67m €. Together with the product component costs of 12.01m € – 40.6% of the total costs – the total actual product costs add up to 29.44m €.

12.6.2 Potential Impact of QoE Technologies

The costs described above are based on state of the art technologies and current network deployment without QoE technologies. Afterwards, the influence of QoE technologies on cost distribution for each product component was developed qualitatively with an expert panel in Table 12.3. In case of empty cells no technical and therefore no economical impact is expected for the corresponding component. In particular pre-congestion notification and policy-based network management qualify for cost reductions. On the other hand, technologies like caching or the Shared Content Addressing Protocol (SCAP) cause considerable additional costs. Overall cost reduction potential was estimated to be a few percent only, at best 5–10% of the actual cost.

12.6.3 Optimal Allocation of Component Costs

The optimal product costs are defined by the enforceable market price, i.e. the product will not prevail on the market if costs are higher than this price.

The market price was deduced from studies about customer needs and their willingness to pay [11, 25]. The market price taken from these studies was compared to real-world Triple Play products in Europe.[8] Combining and averaging the values results in a market price of 13 € per month for an ordinary IPTV product without additional VoD fees. Multiplying this with the estimations of connected IPTV customers described before results in expected discounted revenues of 1.90m € for the seven year period. This marks

---

[8] See Table 12.8 in the Appendix for details.

Table 12.2 Actual costs in all three scenarios [22, 23].

Node	Product-comp.	Double-Play-Scenario			Triple-Play-Scenario			IPTV-Scenario		
		CAPEX in k€	OPEX in k€	TOTAL in k€	CAPEX in k€	OPEX in k€	TOTAL in k€	CAPEX in k€	OPEX in k€	TOTAL in k€
CPE	CPE	n/a[a]	n/a	n/a	433	0	433	433	0	433
NT	RGW	484	2,992	3,477	560	4,361	4,921	75	1,369	1,444
CAB	DSLAM	1,250	4,987	6,237	3,874	7,269	11,143	2,624	2,282	4,906
LE	AGS1	153	1,975	2,127	351	2,878	3,229	199	903	1,102
CE	AGS2	131	1,975	2,106	168	2,878	3,046	37	903	940
	BRAS	492	3,103	3,595	722	4,523	5,245	230	1,420	1,650
SP	C&A S	n/a	n/a	n/a	12	294	306	12	294	306
	Caches	n/a	n/a	n/a	12	294	306	12	294	306
	SR	n/a	n/a	n/a	12	294	306	12	294	306
	HE	n/a	n/a	n/a	12	294	306	12	294	306
	MS	n/a	n/a	n/a	12	294	306	12	294	306
WAN	CR	n/c[a]	n/c	n/c	n/c	n/c	n/c	0	0	0
	LER	n/c	n/c	n/c	n/c	n/c	n/c	0	0	0
	Total	2,509	15,032	17,541	6,170	23,378	29,548	3,661	8,346	12,006

[a] n/a: not available; n/c: not calculated

Table 12.3 Influence of QoE technologies on product components' costs.

QoE technologies	CPE	RGW	DSLAM	AGS1	AGS2	BRAS	C&AS	Caches	SR	HE	MS	CR	LER
Video Quality Management													
Scalable Video Coding	+		-	-	o	o			o	+			
Equal Quality Video Streaming							+			o			
Monitoring and traffic estimation mechanisms													
Pre-Congestion Notification				-	-	-		+	-			o	o
QoS Monitoring	o	o	o	o	o	o	o	+	o	o	o		
QoE Estimation							+	+					
Transport, routing and admission control mechanisms													
Caching								+			++	-	-
Shared Content Addressing Protocol	o						+	+	o	+	++	-	-
Call Admission Control			+			+	+	+	o				
Resource Management			o			o	+	+	o				
Segmentation & Play Map							+			+			
Application Function "QoE EPG"	o						+						
Policy-Based Network Management			-	-	-	-	-	-	+	-		-	-

++: strong cost increases; +: cost increases; o: cost neutrality; −: cost decreases; −−: strong cost decreases

the upper limit for all costs for an IPTV system within our reference model, including product component costs, overhead and target profit.

For the compilation of the optimal cost allocation we assume the shares of target profit and overhead costs to be the same as in the actual costs calculation shown above: 15.9% (0.30 €) respectively 43.34% (0.82 €), see Table 12.4. The remaining 0.78m € of the total revenues can be allocated to the product component costs according to their relevance for the customer.

The degrees of importance of each product component was derived from the customer needs. This was conducted in a two-step process. First, the relevance of each product function for each customer need was determined in a two-dimensional matrix (customer needs × product functions). Thus, the degree of importance for a certain product function is the sum of its relevances for all customer needs. Second, the relative importance of each product component for each product function was calculated the same way in another matrix (product functions × product components). The relationship between

Table 12.4 Actual and optimal costs on product level.

	Share of total costs in %	Optimal costs in k€	Actual costs in k€
Expected / required revenues	100.0	1,904	29,435
Target Profit	15.9	302	4,672
General administration	7.9	149	2,301
Marketing, distribution, logistics	28.3	539	8,333
R&D	2.3	44	677
Service & support	2.6	50	765
Infrastructure	2.3	44	680
Product component costs	40.6	777	12,007

product functions and customer needs and between product components and product functions was established based on expert estimations.

The CPE and the service platform are identified as most important product components. The service platform with its five components (control & application server, caches, service router, head-end, management server) adds up to over 50% of total importance. In other words, about half of the use for customers is provided by the service platform. The reason is that services are the most valuable components that at most value in an IPTV system.

The CPE with its 25% of total importance was revealed to be very important as well. Here, the reason is that many desired functions can be realized within the CPE. The complete access network (RGW, DSLAM, AGS1, AGS2, BRAS) adds up to a quarter of overall importance only. However, Internet connectivity is a precondition for all services and might have been taken for granted by the experts.

12.6.4 Comparison of Current and Optimal Cost Allocation

In total the actual product component costs are about 15 times higher than the optimal costs. The same holds for the required and expected revenues. The latter, however, is in line with the target costing approach: the optimal costs are directly derived from the estimated revenues and the actual costs are the basis for the required revenues. The expected revenues are not even

Table 12.5 Actual and optimal costs on component level.

Product component	DOI in %	Optimal costs in k€	Actual costs in k€	AC / total OC %
CPE	24.6%	191	433	55.8%
RGW	9.5%	74	1,444	186.0%
DSLAM	4.7%	36	4,906	631.6%
AGS1	0.9%	7	1,102	141.9%
AGS2	0.9%	7	941	121.1%
BRAS	7.3%	57	1,650	212.4%
C&A S	23.2%	180	306	39.4%
Caches	5.6%	43	306	39.4%
SR	4.0%	31	306	39.4%
HE	6.3%	49	306	39.4%
MS	11.5%	90	306	39.4%
CR	0.8%	6	0	0.0%
LER	0.6%	5	0	0.0%
Total	**100.0%**	**777**	**12,007**	**1545.7%**

sufficient for the actual product component costs only (without target profit and overhead), see Table 12.4 for details.

This becomes particularly apparent when analyzing single product components in Table 12.5. The costs of some components exceed the total optimal product component costs, e.g. the DSLAM which costs more than six times the amount that all components are allowed to cost.

To visualize the results a value control chart as shown in Figure 12.1 can be used [29]. The x-axis shows the degrees of importance of the product components. The y-axis displays the ratio of actual product component costs to total optimal product component costs. Each product component is one data point. Ideally each data point is on the diagonal line. In that case, the used resources for a product component would be consistent with the importance of the component. Within an optimal zone around the diagonal line no changes are necessary immediately. For data points outside of the optimal zone a re-allocation of resources should be considered: product components above the optimal zone are too cost intensive. Their costs have to be reduced. Product components below the optimal zone can be improved further, additional resources should be allocated to the component to fulfill customer needs optimally.

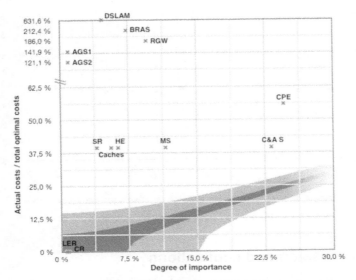

Figure 12.1   Value control chart.

Our analysis reveals that except for the core network components, which cause no costs by assumption, all product components are above the optimal zone. Especially the access network components, DSLAM, BRAS, RGW, AGS1, AGS2, exceed their optimal cost allocation by far. While the CPE is indeed relatively important with its 25% it is not as important as its actual costs imply. These sum up to over half of the total optimal component costs. The same is valid for the service platform. It is the most important component with a combined degree of importance of over 50% but its costs are about twice the total optimal component costs. Thus, according to our analysis there is no room for investments into any product component improvements: there is a strong need for cost optimization for all product components.[9]

Additionally, there is a strong mismatch in relative cost allocation. Thus, resources need not only to be reduced but also to be re-allocated differently. To serve customer needs in an optimal way the cost allocation should be done as shown in Figure 12.2. Resources should be shifted from the access network to the CPE and the service platform. Nevertheless, a reallocation of presently used resources needs to go along with cost reductions for all parts.

---

[9] The core network will be left out of the analysis due to the underlying assumption and its negligible importance of 1.4%.

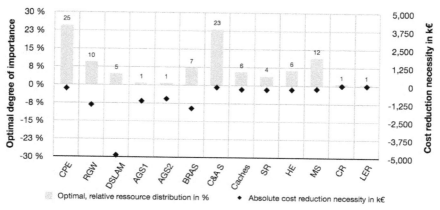

Figure 12.2 Relative costs comparison.

12.7 Discussion and Implications

The implications resulting from our analysis can be grouped in three categories: customers, network and technology and business model.

Customer-related implications center around new customer needs and the changes in their behavior. In the network and technology section related implications and necessary changes are discussed. Afterwards, the resulting effects of the previous two sections on the business model of an IPTV system operator are covered.

12.7.1 Customers

Most important for end users are the ease of use, high quality content and transmission quality. In order to discuss our results, the classical use of TV has to be reconsidered first. It commonly serves for entertainment, recreation, relaxing, information retrieval and quite often for social interaction as well since it provides topics for conversations. Traditionally it is used in a "lean back" way, i.e. passive consumption of content. In this context users need to recognize IPTV as a comparable alternative to conventional TV over satellite, cable or via terrestrial broadcasting. Therefore, a basic feature set of high quality content, appealing signal transmission and pleasant ease of use should be implemented to provide an comparable experience to conventional TV.

To justify premiums for IPTV, additional features that enrich the experience for end users have to be implemented. These can be provided via interactive features. When taking the customer needs into consideration, interactivity in terms of flexibility regarding time and place of consumption emerge as most interesting. However, these features should not contradict the most important customer needs such as ease of use. In that case, we expect interferences with the "lean back" consumption behavior. So far it is not clear to what extent the customers' demand and willingness to use "lean forward" services exists, i.e. services that require active participation. Additionally, data security and customer support are value adding features that might be utilized to differentiate from conventional TV offerings.

Summarized, the features of IPTV systems can be grouped in two categories as can be seen in Figure 12.3: (1) basic features that have to be met to be accepted as alternative to conventional TV, i.e. high quality content, ease of use and high quality of transmission, and (2) value adding features that need to be implemented to gain significant market shares because IPTV is conceived as superior to conventional TV, i.e. interactivity, time, place and device sovereignty, data security and customer support.

Furthermore during our analysis of customer needs it became apparent that customers are not willing to pay for pure Internet connectivity anymore. Both, the access network components and the product functions which are responsible for transport add only little direct value for customer. The current willingness to pay results from the possibility to use services – free and paid, e.g. Google, YouTube, Facebook and many others. Thus, network operators have to re-think their strategies. Most likely, they will need to become much more service-oriented companies because margins for the connectivity will decrease or remain low at least.

12.7.2 Network and Technology

Within the network, the highest costs have their origin in the access and aggregation network. This is mostly due to the cabinet upgrade and the common overdimensioning strategy. With the cabinet upgrade every DSLAM is connected with fiber. This allows much higher bandwidth for end users. Hence, SD and HD video services are realizable.

In our analysis we assumed that the cabinet upgrade is done for the IPTV system only. The analysis clearly shows that this would not be sensible economically. However, extended and new value adding services have the potential to turn the cabinet upgrade into an economically feasible step. On the other hand, the customer needs clearly show the necessity of the cabinet upgrade as this is the precondition for the primary demand for IPTV services: high quality content.

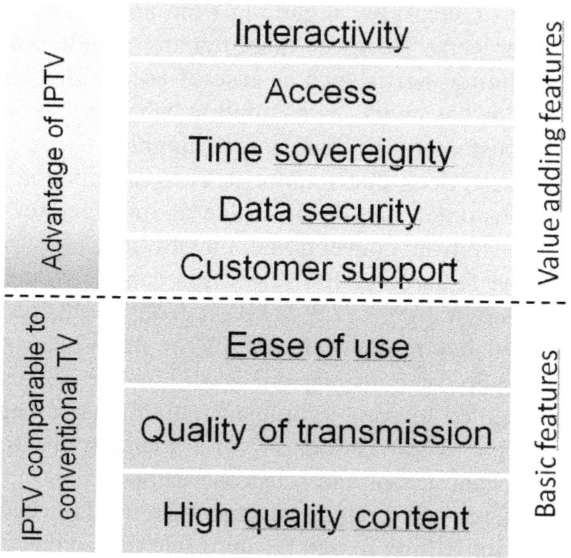

Figure 12.3 Requirements for an IPTV system in terms of customers' needs.

Additionally, operators have different roles today. Quite often they are simultaneously network operators (fixed line and mobile), service providers (VoIP, IPTV and ISP) and equipment vendor (mobile devices, routers, telephones, etc.). Consequently, the cabinet upgrade might be required for other business areas than IPTV as well, for example to avoid loosing customers to other actors such as cable operators.

Nevertheless, cost reductions are inevitable for the cabinet upgrade. At the same time, the actual costs must not be regarded as given and incontestable. To the contrary, further research might reveal alternative cost reduction potential and new revenue streams.

The overdimensioning strategy that network operators commonly pursue today will become even more resource intensive in the future. The main reasons for overdimensioning are (1) the difficulty of forecasting network usage, and (2) the guaranteed bandwidth that network operators need to provide to end users even at times of high network traffic. Through the expected increase of usage of personalized services more unicasts instead of broadcasts will be rendered necessary. Consequently, efficiency increases through smart traffic management in the access and aggregation network have the potential for significant cost savings.

However, the economic feasibility of the introduction of more intelligent network components remains arguable. The feature range of CPEs and service platforms grows steadily due to increasing computing power, new software and new technologies. These features include functions that are more visible and relevant to end users. To date, many of them were realizable with network mechanisms only. Additionally, the costs to integrate additional features into CPEs or service platforms are relatively low. The result is uncertainty in regard to implementation feasibility: should intelligence be implemented within the network or in CPE and service platforms? An answer to this question requires further research and is likely to vary from case to case.

QoE technologies are not a revolution for networks. However, they are expected to represent the next evolutionary level of telecommunication networks. Our analysis revealed that QoE technologies can boost efficiency of networks, particularly in terms of personalization. The anticipated problem with the overdimensioning strategy can be reduced and quality perceived by the end users can be increased. The possibilities of QoE technologies are expected to make it possible to keep prices for network-related products constant. Integration of value-adding features either keeps the perceived quality constant or increases value for the customers. Thus, they are expected to avoid customer migration to other operators.

Summarized, QoE technologies are not expected to provide significant cost improvements. However, they enable new services, are expected to keep customer satisfaction at a constant level and allow to increase network efficiency.

12.7.3 Business Model

The conducted sensitivity analysis reveals that costs are relatively independent of the number of connected customers. The main cost driver is the potential number of customers and the associated sunk costs for network equipment. However, revenues increase proportional with the connected customers. Therefore, IPTV will be increasingly economical sensible provided that the number of connected customers rises. Consequently, the product line-up should aim to realize scale effects by aiming for the mass market.

In this analysis the private market was considered only. Revenue streams from other products and services that potentially benefit from a cabinet upgrade were neglected. In reality, however, operators should seek to realize synergies by offering multiple products on different levels, e.g. on the business market. The 3D TV market is also expected to emerge as potential market, since 3D movie-data requires high bandwidth.

Summarized, a great number of connected customers is needed to refinance high sunk costs for the network and new service offerings and product differentiation are necessary to distribute costs among a larger variety of products and realize scale effects.

12.8 Conclusions and Outlook

Our analysis is a first attempt to apply the target costing approach within the telecommunication industry, particularly to an IPTV system. Due to the complexity of the network deployment and the telecommunication sector many assumptions became necessary. The customer needs were only indirectly evaluated by an expert survey which were possibly biased. Within the degree of importance break down of customer needs to product functions and product components, Internet connectivity was possibly undervalued. This resulted in low degrees of importance for the access and aggregation network. The assumed cabinet upgrade for IPTV only was another assumption with high impact on costs and revenues. Further research should focus on these issues and create models that try to capture the IPTV system more detailed.

In the future, personalization and bandwidth requirements will increase strongly. This is expected to become a problem particularly for IPTV systems. We show that the access and aggregation network is the main cost driver. The large part of customer value, however, is provided by the service platform and the CPE. QoE technologies are revealed to have a rather low potential for cost reduction but should be used to provide enablers for new services.

For now, IPTV service providers should concentrate on delivering high quality content and satisfy customers' needs in terms of ease of use. Network operators should optimize the efficiency of the access and aggregation network and seek alternative cost reduction potential apart from QoE technologies.

Appendix

Table 12.6 Degrees of importance of customer needs, level 1 and 2.

	Customer needs	ϕ, average points	σ, standard deviation	Relative DOI on level 1	Relative DOI on level 2	Absolute DOI
1	Content	4.7	0.562	22%		22.1%
1.1	Selection of channels	4.6	0.502		63%	13.9%
1.2	Integrated services	3.8	0.985		37%	8.2%
2	Quality of transmission	4.5	0.5	16%		16.3%
2.1	Reliability	4.6	0.507		17%	2.8%
2.2	Availability	4.8	0.419		20%	3.3%
2.3	Speed	4.3	0.733		14%	2.2%
2.4	Video quality(HD vs. SD)	4.3	0.82		15%	2.4%
2.5	Audio quality	4.2	0.918		14%	2.3%
2.6	Audio-Video-Synchronization	4.8	0.419		20%	3.3%
3	Time sovereignty	3.7	1.018	10%		9.8%
3.1	PVR (Time-Shift, etc.)	4.1	1.049		61%	6.0%
3.2	Video-on-Demand	3.8	0.787		39%	3.8%
4	Access	3.6	0.684	8%		7.5%
4.1	Place independence	4.1	0.737		29%	2.2%
4.2	Avoidance of discontinuity of media	3.9	1.15		29%	2.2%
4.3	Hardware independence	3.6	0.831		18%	1.4%
4.4	Simultaneous Multi-TV access	3.7	1.108		24%	1.8%
5	Interactivity	3.3	0.82	6%		5.6%
5.1	Participation in TV program (quiz, votings, etc.)	2.5	1.172		17%	0.9%
5.2	Interaction with other users	2.7	1.293		24%	1.3%
5.3	Settings	3.8	0.834		59%	3.3%
6	Ease of use	4.5	0.514	15%		14.9%
6.1	Installation	4.2	0.878		30%	4.5%
6.2	Usability	4.7	0.594		41%	6.0%
6.3	Set-Top-Box-Hardware characteristics	3.1	1.183		11%	1.7%
6.4	Program survey	3.4	1.294		18%	2.6%
7	Data security	3.6	1.165	11%		10.7%
7.1	Security of personal data	4.2	1.015		30%	3.2%
7.2	User behavior profiling	3.6	1.121		19%	2.0%
7.3	Privacy	4.1	0.994		28%	3.0%
7.4	Identity Management	3.8	1.134		23%	2.4%
8	Customer support	4.0	0.943	13%		13.1%
8.1	Service	4.4	0.761		47%	6.2%
8.2	Charging	4.1	0.78		35%	4.6%
8.3	Image / Reputation of vendor	3.3	1.098		17%	2.3%

Table 12.7 Customer development in different scenarios.

Year	Connected customers				
	DP scenario	TP scenario (normal penetration rates)		TP scenario (higher penetration rates)	
	Double Play	Double Play	Triple Play	Double Play	Triple Play
1	16,472	16,118	355	16,093	379
2	20,442	19,749	693	19,052	1,390
3	22,339	22,339	1,200	21,139	2,400
4	24,032	24,032	1,862	22,167	3,722
5	24,998	24,998	2,770	22,223	5,535
6	23,609	23,609	5,462	18,044	10,995
7	24,462	24,462	6,285	17,862	12,835
Average	**22,336**	**22,187**	**2,661**	**19,511**	**5,322**

Table 12.8 Triple Play prices in Europe.

Land	Basis	Normal	Premium
United Kingdom (BT)	33.00 €	44.00 €	53.00 €
France (Orange, FREE, SFR, Numéricâble)	29.92 €	34.92 €	38.67 €
Spain (Telefónica)	43.90 €	59.30 €	85.90 €
Belgium (Belgacom)	49.95 €	60.15 €	70.15 €
Germany (Deutsche Telekom, Alice)	37.45 €	54.88 €	67.43 €
Switzerland (Swisscom)	51.50 €	68.00 €	73.30 €
Europe, Triple Play	40.95 €	53.54 €	64.74 €
thereof 33% IPTV-Share	13.51 €	17.67 €	21.36 €

Table 12.9 Return on Sales of telecommunication operators, 2006–2008.

	2008	2007	2006	Weighted
Deutsche Telekom	11.4%	8.5%	8.6%	10.1%
France Telecom	19.1%	20.2%	13.4%	18.8%
KPN	17.8%	19.8%	18.4%	18.5%
Telefónica	23.7%	23.5%	17.6%	22.9%
BT Group PLC	1.9%	11.4%	12.6%	6.3%
AT&T	18.2%	17.2%	16.3%	17.7%
Vodafone Group PLC	14.3%	28.3%	−5.0%	16.8%
Average	**15.2%**	**18.4%**	**11.7%**	**15.9%**

Acknowledgements

The research was performed partially within the framework of the EUREKA CELTIC RUBENS project. The authors would like to thank all involved partners.

Financial support was granted by Deutsche Telekom AG.

References

[1] Shahid L. Ansari, J.E. Bell, and the CAM-I Target Cost Core Group. *Target Costing: The Next Frontier in Strategic Cost Management*. McGraw Hill, New York, 1997.

[2] Ali Arnaout. Target costing in der deutschen Unternehmenspraxis: Eine empirische Untersuchung. PhD Thesis, Stuttgart Universität, 2001.

[3] J. Bryant and M.B. Oliver. *Media Effects – Advances in Theory and Research*, 3rd edition. Routledge, New York, 2008.

[4] Willi Buggert and Axel Wielpütz. *Target Costing*. Hanser, München, 1995.

[5] Commission of the European Communities. Communication on future networks and the internet. *COM(2008)*, 594, 2008.

[6] R. Cooper and R.S. Kaplan. Profit priorities from activity-based costing. *Harvard Business Review*, 130–135, May–June 1991.

[7] Robin Cooper and W. Bruce Chew. Control tomorrow's costs through today's designs. *Harvard Business Review*, 74(1):88–97, 1996.

[8] Robin Cooper and Regine Slagmulder. Develop profitable new products with target costing (cover story). *Sloan Management Review*, 40(4):23–33, 1999.

[9] Christian Esteve Rothenberg and Andreas Roos. A review of policy-based resource and admission control functions in evolving access and next generation networks. *Journal of Network and Systems Management*, 16(1):14–45, 2008.

[10] M. Fiedler, K. Kilkki, and P. Reichl. 09192 Executive summary – From quality of service to quality of experience. In M. Fiedler, K. Kilkki, and P. Reichl (Eds.), *From Quality of Service to Quality of Experience*. 2009.

[11] GfK Consumer Tracking. Bekanntheit, Nutzungsinteresse und Nutzerpotenzial von IPTV, http://www.bitkom.org/files/documents/3TVundUserforschungLechnerGfK.pdf, 2008.

[12] Toshiro Hiromoto. Management accounting in Japan: Ein Vergleich zwischen japanischen und westlichen Systemen des Management Accounting. *Controlling*, 1(6):316–322, 1989.

[13] P. Horváth and W. Seidenschwarz. Zielkostenmanagement. *Controlling*, 4(3):142–150, 1992.

[14] Péter Horváth, Stefan Niemand, and Markus Wolbold. Target costing – State of the art. In Péter Horváth (Ed.), *Target Costing – Marktorientierte Zielkosten in der deutschen Praxis*, pages 1–27. Schäffer-Poeschel, Stuttgart, 1993.

[15] M. Inouyem. The "audience of one": Long-form mobile and portable content slowly emerges. *In-Stat*, 2006.

[16] International Telecommunication Union. ITU-T Recommendation G.1081, 2007.

[17] S. Latré, B. De Vleeschauwer, W. Van de Meerssche, S. Perrault, F. De Turck, P. Demeester, K. De Schepper, C. Hublet, W. Rogiest, S. Custers, and W. Van Leekwijck. An autonomic PCN based admission control mechanism for video services in access networks. In *Proceedings of IM'09, IFIP/IEEE International Symposium on Integrated Network Management*, pages 161–168. Springer, Berlin/Heidelberg, 2009.

[18] Steven Latré, Filip de Turck, Bart Dhoedt, and Piet Demeester. Scalable simulation of qoe optimization for multimedia services over access networks. In *ICOMP2007, the 2007 International Conference on Internet Computing (part of the 2007 World Congress in Computer Science, Computer Engineering, and Applied Computing)*, 2007.

[19] Steven Latré, Bart de Vleeschauwer, W. van de Meerssche, F. de Turck, and P. Demeester. Design and configuration of pcn based admission control in multimedia aggregation networks. In *Proceedings of Global Communications Conference. Exhibition & Industry Forum (Globecom 2009)*. Honululu, Hawaii, 2009.

[20] Michael Menth and Matthias Hartmann. Threshold configuration and routing optimization for pcn-based resilient admission control. *Comput. Netw.*, 53(11):1771–1783, 2009.

[21] Thomas Monath. Economics of fixed broadband access network strategies. *IEEE Communication Magazine*, 41(9):132–139, 2003.

[22] MUSE Consortium. Techno-economic evaluations: Results from initial use cases. Deliverable DA3.1, 2004.

[23] MUSE Consortium. Techno-economics for fixed access network evolution scenarios. Deliverable DA3.2, 2006.

[24] Stefan Niemand. Target Costing für industrielle Dienstleistungen. PhD Thesis, Stuttgart Universität, 1996.

[25] PriceWaterhouseCoopers. IPTV – Das neue Fernsehen?, 2008.

[26] M. Sakurai. Target costing and how to use it. *Journal of Cost Management*, 3(2):39–50, 1989.

[27] José Ma Saldaña, José I. Aznar, Eduardo Viruete, Julián Fernández-Navajas, and José Ruiz. Qos measurement-based cac for an ip telephony system. In *Quality of Service in Heterogeneous Networks*, Lecture Notes of the Institute for Computer Sciences, Social Informatics and Telecommunications Engineering, pages 3–19. Springer, Berlin/Heidelberg, 2009.

[28] Julie Schlembach, Anders Skoe, Ping Yuan, and Edward Knightly. Design and implementation of scalable admission control. In G. Goos, J. Hartmanis, and J. van Leeuwen (Eds.), *Quality of Service in Multiservice IP Networks*, pages 1–15. Springer, Heidelberg, 2001.

[29] Claus Schulte-Henke. *Kundenorientiertes Target Costing und Zuliefererintegration für komplexe Produkte: Entwicklung eines Konzepts für die Automobilindustrie*, 1st ed. Gabler, Wiesbaden, 2008.

[30] Werner Seidenschwarz. *Target Costing*. Vahlen, München, 1993.

[31] D. Soldani, M. Li, and R. Cuny. *QoS and QoE Management in UMTS Cellular Systems*. Wiley, 2006.

[32] Martin Stirzel and Stefan Zeibig. Target costing. *Controlling*, 21(6):322–325, 2009.

[33] M. Tanaka. Cost planning and control systems in the design phase of a new product. In Y. Monden and M. Sakurai (Eds.), *Japanse Management Accounting: A World Class Approach to Profit Management*, pages 49–71. Productivity Press, Cambridge, MA, 1989.

[34] S. Van den Berghe, P. Nooren, S. Latré, B. Crabtree, M. Kind, E. Viruete, and C. Pons. Qoe-driven broadband access. In *Conference Proceedings: NEM Summit*, Saint-Malo, October 13–15, pages 71–80, 2008.

[35] M. Varela. Évaluation pseudo-subjective de la qualité d'un flux multimédia. PhD Thesis, INRIA/IRISA – University of Rennes, Rennes, France, 2005.

[36] Zheng Wang. *Internet QoS: Architectures and Mechanisms for Quality of Service*. Morgan Kaufmann Publishers, San Diego, 2001.

[37] Ruediger Zarnekow and Walter Brenner. Quality of service business models for the broadband internet. In *Proceedings of 18th European Regional ITS Conference*, Istanbul, 2007.

PART 3

CONTEXT AWARENESS

13

Context Discovery for Autonomic Service Adaptation in Intelligent Space

Yazid Benazzouz

ENSM-SE & Orange-Labs, Centre G2I, 158 cours Fauriel,
F-42023 St. Martin-d'Hères, France; e-mail: benazzouz@emse.fr

Abstract

Recent advances in pervasive computing have led to the emergence of a number of interesting challenges. Services in particular require to overcome higher costs of development resulting from human intervention, redevelopment of services and dedicated design solutions. Moreover, the various practices and habits of different users of such monitored services increase the difficulty in interpreting generic usage.

We present an approach that tackles the problem of determining relevant situations for service, and the ability of this approach to carry out autonomic service adaptation to design an intelligent space. Several approaches have been presented as candidates to address service adaptation problems by taking into account the context of the user and his environment. These approaches include context reasoning, pattern recognition and context prediction. These research efforts, despite their high quality, do not resolve the requirements of autonomic service adaptation. Making context-aware services adaptive requires an autonomic determination of relevant situations for services to adapt, compose or configure these services.

The goal of this chapter is to motivate a research direction which aims to make pervasive spaces intelligent through services delivery. This study promotes context-aware adaptive services and presents context discovery as challenge to help in their advancement. Context discovery is the process of

Anand R. Prasad et al. (Eds.), Advances in Next Generation Services and Service Architectures, 281–305.

discovering relevant patterns, intention and correlations for services from context data stored in databases. Most influential algorithms are those widely used in data mining and knowledge discovery community. Among the numerous scenarios that can be derived from context data, this chapter presents a case study of autonomic determination of service usage.

Keywords: intelligent space, context-awareness, autonomic adaptation, service computing, data mining, context discovery, adaptive services, pervasive computing.

13.1 Introduction

An intelligent space is a space with features that can provide appropriate services to humans by capturing events in space and using information intelligently with computers and robots [14]. In such a space, life can be more convenient and satisfactory. Intelligent space is viewed as confined space (room, street, building, or a city), equipped with distributed sensory intelligence. The various devices of sensory intelligence cooperate with each other in autonomic fashion or by a centralized control in order to truly represent the conditions or environments that exist within the intelligent spaces. The concept of intelligent spaces offers a new paradigm in the area of human, or environment, to machine communication and interaction.

Nowadays, pervasive computing has helped make our living spaces more intelligent. These spaces tend to provide a set of applications and services sensitive to the user context and his environment in order to customize, adapt, anticipate actions or to detect deviation from the pre-known situations. This continuing growth in the deployment of context-aware services has been driven by the continuous decrease in price of electronics, miniaturization and multi-channel access. Examples of these services include electronic equipment control, user interfaces and assisted living.

Context-aware services are defined by Kirsch-Pinheiro et al. [21] as services that their description is enriched with context information relating to the execution environment of services and their adaptation. This definition is broader compared to that in [31] which focuses on user services. The author appoints context-aware services those that exploit context information in the provision of user services.

We distinguish two categories of services, user services and software services. The user services offer features that facilitate their use or assist in carrying out its tasks. This concept necessarily covers software aspects. Software services are computer programs able to communicate and exchange data between applications in distributed environments. In this chapter, the primary concern is that of user services where the interaction Person-Environment (PE) and Person-Applications (PA) is done through context-aware services. In Persons-Environments interaction, the context is used to distribute personalized information, to provide aid or assistance, to avoid situations of risk, to automate repetitive procedures, to automatically trigger actions, etc. In Persons-Applications interactions, the context is used to customize the user interface, recommending certain products, autonomic changing of communication mode, etc. These services must overcome the higher costs of development resulting from human intervention, redevelopment of services and dedicated design solutions. In addition, they should be able to adapt autonomously to meet the specifications of intelligent spaces.

In this chapter, autonomous services refers to services able to adapt automatically in case of deviation from usual situations or encountering an unexpected situation relating to the use of these services. We present an approach that tackles the problem of determining relevant situations for service. Thus, it is possible to achieve future adaptation of services by control or anticipation, according to these usage or in case of deviation. For example, consistently opening the windows/shutters at certain angle, selecting a particular temperature, etc. However, the various practices and habits of different users of such monitored services increase the difficulty in interpreting generic usage.

Chaari et al. have expressed this challenge in [6], when saying that one of the creative challenges of intelligence is to analyze the context information and deduce the meaning or understanding and to integrate this knowledge to adapt applications. Several works of research have been involved in this. The main approaches are context reasoning [9], context recognition [4], context prediction [27] and episode discovery [15]. These research efforts, despite their high quality, do not resolve the requirements for autonomic service adaptation. Making context-aware services adaptive requires an autonomic determination of relevant situations for services to adapt, compose or configure these services. Their limitations are, in part, due to:

- Pattern recognition process ignore information about services which is an important parameter for the success of service usage patterns recognition.
- Service adaptation can be achieved using rules engines and pattern recognition methods. For example, a rule of the form: *if a pattern X is*

recognized then activate service Y. These solutions are unable to learn new living situations of people nor to follow the changes in their habits, because in most cases rules are predefined by the system developer.

- The approaches based on context models need to be fixed in advance, for example an entity-relation model of context should be established before implementing. The proposal to learn models is limited to particular models but still a big challenge.
- Context data is heterogeneous as they can come from different source such as Web services, applications, sensors, devices and appliances. The advances in communication technology enable richer readings, the semantics of which can be utilized to learn context.

The successful resolution of this challenge is advantageous to the development of innovative services capable of operating independent of the user with no prior information or definition of potential patterns. Notwithstanding, existing approaches differ from the objective of this research, which is focused on a pattern discovery and learning method which accounts for the use of services in the derivation of the pattern. Existing mechanisms are only capable of recognizing known patterns as opposed to generating them based on human-service interaction.

The remainder of this chapter is organized as follows: Section 13.2, adaptation in context-aware services is introduced. Section 13.3 presents challenges for context-aware adaptive services. Section 13.4 contains a brief overview on context discovery methods and applications followed, an explanation on how context discovery can improve service adaptation and an example of architecture for autonomic service adaptation. Section 13.5 presents a case study of context discovery. It aims to identify usage situations of services. The method proposed and implemented in this work employs an agglomerative clustering algorithm considering semantic data and specific similarity. The success of the method is supported through experimental results obtained through real-world deployment in the home environment; incorporating live pervasive sensors. Finally, we present a conclusion in Section 13.6.

13.2 Adaptation in Context-Aware Services

Mayrhofer [28] considers adaptation as one of the main motivations for the use of context. In several studies adaptation is seen as the action to act in context. Preuveneers et al. [7] even consider that context-awareness has been introduced to support non-intrusive adaptability of application without a need for an intervention from the user or the application administrator. Indulska et al. [17] mention that a major factor in ubiquitous systems is their ability to adapt to context change. According to Lombardi et al. [25], context adaptation concerns the detection and the positive response to context change. De Virgilio and Torlone [8] go further by considering that a well-developed model of context is a fundamental ingredient of any adaptive system.

Saowanee [37] describes two types of adaptation: *self-adaptation* and *controlled adaptation by the user*. In a self-adaptation process, the system adapts without any interaction between the user and the system, whereas controlled adaptation is kind of system where the user makes the decision, but the system automates the rest. Oppermann [32] distinguishes *adaptive systems* from *adaptable systems*. Systems that allow the user to modify certain parameters and to adjust their behavior accordingly are called adaptable. In contrast, systems that adapt to users automatically based on assumptions of the system needs are called adaptive. Personalization is considered by McBurney et al. [29] as a form of adaptation. The authors mentioned that the purpose of personalization is to tailor the functionality and behavior of the system to respond differently depending on the available resources, the user context and his preferences. Recommender systems are particular tools of personalization which aim to help users find information or services of interest [39]. Pignotti et al. [10] add that recommendations can not be made without a full understanding of the context. Additionally, Grondin [12] presents adaptation from the software reconfiguration perspective, i.e. the ability to comply with new or different conditions. Giovanni et al. [1] has shown that adaptive strategy can be applied to dynamically reconfigure services in home automation according to new habits of users and sensors states.

Previous works have argued for a high dependence between context awareness and adaptation concepts. This appears particularly in research works focusing on context-aware services. Figure 13.1 covers known works according to a user-centric vision. This analysis confirms the two classes of services interaction Persons-Environments and Persons-Applications in the development of context-aware techniques for personalization, recommendation and content adaptation. Contrariwise, context history was limited to users' profiles and previous transactions.

In addition, these approaches aim to make services context-aware by triggering appropriate responses to the current user context but their two main

service	adaptation	context	used technique	comment	reference
wireless Web services	content adaptation	devices, time, date, location, physical environment (weather, ..)	XML based technology	no use of historical information	32
home services (TV services, lighting,..)	personalization	user control preferences	preferences management based on applications API	a previous study on users preferences	11
Internet radio service	recommendation	current users choices	similarity computation	no use of historical information	31
booking Web services	content adaptation	location, travel destination, ...	development platform	no use of historical information	20
VoIP and information Web service	personalization of services composition	users preferences (states (busy, reading), language, ..)	rules based setting	no use of historical information	40
e-commerce Web services	recommendation	user profile, location, ..	rules generating	no use of historical information	22
Web meal services	personalization	users personal information and activities	decision tree	historical of users information	16
shopping Web services	personalization of recommendation	users profiles and interest	similarity analysis	historical of transactions	39

Figure 13.1 Analysis of context-aware services [11, 16, 20, 22, 33, 34, 40, 41].

limitations are the lack of adaptive mechanism for services and the restricted bound of context interpretation.

13.3 Challenges for Context-Aware Adaptive Services

Notwithstanding, context-aware services remain largely unable to handle autonomic adaptation when providing services. The following provides a summary of the principal requirements for context-aware adaptive services:

- A method that takes into account the action and the situation in the adaptation process. This means that neither the action, nor the situation, is predefined in the system. The service or system is autonomously capable of determining the type of situation, and it must act based upon the type of service. By analyzing historical data on the context, the system determines the links between situations and service users, and the relevance of a service for a particular type of situation.
- The method should have the ability to include other context data not considered initially in the construction of the adaptation model.
- A service depends on its use and the system perceptual abilities. It should not carry on specific well-designed services.
- A robust method with respect to the nature of the environment such as multiple houses configuration, and different types of sensors. Providing

a system that works only for a single sensor configuration and the same type of space structure is not very advantageous.

To better understand the limitation of context-aware service adaptation, we should understand that adaptation models commonly include two main components. The first is to identify the situation of interest and the second is to select the correct action to be taken accordingly. The action may be associated prior to the situation, governed by rules, or be selected automatically according to a well-defined algorithm.

Autonomic adaptation is the fact that the system taught itself to build a model of adaptation which makes it totally adaptive. An adaptive system uses the models determined automatically to act properly or adjust to unexpected or unusual situations. Services are then used automatically in case of deviation from the usual situations or encountering an unexpected situation. Usual situations are two types, usage situations of services and normal states of the environments. A usage situation of a service is the usual situation where a service is used (enable, trigger, etc.). If the challenge for unusual situations determination is the autonomic identification of usual situations and related services first, the challenge for unexpected situations is to determine what action, if any, should be taken (what services should be invoked, who should be informed, etc.) because the system has no prior knowledge of the current state, which may, or may not, be indicative of an alarm-type condition.

In the following section, context discovery is presented and its applicability to the resolution of the problem of delivering situation awareness, eventually extending towards adaptive services.

13.4 Context Discovery for Autonomic Service Adaptation

Context discovery is the process of autonomic extraction of interesting and unknown knowledge from context data. It is a particular research topic that is part of a broader area which is knowledge discovery. This knowledge is mostly represented in the form of patterns. A pattern embodies both the consistency and completeness of existing relations between context data. They look like an archetype representing a situation with both a notion of repetition and prediction. Context discovery is based on methods and algorithms developed in statistics, data mining, machine learning knowledge discovery and pattern recognition. However, it is not the purpose of this chapter to describe all the available algorithms and derived techniques. The next section presents some methods and applications that we consider under this topic.

13.4.1 Context Discovery Methods and Applications

The solutions presented in [2,24,30] explore log files for the improvement of service delivery process by considering service usage patterns. For example patterns of execution sequences in the composition of services, detection of similar use of services for eventual recommendation. These works do not integrate environmental and user real-time information in the determination of services usage.

The solution presented in [15] proposes an autonomic interaction of home devices based on the history of interactions. The goal is to find regular uses in the interactions with the device. The data is stored and used by an episode discovery algorithm called *ED*. These efforts tend to propose dedicated solutions to a particular type of device or apparatus in which each device or apparatus is treated separately. In addition, these solutions are dedicated to extract a particular type of pattern from a specific device data to satisfy a specific need of the application.

Bridiczka et al. [5] employ supervised learning to learn situation models by indicating the corresponding service to each situation. The user is included in the learning loop by associating services to situations. Situations are recognizable on the basis of the model already tested; thus supervised by an expert. This approach attempts to provide context-aware services, however, it makes assumptions on the expert labeling and relationships between services and situations. Nevertheless, for commercial applications, it seems inappropriate to introduce the user in the learning loop,

After this brief overview on related works to context discovery, we put forward the citation of Truong and Dustdar who specify in [38] that adaptation to context is generally specific to the application. Many context-aware middleware allows developers to specify actions that should be performed, given specific contexts. In most cases, the middleware can just support the management and exchange of context information. Adapting to the context in these applications, from our point of view, is synonymous to the autonomic selection of the action to be executed, regarding the reason defined in the system but the reason itself is not determined autonomously. It also shows the limitation of current solutions in their exploration of context data to design mechanisms for autonomic services adaptation. This limitation hinders the development of more intelligent spaces. The next sections proposes context discovery as one of the important steps for the autonomic services adaptation.

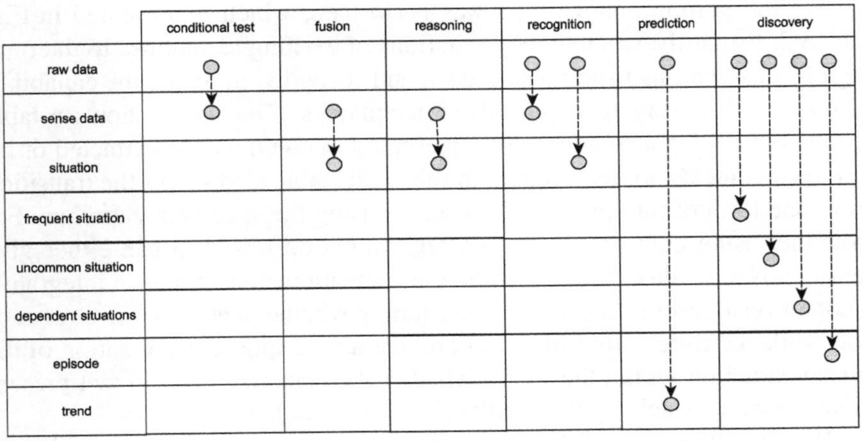

	conditional test	fusion	reasoning	recognition	prediction	discovery
raw data	○			○ ○	○	○ ○ ○ ○
sense data	○	○	○	○		
situation		○	○	○		
frequent situation						○
uncommon situation						○
dependent situations						○
episode						○
trend					○	

Figure 13.2 Relationship between knowledge and context data.

13.4.2 How Context Discovery Can Improve Service Adaptation

Context discovery helps in the autonomic adaptation of services by its ability to discover more significant situations for services without supervised techniques, and without pre-defined situations during software development. It is important to consider that each instance, or description, of a discovered situation refers to at least one service that is used, or may be used, in that particular situation. The link between situation descriptions and services is vitally important for future adaptation.

The Web is a better example of the improvement of service using high-level situations discovery. In the case of the Web, data mining and knowledge discovery have experienced a rapid increase in interest both research and industrial development. Web data mining is used specially to discover Web usage patterns or trends from Web data, to understand and better meet the needs of Web applications and users. WebSIFT [36] is an example of a development framework that operates the content and structure of information from a Web site to identify interesting patterns. It is used, for example, in recommendation, Web site personalization and user behavior discovery. The particular technique used for Web data mining is *clustering*. Clustering is a technique used to group a set of objects with similar characteristics. Drawing on research efforts in knowledge discovery from the Web, we obtain a better understanding of the more significant situations that can be discovered from context data.

For the purpose, an analysis was performed; which is presented in Figure 13.2. Primarily, to show the limitations of existing techniques to interpret high level situations from context data, and secondly, to image the capability of context discovery to bypass these techniques. The figure shows a table with two axes, the techniques are represented in columns and extracted or interpreted contexts are represented in rows. The table also shows the transition from one level of interpretation to another using the specified technique. For example, using context recognition algorithms on raw data can either give sense to data, or allow the determination a situation. A camera can integrate a recognition algorithm capable of determining whether a person is in a room or not; while a different algorithm can provide a description/visualization of the current situation within the room (whether the person, who is in fact present in the room, is watching TV, reading, sleeping, etc.).

We nevertheless hold that the empty boxes correspond to a logical intuition. For example, applying recognition techniques to symbolic data cannot help in the process because we already have raw data. On the other side, symbolic data can be used by reasoning algorithms. Therefore, the realization of certain transitions seems inappropriate because of the nature of data. Notwithstanding, the table shows clearly the potential for the discovery of context on raw data in the determination of more significant situations that can be explored during autonomic services adaptation on the one hand, because of the large volume of data and on the other hand because of its spatial-temporal characteristics. We note that only very few studies were interested in unusual situations. The discovery of the context is not only a potential, but also a challenge in itself and a promising research topic for the realization of intelligent spaces.

13.4.3 A New Architecture for Service Adaptation

The type of architecture that we propose aims to adapt the service automatically in case of deviation from usual situations. This architecture can be easily used for other purpose. The architecture presented in Figure 13.3 includes various components mainly:

1. Data acquisition in a common format to support multiple sources and heterogeneous data.
2. Context discovery of user and environment situations.
3. Determination of relevant situations for services by exploring data that form these situations and their relationship with the environment onto-

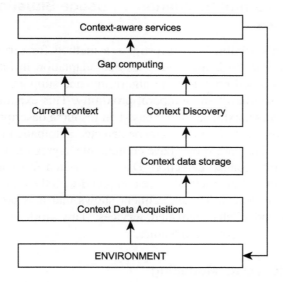

Figure 13.3 An architecture for autonomic services adaptation/

logy. The environment ontology describes the environment and services data.

4. The real-time detection of a gap between the current situation and any of the already discovered situations.
5. Activate, notify or alert depending on the specific service.

Acquisition of data is achieved by a software interface. Sensors, applications and web services communicate with the storage module through this interface according to a subscription mode (subscription, i.e., if the value changes prevents me). In this way only changes are stored in the database.

The data from different sources at time t constitute a situation. The situations over time are compared and grouped using a clustering algorithm. The algorithm employs specific similarities measures to form situations. If a situation presents one or more links with services that are defined on the environment ontology, the situation is considered relevant for these services. A gap is calculated in term of thresholds, between context data belonging in the current and previous situations. However, this architecture needs some modifications to take into account changes in user habits or practices.

13.5 Experimental Evaluation on Usage Situations of Services

The goal of this section is to investigate a method for service usage situations discovery towards autonomic service adaptation in intelligent space. This method is based on clustering algorithms and supported by conceptual and relational similarity and temporal proximity. This experiment argues in favour of a new context data model based on resource description framework *RDF* so that context and services can be detected automatically by looking at the data values. The data values have computable aggregation so that the data values merge into context detection. Data is collected from an instrumented home. The results of the method were compared with the real data previously recorded and annotated. The experiment shows that it is possible to detect user contexts and related activated services. This work is destined to the development of the above architecture.

13.5.1 Context Data Modelling

Context data includes physical data (e.g. temperature, pressure and other environmental information), system data and network information (e.g. devices and machines used, sensors data and electrical shareholder states). We propose a semantic context data model motivated firstly by the need to take into consideration a variety of distributed context sources such as sensors, applications and services to ensure interoperability via a common format of context at its data level.

The context data model is a collection of Resource Description Framework (RDF) triples.[1] RDF is a language that was originally designed for representing information about resources in the World Wide Web. An RDF triple consists of a *subject*, a *predicate*, and an *object*. The subject identifies what object the triple is describing. The predicate defines the piece of data in the object we are giving a value to. The object is the actual value. The choice of RDF as a representation model of context data is motivated by a number of considerations. RDF is a model that supports integration and uniform access to context sources and services as well as applications. RDF principles of representation allow data interchange and data semantics handling. RDF data model can be visualized graphically and is extremely flexible, allowing for future change. The simplicity of the RDF data model complements existing

---

[1] http://www.w3.org/RDF

Table 13.1 RDF triples of the context data model.

Subject	Predicate	Object
contextParameter	cxt:isContextOf	cxt:entity
contextParameter	cxt:hasContext	cxt:context
contextParameter	cxt:hasSource	cxt:source
contextParameter	cxt:hasConfidence	rdf:int
contextParameter	cxt:hasTimestamp	rdf:date
cxt:entity	cxt:hasIdentifier	rdf:string
cxt:context	cxt:hasIdentifier	rdf:string

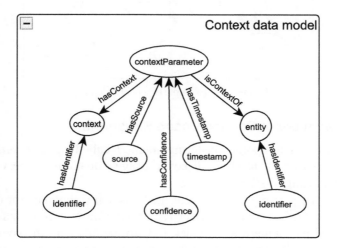

Figure 13.4 The context data model.

services architectures and makes the development of autonomic adaptability of services more accessible.

The set of triples of the context data model constitutes an RDF graph to which they give a collective meaning. This set is listed in Table 13.1 and is illustrated in Figure 13.4.

"Context parameter" is an identifier of the context data. It is linked to the "entity" and "context" nodes by the predicates *isContextOf* and *hasContext* respectively. Context characterizes a situation of an entity (person, object or a measure). Additionally, the *source, confidence* and *timestamp* are literals that indicate the source providing the context and the confidence and timestamp attributed to this context.

Figure 13.5 illustrates a working model of context data according to the context data model.

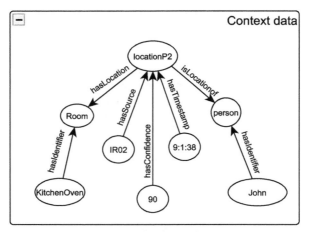

Figure 13.5 Example of a context data.

13.5.2 Discovery of Service Usage Situations

In this section, a temporal clustering method called *agglomerative hierarchical clustering* is used to discover usage situations of services and implemented using *Matlab*[2] software.

Among a large variety of different methods and techniques for data analysis in the smart-home environment [26] *clustering* is a popular technique and was used by Giguere and Dudek [23] to identify terrain types for unmanned ground vehicles, where sensors data is represented by sequences of consecutive measurement from the vehicles as it travels.

Clustering is an unsupervised learning method; a characteristic that makes it preferred to other methods, such as supervised classification, but it is a difficult task because the classes of data are not already identified, and thus, a system can not be adequately trained. Moreover, in the case of this work, it is necessary take into account the semantic nature of data which necessitates particular similarity. Appropriate algorithms are presented in the following sections.

13.5.2.1 Algorithm

In these experiments, an agglomerative hierarchical clustering mechanism is employed [19]. The distance between two clusters is computed by using the minimum distance *Single Link* (13.1). Single linkage is one of the simplest agglomerative hierarchical clustering method. It defines the distance

---

[2] http://www.mathworks.fr/

between context data groups as the distance between the closest pair of context data, where only pairs consisting of one context data from each group are considered.

$$d_{\min}(C_i, C_j) = \min_{a \in C_i, b \in C_j} d(a, b) \tag{13.1}$$

where d is a distance function, C_i is the i-th cluster, C_j is the j-th cluster, a is a member of C_i, and b is a member of C_j.

The agglomerative hierarchical clustering initially takes each context datum as a single cluster and then constructs progressively more expressive clusters by grouping similar context data together until the entire dataset is assembled in one cluster. Thus, the algorithm is able to cluster new data without resuming the treatment. Additionally, this algorithm is more representative of the original data structure at the bottom levels than at the top levels of the dendrogram. A dendrogram is a tree for visual classification of similarity, commonly used for grouping data. Elements of a dendrogram are clusters and the height of the dendrogram corresponds to the similarity between clusters. An association of two clusters forms a new cluster. In a dendrogram, two elements are grouped in one cluster when they have the closest values of all elements available. These characteristics make it possible to identify multiple abstraction levels of service usage situations.

The input to the program must be a *one space separated* value of $M \times M$ distance matrix. This matrix is constructed based on two similarities (conceptual and relational) and a temporal proximity. Each one takes values between $[0 \dots 1]$ and are presented in the next sections. These measures are defined to allow the discovery of two types of situations: *non-sequential situations* and *sequential situations*. The first one involves relationships among events in the same zone, ignoring the temporal aspects of the data and, the second involves temporal relationships among events occurring in different zones. Using these measures, we define the similarity between two context data as

$$\mathcal{S}_1 = \sum_{c_m \in cxt_i, c_n \in cxt_j} \text{sim}_c(c_m, c_n) \tag{13.2}$$

$$\mathcal{S}_2 = \sum_{p_k \in cxt_i, p_L \in cxt_j} \text{sim}_r(p_k, p_L) \tag{13.3}$$

$$\mathcal{S}_3 = t_p(cxt_i, cxt_j) \tag{13.4}$$

$$\text{sim}(cxt_i, cxt_j) = \frac{\alpha \mathcal{S}_1 + \beta \mathcal{S}_2 + \gamma \mathcal{S}_3}{\alpha + \beta + \gamma} \tag{13.5}$$

Formula (13.5) must verify the condition below which ensures the temporal order of context data: $\gamma > \alpha + \beta$, where α, γ, β are weights verifying the above condition.

Then, the distance matrix values are computed using

$$d(cxt_i, cxt_j) = 1 - \text{sim}(cxt_i, cxt_j) \qquad (13.6)$$

13.5.2.2 Conceptual Similarity

The conceptual similarity calculates the proximity in terms of distance between concepts in two distinct context data graphs. It is calculated by their respective positions in the hierarchy of the ontology defined in [3]. An ontology is a "formal, explicit specification of a shared conceptualisation" [13]. In pervasive computing environment ontologies are key requirements for building ubiquitous systems, and describing the environments. Through ontologies, ubiquitous systems can easily share knowledge and provide relevant services and information to users. The use of ontologies is not limited to reasoning and information sharing and reuse, their use can be extended to include other important aspects by providing support to other types of reasoning than logical inference for example, the hierarchy of concepts contained in ontology can be seen as a semantic space for the concepts proximity computing.

The conceptual similarity defined in [18] is adopted in this work and used by the agglomerative hierarchical clustering algorithm presented in the above section.

Let c_1, c_2 be two concepts of the context data cxt_1 and cxt_2 respectively. The similarity between these two concepts is defined by

$$\text{sim}_c(c_1, c_2) = 1 - d_c(c_1, c_2) \qquad (13.7)$$

Every concept in the ontology has a value called *milestone* which is obtained by the formula $nv = \frac{1/2}{k^{l(n)}}$, where k is a predefined factor larger than 1, that indicates the rate at which the value decreases along the hierarchy. $l(n)$ is the depth of the node n in hierarchy such as $l(root) = 0$. In this work k is fixed to the value 2. The distance between two concepts will be determined by their milestone nv and their closest common parent as follows:

$$d_c(c_1, c_2) = d_c(c_1, ccp) + d_c(c_2, ccp) \qquad (13.8)$$

such as

$$d_c(c, ccp) = nv(ccp) - nv(c) \qquad (13.9)$$

13.5.2.3 Relational Similarity

The relational similarity is stronger than the conceptual similarity as it is calculated on the base of syntactic equality. Syntactic equality, in the context of this chapter, refers to a sequential comparison of the letters of two predicates, determining membership of two distinct context datasets.

Let p_1 and p_2 be two predicates of the context data cxt_1 and cxt_2 respectively. The relational similarity between these two predicates is defined by $sim_r(p_1, p_2) = 1/np$ iff $p_1 \underset{syn}{=} p_2$ else 0. np is the number of relationship between concepts.

13.5.3 Temporal Proximity

The temporal proximity between two instances of context data is calculated using their individual timestamps. Intuitively, the proximity between two context data instances far away from each other is less likely to those context data instances much closer to each other. The temporal proximity between two context data is given by

$$tp(cxt_1, cxt_2) = e^{-a\frac{|t_1 . t_2|}{T}} \tag{13.10}$$

where T is the duration of time series partition as defined in Section 13.5.4.4, and a is the time decaying weight which is between 0 and 1. For example, considering $a = 86$ ensures that the decadence is stable if the difference between context data instances is less than 7 seconds.

13.5.4 Empirical Study

This section describes the experimentation environment and the used scenario. The scenario was carried out for multiple use by R. Kadouche, member of the Domus Laboratory (Canada).

13.5.4.1 Sensors

The Domus Laboratory is a multidisciplinary lab aimed at research in demotic and mobile computer science. The laboratory is equipped with a fixed infrastructure sensing system. For the need of this experiment, 36 fixed sensors are used. Also, the laboratory is divided in six logical zones corresponding to the six areas covered by infrared sensors (see Figure 13.6). The Domus Laboratory is also equipped as any apartment for living.

The following gives details about sensors types used in the experiment:

Figure 13.6 Domus Laboratory.

- Infrared (IR) movement detectors provide information on the users' location in a zone. For example, there is only one IR detector that covers the entire zone of the dining room, whereas there are three IR installed in the kitchen.
- Pressure detectors, in the form of tactile carpets, are placed on the floor of entrance hall. This allows for the detection of the user moving between the bedroom and living room. There are two paths available to move between these two zones: through the kitchen, or the entrance hall.
- Light switches. These sensors send an event every time the occupant turns the lights on or off.
- Door contacts. These sensors are placed on the doors. They send an event related to the door state (open or closed).
- Switch contacts are the same as the door contacts. They are placed on the lockers and fridge. They provide an event when their state is changed: either open or closed.
- Flow meters provide the taps and the toilet flush statistics; two on the cold and hot water taps of the kitchen sink, one on the washbasin's cold water tap (because of setup constraints) and another in the toilet flush mechanism. They send an event when the tap is turned on or off and the toilet is flushed.

Table 13.2 defines the sensors per zone in the apartment.

13.5.4.2 Experimental Scenario

In this work, the focus of interest is on personal habits (personal grooming, making/eating breakfast). There are two versions of the scenario. The first one is called *series 1*. The user is asked to perform his early morning routines without constraints during the execution of his tasks. In the second *series 2* a constraint is introduced and it consists of preparing a tea for breakfast. The tea recipe is displayed in the home interface. The preparation of the tea recipe takes at most 10 minutes.

Six persons participated in the two versions of the scenario during the experiment. Each person participated 10 times in series 1 during two consecutive weeks, and one week for series 2 after a break of two weeks.

The experiment time was about 45 minutes. It started with the same apartment condition for all the users, all doors closed and lights switched off. The user was asked to stay in the bedroom for one minute (time required to start the data recording process). Each experiment returns sample data formatted according to the context data model presented in Section 13.5.1. The total of data is 60 samples for series 1 and 30 samples for series 2.

13.5.4.3 Data Collection and Preprocessing

The Context Management System (CMS) [35] is used for the collection and delivery of context data. The CMS abstract information coming from various sources, such as physical sensors and applications into a semantic model. The contribution of this design is to be able to represent complex and heterogeneous information.

The CMS is implemented in conformity with the Web service architecture and support Resource Description Framework (RDF). In this work, the information coming from the sensors described in Table 13.2 are abstract according to the context data described in Section 13.5.1.

Table 13.2 List of sensors used per zones.

	Entrance hall	Living room	Dining room	Kitchen	Bath-room	Bed-room	Total
IR	0	1	1	3	0	0	5
Pressure Detector	1	0	0	0	0	0	1
Lamps	0	1	1	1	1	1	5
Door contacts	0	0	0	0	1	1	2
Switch contacts	0	0	0	19	0	0	18
Flow meter	0	0	0	0	2	2	4

Table 13.3 Context data.

ID	Context data
cxt1SS	status device hasStatus isStatusOf Open Lampe1 1970-01-01T09:39:00 5105 100
cxt1SD	status door hasStatus isStatusOf Open Door2 1970-01-01T09:39:03 0215 100
cxt2SD	status door hasStatus isStatusOf Open Door1 1970-01-01T09:39:08 0216 100
cxt2SS	status device hasStatus isStatusOf Open Lampe2 1970-01-01T09:39:09 5106 100
cxt3SD	status door hasStatus isStatusOf Close Door1 1970-01-01T09:39:14 0216 100
cxt1ST	status tap hasStatus isStatusOf Open TapeColdWaterWashbasin 1970-01-01T09:39:43 Fl01 100
...	...

Table 13.3 illustrates a portion of the context data used/obtained in our experiment. This data is represented according to the model presented in Section 13.5.1. Each line in the table corresponds to an RDF graph.

13.5.4.4 Time Series Segmentation

The experiment time was about 45 minutes, an arbitrary value to cover the time needed by the above scenario. Context data is partitioned based on fixed time period (10 minutes). This time period is fixed to allow proper analysis and visualization of the dendrogram and also to permit the presence of noise data. Then, each partition is clustered using the agglomerative algorithm.

13.5.5 Results and Discussion

Clustering analysis is faced with two problems: determination of the number of true clusters and how to evaluate the samples assigned to those clusters. In this case, the number of clusters is determined by the level of abstraction that we want to generate. By moving up from the bottom layer to the top node, a dendrogram allows us to reconstruct multiple levels of abstraction of situations. The agglomerative clustering algorithm does not require a prescribed number of clusters. The hierarchy needs to be cut at some prespecified level of similarity.

A number of criteria can be found in the literature to determine the cutting point. In this study, the cut is done incrementally from the bottom to top of the dendrogram. This aims to construct and compare the sets of clusters with annotated data. A method for determining the cut value is required in the future for determining the best clusters; those which represent significant service usage situations.

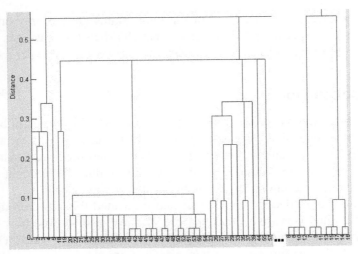

Figure 13.7 Agglomerative clustering result.

Two possible cuts of the dendrogram are shown in Figure 13.7: at 0.3 and at 0.4. Table 13.4 corresponds to a dendrogram cut at a minimum similarity of 0.3. It can be seen that the clusters (6, 8, 10, 12, 7, 9, 11) and (13, 15, 17, 14, 16) are at the right of the dendrogram; this is for visualization reasons to allow adequate interpretation. Another concern relates to conflicts between the status open and closed of the cold tap whilst the subjects were washing. This was found to be the result of an issue relating to the flow-meter sensor's sensitivity.

The clusters forming a situation are RDF graphs according the context data model presented in this chapter. These graphs are fused to form a unique graph representing the service usage situation.

Table 13.4 Clusters.

Clusters	Cluster samples	Contexts or situation	Service
1,2,3,4	lamp bedroom open door bedroom open door bathroom open lamp bathroom open	wake up and going to bathroom	lamp bedroom open lamp bathroom open
6,8,10,12 7,9,11	open tape col water washbasin bathroom close tape col water washbasin bathroom	washing	no service here
13,15,17 14,16	close tape col water washbasin bathroom open tape col water washbasin bathroom	washing	no service here
18,19	open the door bathroom and put the kitchen lamp on	finish washing and go to kitchen	lamp kitchen on
...	...	...	...

The services *open* and *close* of the lights status are the only ones considered in the experiment for the reason that DOMUS Laboratory is capable of light control. Therefore, it was possible to carry out a full experiment.

13.6 Conclusion

Context discovery is an important step in a process which aims to conceive intelligent systems that can assist people in their everyday lives through service delivery. For example, ensuring less energy consumption, assistance or alerting. Also, context discovery can be used to reduce the amount of time users spend interacting physically with devices or appliances.

We have defended in this chapter and motivated this research direction in the areas and pervasive smart spaces. We have also showed how context discovery can be achieved on real scenarios. We have investigated a method capable of discovering service usage situations to carry out autonomic services adaptation. The experimental study was conducted on a real scenario in an instrumented space.

Acknowledgements

The research leading to these results has been supported by the "Region Rhones-Alpes France" through the Explora'doc programme 2008. This programme provides partial funding for this work. Also, many thanks to the DOMUS Laboratory Team for their help in the realisation of the scenario.

References

[1] Giovanni Acampora, Vincenzo Loia, Michele Nappi, and Stefano Ricciardi. Ambient intelligence framework for context aware adaptive applications. In *Proceedings of the Seventh International Workshop on Computer Architecture for Machine Perception*, pages 327–332, IEEE Computer Society, Washington, DC, 2005.

[2] Mohsen Jafari Asbagh and Hassan Abolhassani. Web service usage mining: Mining for executable sequences. In *Proceedings of the International Conference on Applied Computer Science*, pages 266–271, World Scientific and Engineering Academy and Society (WSEAS), Stevens Point, Wisconsin, 2007.

[3] Chikhaoui Belkacem, Benazzouz Yazid, and Abdulrazak Bessam. Towards a universal ontology for smart environments. In *Proceedings of the 11th International Conference on Information Integration and Web-based Applications and Services*, Kuala Lumpur (Malaysia), 2009.

[4] Mark Lawrence Blum. Real-time context recognition. PhD Thesis, Swiss Federal Institute of Technology Zurich (ETH), 2005.

[5] Oliver Brdiczka, James L. Crowley, and Patrick Reignier. Learning situation models for providing context-aware services. In *Proceedings of 4th International Conference on Universal Access in Human-Computer Interaction*, Lecture Notes in Computer Science, Vol. 4555, pages 23–32, Springer, 2007.

[6] Tarak Chaari, Dejene Ejigu, Frédérique Laforest, and Vasile-Marian Scuturici. A comprehensive approach to model and use context for adapting applications in pervasive environments. *Journal of Systems and Software*, 80(12):1973–1992, 2007.

[7] Preuveneers Davy, Victor Koen, Vanrompay Yves, Rigole Peter, Kirsch Manuele, and Berbers Yolande. Context-aware adaptation in an ecology of applications. In *Context-Aware Mobile and Ubiquitous Computing for Enhanced Usability: Adaptive Technologies and Applications*, chapter 1, pages 1–25, Information Science Reference, Hershey, New York, 2009.

[8] Roberto De Virgilio and Riccardo Torlone. Modeling heterogeneous context information in adaptive web based applications. In *Proceedings of the 6th international conference on Web engineering*, pages 56–63, ACM, New York, NY, 2006.

[9] Dejene Ejigu Dedefa. Context modeling and collaborative context-aware services for pervasive computing. PhD Thesis, INSA de Lyon, France, 2007.

[10] Pignotti Edoardo, Edwards Peter, and Aastrand Grimnes Gunnar. Context-aware personalised service delivery. In *Proceedings of the European Conference on Artificial Intelligence*, pages 1077–1078, 2004.

[11] Kawsar Fahim, Fujinami Kaori, Pirttikangas Susanna, and Nakajima Tatsuo. Personalization and context-aware services: A middleware perspective. In *Proceedings of 2nd International Workshop Proceedings on Personalized Context Modeling and Management for UbiComp Applications*, 2006.

[12] Guillaume Grondin. Un modèle d'agents auto-adaptables à base de composants. PhD Thesis, Ecole nationale supérieure des Mines de Saint-Etienne, Saint-Etienne, France, 2009.

[13] Thomas R. Gruber. A translation approach to portable ontology specifications. *Knowledge Acquisition*, 5(2):199–220, 1993.

[14] H. Hashimoto. Intelligent space: Interaction and intelligence. *Artificial Life and Robotics*, 7(3):79–85, 2003.

[15] Edwin O. Heierman and Diane J. Cook. Improving home automation by discovering regularly occurring device usage patterns. In *Proceedings of the Third IEEE International Conference on Data Mining*, pages 537–540, IEEE Computer Society, Washington, DC, 2003.

[16] Jongyi Hong, Eui-Ho Suh, Junyoung Kim, and SuYeon Kim. Context-aware system for proactive personalized service based on context history. *Expert Systems with Applications*, 36(4):7448–7457, 2009.

[17] Jaga Indulska, Seng Wai Loke, Andry Rakotonirainy, Varuni Witana, and Arkady B. Zaslavsky. An open architecture for pervasive systems. In *Proceedings of the Third International Working Conference on New Developments in Distributed Applications and Interoperable Systems*, pages 175–188, Kluwer, Deventer, the Netherlands, 2001.

[18] Zhong Jiwei, Zhu Haiping, Li Jianming, and Yu Yong. Conceptual graph matching for semantic search. In *Proceedings of the 10th International Conference on Conceptual Structures (ICCS)*, pages 92–196, Springer, London, UK, 2002.

[19] Stephen Johnson. Hierarchical clustering schemes. *Psychometrika*, 32(3):241–254, September 1967.

[20] Markus Keidl and Alfons Kemper. Towards context-aware adaptable web services. In *Proceedings of the 13th international World Wide Web Conference on Alternate Track Papers & Posters*, pages 55–65, ACM, New York, NY, 2004.

[21] Y. Berbers, M. Kirsch-Pinheiro, and Y. Vanrompay. Context-aware service selection using graph matching. In *Proceedings of 2nd Workshop on Non Functional Properties and Service Level Agreements in Service Oriented Computing*, Lecture Notes in Computer Science, pages 199–205, Springer, 2008.

[22] Ohbyung Kwon and Jihoon Kim. Concept lattices for visualizing and generating user profiles for context-aware service recommendations. *Expert Systems with Applications*, 36(2):1893–1902, 2009.

[23] Kristof Van Laerhoven. Combining the self organizing map and k-means clustering for online classification of sensor data. In *Proceedings International Conference on Artificial Neural Networks*, pages 464–469, 2001.

[24] Qianhui Liang and J.Y. Chung. Analyzing service usage patterns: Methodology and simulation. In *Proceedings of the 7th IEEE International Conference on e-Business Engineering*, pages 359–362, IEEE Computer Society, Washington, DC, 2007.

[25] Paolo Lombardi, Bertrand Zavidovique, and Michael Talbert. On the importance of being contextual. *Computer magazine*, 39(12):57–61, 2006.

[26] Jani Mantyjarvi, Johan Himberg, Petri Kangas, Urpo Tuomela, and Pertti Huuskonen. Sensor signal data set for exploring context recognition of mobile devices. In *Workshop on Benchmarks and a Database for Context Recognition of 2nd International Conference on Pervasive Computing*, 2004.

[27] R. Mayrhofer. *An Architecture for Context Prediction*. Schriften der Johannes-Kepler-Universität Linz. Trauner Verlag, 2005.

[28] R. Mayrhofer. *An Architecture for Context Prediction*, Schriften der Johannes-Kepler-Universität Linz, Vol. C45, Trauner Verlag, April 2005.

[29] Sarah McBurney, M. Howard Williams, Nick K. Taylor, and Eliza Papadopoulou. Managing user preferences for personalization in a pervasive service environment. In *Proceedings of Advanced International Conference on Telecommunications*, pages 32–32, 2007.

[30] Richi Nayak and Cindy Tong. Applications of data mining in web services. In *Web Information Systems - WISE*, Lecture Notes in Computer Science, pages 199–205, Springer, 2004.

[31] Katsumi Nihel. Context sharing platform. In *Advanced Technologies and Solutions toward Ubiquitous Network Society*, pages 200–204, 2004.

[32] Reinhard Oppermann (Ed.). *Adaptive User Support: Ergonomic Design of Manually and Automatically Adaptable Software*. L. Erlbaum Associates, Hillsdale, NJ, 1994.

[33] Cunningham Padraig and Hayes Conor. Recommendation in context. In *Workshop Proceedings of the 6th European Conference on Case Based Reasoning*, pages 18–19, 2002.

[34] Ariel Pashtan, Shriram Kollipara, and Michael Pearce. Adapting content for wireless web services. *IEEE Internet Computing*, 7(5):79–85, 2003.

[35] Fano Ramparany, Remco Poortinga, Maja Stikic, Jorg Schmalenstroer, and Thorsten Prante. An open context information management infrastructure the IST-Amigo project. In *Proceedings of the 3rd International Conference on Intelligent Environments*, pages 398–403, 2007.

[36] Cooley Robert, Tan Pang-Ning, and Srivastava Jaideep. Websift: The web site information filter system. In *Proceedings of Workshop on Web Usage Analysis and User Profiling WEBKDD in Conjunction with ACM SIGKDD International Conference on Knowledge Discovery and Data Mining*, San Diego, CA, 2009.

[37] Saowanee Schou. Context-based service adaptation platform: Improving the user experience towards mobile location services. In *Proceedings of the International Conference on Information Networking*, pages 1–5, 2008.

[38] Hong-Linh Truong and Schahram Dustdar. A survey on context-aware web service systems. *International Journal of Web Information Systems*, 5(1):5–31, 2009.

[39] Quan Wen and Jianmin He. Personalized recommendation services based on service-oriented architecture. In *Proceedings of the IEEE Asia-Pacific Conference on Services Computing*, pages 356–361, IEEE Computer Society, Washington, DC, 2006.

[40] Quan Wen and Jianmin He. Personalized recommendation services based on service-oriented architecture. In *Proceedings of the IEEE Asia-Pacific Conference on Services Computing*, pages 356–361, IEEE Computer Society, Washington, DC, 2006.

[41] Yang Yuping, Mahon Fiona, Williams M. Howard, and Pfeifer Tom. Context-aware dynamic personalised service re-composition in a pervasive service environment. In *Proceedings of the Third International Conference Ubiquitous Intelligence and Computing*, Lecture Notes in Computer Science, Vol. 4159, pages 724–735, Springer, 2006.

14

Modelling Context-Aware Smart Spaces, Services and Context Management Frameworks

Ahsan Ikram, Saad Liaquat and Madiha Zafar

Center for Complex and Cooperatives Systems, Bristol Institute of Technology, UWE, Bristol BS16 1QY, UK; e-mail: ahsan2.ikram@uwe.ac.uk

Abstract

Pervasive devices today have evolved as a cognitive device that has been woven into the daily fabric of every aspect of every day life. Nowadays, pervasive devices are used for navigation, paying bills, used as credit cards, surfing Internet, managing diaries, taking pictures, recording movies and much more. With sensors like the accelerometer, gyrometer, compass and so on becoming a default part of the next generation devices a plethora of new possibilities have opened in making the services and physical environments more cognitive, intelligent and context aware. At the same time, users want to consume services designed for their preferences and adaptable to their context and situation. In this chapter we highlight challenges in modelling, evolving and managing context and smart adaptive environments. In the end we present a novel approach to model context and services dynamically.

Keywords: IMS, context awareness, context management, smart space.

14.1 Introduction

In the last decade mobile devices have evolved as smart gadgets with sensors offering, multimedia recording, positioning systems so on and forth. These

Anand R. Prasad et al. (Eds.), Advances in Next Generation Services and Service Architectures, 307–327.

enhancements enable mobile devices to develop dynamic understanding of device, user and surroundings. This information is useful in adapting services, surroundings and usage of pervasive technology to an individual's needs and preferences. These developments have led to the realization of services and concepts such as smart living, context-aware navigation systems and augmented reality browsers. Similarly, recent success of community and social networking platforms that offer personalized user and community spaces have invoked the social dimension of pervasive computing where social networks and pervasive computing are now evolving as a means of social communication, awareness and adaptation. Context awareness about communities and groups of users is one of the main objectives of social dimension of pervasive computing and context awareness. This trend is supported with the increasing weaving of smart artefacts and sensors into daily life, vending machines, context-aware advertisements, digital displays and parking sensors, where every service and device is capable of being aware and communicating with the users and vice versa.

Context and context-aware systems have been widely studied and there are well known challenges to achieving context awareness including validating correctness of context, reasoning on context and creating a suitable response to the situation of a user. However, context-aware communities, an area of emphasis in this research, are a recent aspect of context-aware computing and it adds more challenges to the domain.

Loke [25] highlighted research challenges in the domain of task-based adaptive systems and services. One of the main challenges presented by the domain is to model possible scenarios and situations. Most of the existing programming models are event based and/or service orientated. The problem with these techniques is the difficulty in listing the possible events and resulting services. Ontologies have been proposed as a solution to the problem, where ontologies can surely serve as a higher-level abstraction the problem remains the same as to how many ontologies would be needed to model every possible environment? On the other hand the problem is not new for the artificial intelligence community and there have been various theorems and algorithms for managing such systems based on evolution. Where evolutionary algorithms are surely a possible solution, the specific domain for the application of this research is pervasive environments where situations are very short lived and context quickly changing over a broad spectrum.

In this chapter a context and service modelling framework is proposed which uses a bio-inspired novel context representation mechanism in conjunction with graph theory.

14.2 Smart Spaces: Some Scenarios

In this section a few scenarios are discussed that present smart environments and spaces using storylines. This will give an overview and understanding of the possible applications of smart spaces in context of pervasive environments. Assume environment of a city centre enriched with smart entities like RFID readers, smart card readers, weather/light/temperature sensors, etc.

Consider a set of users arrive near a Carnival in progress; some of these users have subscriptions to services such as 'Cultural Podcast' and 'Events-Around-Me'. Some of the users have bought tickets for the carnival online and stored them on smart cards on their phones. Two of the users (A and B) also happen to be friends and are subscribed to each other's presence service. When the users A and B arrive in the vicinity of carnival, the spaces interact to deliver the carnival's history and today's offer and also lets A and B know of each other's presence.

Now consider A and B are two tourists from, say, Germany and France respectively. When they arrive in the city centre, their spaces interact with that of the city centre and offer a French and German translated presentation of facilities, which might include lunch menus, retrieved by interacting with a restaurant smart space. Another scenario could be that A and B arrive at the airport. On their arrival at the airport A's space interacts with the airport's space which presents a list of services offering travel and hotel options. However, in the same space B's space tries to find and connect to the space of his host who had to come to pick him up. This was scheduled in the user's device using calendar, planner or timed task.

Finally, consider D and E are two shoppers in a shopping mall. Half way through the shopping they start to feel hungry and change their mood/activity status to 'need lunch'. D and E are both subscribed to 'smart services'. Then, as they happen to pass nearby a Mexican restaurant, both D and E are sent a 'menu of the day' message. In the remainder of this chapter this scenario is used to explain and trace working of various components. Another user F is also involved in this scenario, F is at a different location but shares partial context with D and E.

14.3 Smart Spaces Modelling Requirements

We have previously highlighted [17] some of the challenges involved in modelling and management smart spaces. The dynamic, distributed and short lived nature of pervasive smart spaces pose distributed management, partial

Table 14.1 Smart space types.

	User Centric	**Community Centric**
Active	A user connecting to the host smart space in airport scenario	Friends connecting to the carnival space and using presence
Proactive	Tourist at airport customizes and connects to hotel and travel spaces in the vicinity	Users in a pervasive environment playing a game, such as treasure hunt, forming a smart community

validation, uncertain information handling and monitoring challenges to the environment. These issues and the complexity of events along with relationships between individual context elements are difficult and complex to model. Extending earlier observations, the main modelling challenges are listed in the following subsections.

14.3.1 Event Processing: What Starts the Process of Modelling Smart Spaces

The process in which a situation has to be modelled can be triggered in more than one way and at more than one time. A change in context A could trigger a different response as compared to a change in context B, and a change in contexts A and B together result in different responses again. This is valid for a large number of contexts and constraints. Therefore, the nature, time and sequence of these events are of vital importance in understanding the needs of the respective situation. Also the large amount of such possible combinations makes it difficult to enlist all possible event combinations. In terms of what event or trigger starts the process of forming a smart space, it is assumed that a situation modelling process can be triggered in two ways (summarized in Table 14.1):

1. Pre-defined set of conditions that start to hold true, e.g., remind me to buy shoes next time I visit mall. These situations can be handled by traditional rule-based systems.
2. Pro-Active; automated reaction where change in value of a contextual entity, e.g., location co-ordinates, which when fed to a smart space trigger a response/service. Some contexts can have the feature of being triggering catalysts while others may not and serve as bridging context or informational context, e.g., location change can have the feature to

trigger a situation and service subscriptions may not. We discuss this aspect later while introducing context model.

14.3.2 Scope: Smart Space Boundary Marking

From Section 14.2 it can be observed that a smart space scope might be restricted to a small area such as a restaurant or a cafe, or it might encompass a broader physical scope such as the whole city. The rich relations between contextual entities can continue iteratively till all the entities have been addressed. It is, therefore, vital for formal representation and modelling of a situation that the underlying model understands how to mark the scope/boundary of a situation, otherwise the whole system would recursively grow to represent a single smart space. This could be achieved by limiting the scope of a smart space to a physical location, service subscription, presence or any other classifiable context source. Having talked about shaping a space scope and avoiding a single large space it is also observed that logically the concept of a single parent space is possible with nested sub spaces but it is important to mention here that this concept renders complex modelling and reasoning scenarios that might affect the efficiency in a highly dynamic environment such as pervasive smart spaces. In a nested smart space hierarchy a change in a smart space renders a re-evaluation or update to all the space underneath which can be expensive on resources and efficiency. The model proposed in this research will partially address this boundary marking issue. Figure 14.1 gives a visual overview of the difference between two hierarchies. Where structured spaces are easier to synchronize, they can be expensive on time consumption when reorganizing on context updates. Similarly, exclusive spaces are easier to update yet they need a separate monitoring and update thread for each smart space.

14.3.3 Complexity: How Much Information Is Minimum Requirement?

For the reactive models that have been predefined, the participants and entities of the situation are clearly defined and documented. In these cases the triggers happen when predefined values, criteria and thresholds start to appear and/or match. For proactive reactions the entities involved must react and collaborate, in real time, to decide if a context-aware reaction is possible or necessary and this precisely makes the problem at hands complex. Certain approaches exist to achieve this goal such as rule-based engines, Bayesian classification,

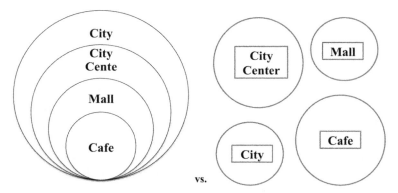

Figure 14.1 Overview of two smart spaces.

ontologies, etc. However the basics of all these approaches rely on the relationship between entities. For example, for the weather context to update itself with a location change it needs a handle/subscription to changes in location value. We discuss the suitability of existing approaches to pervasive smart spaces in later sections.

14.3.4 Uncertainty: Can the Same Situation Be Responded to with More Than One Option?

Having the characteristic of managing uncertainty in both the context model and response selection enriches the context awareness experience to the end users and makes the system more scalable and robust. For more than one option, responses can be ranked with correctness or probability scores. For uncertain context, sources can be structured by reliability. The presentation of multiple response options can (proactive) or cannot (reactive and interactive) be transparent to the user depending on the business model. However, presentation of options tend to make the system more interactive but less pro-active. The proposed method addresses this issue by finding all possible response options and rates them based on the matching to the situation and reliability of individual context source. Smart space representations having unreliable context sources as the main triggers are ranked less and vice versa.

14.3.5 Temporal Representation: Does the Smart Space Keep Track of Time?

Temporal representation support is one the most important requirements of pervasive smart spaces. Temporal support in terms of representation and reasoning is vital as pervasive environments are short-lived and are often triggered by temporal constraints. In context-aware spaces occurrences of events, spatial monitoring, multimedia content delivery, all are dependent on the capability of the underlying model to represent and reason time. Temporal representation in itself can be categorized into two themes, absolute and relative. Whereas absolute time deals with time-stamping events, relative temporal representation defines relations such as before, after, etc. This can be addressed by adding an 'expiry' dimension to the validity of context and at the same time considering context as a context source.

14.4 Smart Spaces Components

Dobson et al. [20] highlighted that any autonomic framework based on context has to fulfill the four phases of autonomic communication lifecycle, collect (data), analyze (data), decide (what to do) and act (on decision). Kjær et al. [21] have characterized context-aware frameworks to have eight aspects namely, environment, storage, reflection, quality, adaptation, migration and composition. In this section we introduce components that serve as building blocks for fulfilling the issues discussed above and enabling smart spaces. It can be said that modelling smart spaces and delivering smart services is deeply dependent on the efficiency of all phases in the context management life-cycle. Context generation, acquisition (representation, storage, retrieval), modelling, reasoning and provisioning response are some of the vital aspects of context management life-cycle within a smart space.

This chapter addresses the management, modelling, partial reasoning aspects of the life cycle. We discussed context management paradigms in the domain of distributed context management architectures in [22]. Bettini et al. [11] discussed various approaches of context modelling and pointed out key characteristics of a formal context model. Figure 14.2 identifies different activities within each phase of context management life-cycle proposed by Dobson et al. [20].

In the following sections three main components are discussed, namely, context modelling component, context brokering component and smart space generation component.

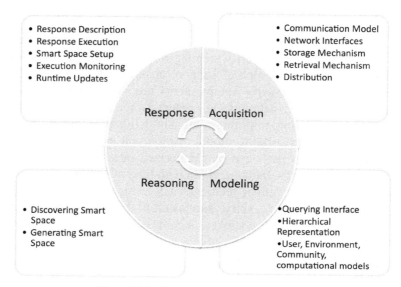

Figure 14.2 Context management activities.

14.4.1 Context Management

Managing context is the backbone of any context-aware system. In order to design a large context-aware system that is able to support emerging services, a large variety of context information needs to be created, maintained and distributed. Although in context-aware systems, domain knowledge is very much tied to the application, building context-aware applications from scratch is not practical. Therefore it is vital to decouple context from application to enable reuse and support for many applications.

A well-known paradigm of combined context-aware service and networked service model is the producer-consumer model. Network components may take the role of a Context Provider (CP) or may take the role of information sinks, i.e. Context Consumers (CC) [9, 15]. These basic entities may interconnect by means of Peer-to-Peer (P2P) techniques [26] or by a Context Broker (CB) providing a directory and lookup service. Broker architecture in its various forms exists as a middleware technology that manages communication and data exchange between objects or entities. Chen et al. [10, 13] presented a Context Broker Architecture (CoBrA), which is an agent-based architecture for supporting context-aware systems in smart spaces.

In addition to inference of context information, which is usually delegated to specialised components, dissemination of context data from context producing entities to context consuming entities is a fundamental functional task of context-aware systems. Broker-based approaches have been successfully demonstrated in a number of prototype systems for context dissemination. With the increase in sensing capabilities of mobile devices, such devices are not merely consumers of context information any more but also have the ability to be providers of context gathered through integrated sensors. In such a provider-consumer model, where context dissemination is aided by a central broker, device broker interaction can become a communication and computation bottleneck in presence of multiple context providers and consumers in mobile devices. To reduce management and communication overheads, it is desirable to have multiple brokers in the system divided into administrative, network, geographic, contextual or load-based domains. Context providers and consumers may be configured to interact only with their nearest, relevant or most convenient broker. But this setup demands inter-broker federation so that providers and consumers attached to different brokers can interact seamlessly. To achieve this, a simple event system can be implemented by an overlay network of distributed brokers for relaying subscriptions and notifications. In [22] we discussed presented such a distributed and federated context brokering system. This brokering model is distributed residing at control layer (the main broker) and federated instances at network and user planes. In principle any entity that handles more than one context (consume or produce) involves a broker. The footprints of brokers might vary and the brokers on mobile devices are called 'Mobile Brokers'.

Figure 14.8 presents an overall view of the proposed architecture designed on top of an IMS (IP Multimedia Subsystem) backbone. Previously, Long et al. [23] and Connell et al. [24] discussed context-aware service delivery in IMS-based environments.

14.4.2 Periodic Table-Based Context Modelling

The proposal for context modelling presented in this paper is inspired by the bonding concepts of chemical reactions and bonding. Context is differentiated into types and then different instances of context are placed in the table according to their type and quality. The context sources giving exact and/or certain information about their parent type are placed higher and the context sources with unreliable information are placed lower. The certainty value is incremented by 1 across each row downwards, e.g. in the context

Table 14.2 Context model periodic table.

Where	When	How	What	What'	Who
LOCATION [1]	TEMPORAL [2]	ACTUATOR [3]	ACTIVITY [4]	CONTENT [5]	PROFILE [6]
GPS Co-ordinates	GMT	Display	Driving	Movie	Subscriptions
Street Address	After	Motion	Meeting	Song	Presence
Room Type	Before	Temperature	Eating	News	Interests

column 'Location', set of GPS co-ordinates provide certain information about location and are put at the top, whereas 'Room Type' (e.g. Dining Room) is placed lower as it gives a higher abstraction of the location type. The table is similar to user profile and can vary for different users, meaning each entity has its own table where the columns stay consistent but the content can vary. Furthermore, the table is organized horizontally to feed information into a smart space based on 'where', 'when', 'how', 'what', '(what)' and 'who', as shown in Table 14.2. It must be pointed out that the context values under each column are placed as samples for understanding.

- Who: User and social profile come under this section and are represent-ation of entities in the pervasive environment for which context is being modelled.
- What': Refers to the content, services and goals of a resulting context model and is based on the columns audio and visual content and service subscriptions element from user profile column.
- What: Represents the activity by a 'Who' in a context model and is modelled using the Activity column. Activity column generally contains elements composed of one or more elements from the rest of the table.
- Where: Refers to the locative component of an urban pervasive environ-ment and is covered under relative and absolute location.
- When: Addresses the temporal and conditional representation of trig-gers and events in a pervasive model of context. Absolute and relative temporal context columns cover it.
- How: Answers the communication interfaces and end-point devices or services in a context model. Actuator context family generally covers this aspect.

Next is the representation of individual context entity. An element of the table can be an atomic representation or a compound, e.g., GPS coordinates is an atomic context where as driving can be a compound of accelerometer

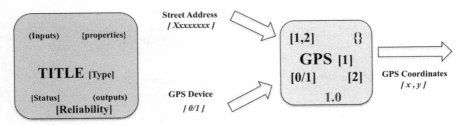

Figure 14.3 Context entity data structure.

and heartbeat sensor. For atomic contexts, the context element data structure is organized into the following fields:

- Title is the name of the context entity e.g. GPS, Address, News Clip, etc.
- Type is the group of periodic table a particular context belongs to, e.g., 1, 2, ..., 6.
- Status is the status of a context entity at a particular instance in time, e.g. Movie Clip might exist in content database but may not become active before a certain date/time.
- Output/s is the set of outputs a context is capable to generate, e.g., 'Street Address' can provide output in different formats.
- Input/s is the set of inputs necessary for the entity to generate context, e.g., pre-requisite for GPS is hardware availability.
- Properties is a set of basic characteristics of the context entity, which can be later used in discovery and matching algorithms.
- Certainty Value gives the reliability of context information, with 1 being the highest and 0 being the lowest.

Figure 14.3 gives an overview of the context entity data structure with an example of GPS location provider. Next, each User, Location and Service keep a context stack of their active context sources. The stack is arranged on FIFO basis and shows the active values of the periodic table. The stacks are locally maintained at user, location and service levels, see Figure 14.4.

On triggers such as registering to a new location or at regular intervals local spaces are updated to centralized SCIM (Service Capability Interaction Manager) using the 'Space Discovery' module. The centralized SCIM module 'Space Modeler' using the concept of dominating sets in graphs models a collective smart space for the individual spaces. The resulting space must fulfill all the six (what, what', who, how, when, where) parameters which serve as the basis for the context table. If the resulting dominating set is in-

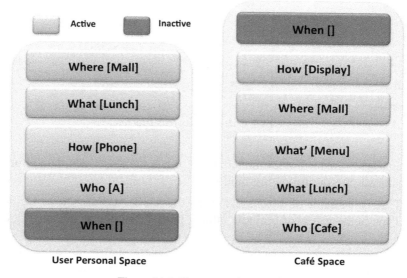

Figure 14.4 Context stack examples.

complete on any of the parameters, the space modeler uses external reasoning engine to fill it. Once the space graph is complete the smart space scripting engine creates a smart space script for execution, the script consists of data corresponding to each of the five (where, when, how, what and who) parameters. The 'what' parameter is of prime importance in a space as it contains the media/services/content/aim of the space. Furthermore, we propose two categories of 'what', one referring to the current activity (what) and the other referring to relevant multimedia content in the smart space (what'). An important aspect of the modelling algorithm is that context graph for an individual or entity is based on attachment to that 'who' and collective outcome of a space is modelled around 'what'. Lastly, as an output of the SCIM we have a smart space structure which is a set as follows:

$$Ss = \{who, where, when, how, what, what'\}$$

Till now we have addressed the model used to represent context, which is vital for the higher-level abstractions to work efficiently. Now we see how the spaces interact to generate higher-level abstractions to represent the current state of a smart space. To generate abstractions we use graph theory to visually represent the state of ecology and trace the possibilities of interaction. In

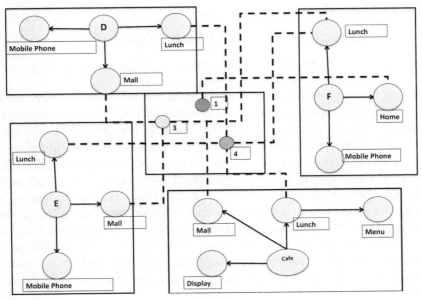

Figure 14.5 Smart space interaction graph.

the following sections we elaborate the terms, theorems and algorithms used for modelling smart ecologies using graphs.

Different graphical models have previously been explored to model context awareness, e.g. Milner's bigraph approach [12] proposed use of bigraphs to model situations, Kelleher et al. [18] proposed use of dominating sets in graphs for deducing patterns and Zimmer [27] has proposed 'Calculus for context awareness' as a formal tool to model context awareness. In this section we introduce our model for modelling smart spaces. In the previous section we defined the data-modelling proposal based on chemical bonding concept and presented a context-aware periodic table, now we define the smart space-modelling layer, which uses the concept of dominating sets in graphs [18]. In using dominating sets for smart spaces we have extended the concept to use different type of edges, vertices shapes and vertices colors. Using this approach the resulting smart space graph is a combination of context stacks from all available players in a space. A representation of 'Lunch Smart Space' scenario is shown in Figure 14.5.

Taking its description as a sample scenario and applying partial dominating sets algorithm on its smart spaces. The resulting dominating set is represented as follows:

$$Ss_{Lunch} = \{[(D, E, Cafe), (F)], [(Mall, Home)], [],$$

$$[(Phone, Display), (Phone)], [(Lunch)], [(Menu)]\}$$

We then move on to discuss the certainty factors of a smart space. As we described earlier in the periodic table of context that context are arranged top to bottom with decreasing certainty. The resulting certainty factor of a smart space is a collective average of the individual contexts involved in the space. The mechanism for allocating weights can vary from table to table (between 0 and 1). For example, in the scenario in discussion, lets assume, 'Mall' is not a certain representation of location (best case being GPS co-ordinates) we assign it a certainty factor of 0.5, similarly, 'Lunch' does not give a certain representation of activity such as eating or walking we assign it 0.5, lacking of temporal constraints make the smart space not trackable in time so we assign it a 0.0 and we assume user profiles are complete and 'menu' content is validated so we give each 1.0, this yields a smart space certainty factor as follows:

$$Ss_{Lunch}Certainty - Factor = (1 + 0.5 + 0 + 1 + 0.5 + 1)/6 = 0.66$$

The certainty factor calculation can be varied based on using different mathematical models for assignments of weight and is an active topic of future directions of this research. This certainty factor is useful in many ways, e.g. application servers can compose services based on the reliability of a smart space and users can prefer in their smart space service subscription the reliability of a space for which they want their presence to be involved or ignore otherwise.

The last requirement discussed in the earlier section is to mark the scope or boundary of a smart space. In the modelling approach proposed in this chapter there are three scenarios for boundary specification of a smart space. The first is when a space is considered as a fully connected graph, any node outside that graph will be outside the space, however, this still does not resolve the issue of a space growing extensively large. The second is to have a snapshot of a smart space for each type (column from context table) of context in the space. The last one is to allow nested smart spaces to the level equivalent to the context categories in the context table (creating a minimum of 2^6 possible spaces at a given instance in time).

14.4.3 Service Selection Component

A service is considered to be an atomic service, meaning it consists of only one service or a service can be a compound service in which case it is composed of two or more atomic services. Each service has a template representing each of the periodic table column and contains the description of service. e.g. the 'how' section can contain the constraints of the service or content in terms of devices or codecs, etc.

The service selection module is a centralized module within 'Space Modeler' that keeps a catalog of services hosted on each of the application servers. Once a smart space is formed the modeler finds a best match for the space by sorting the services in decreasing order of their sub-tasks and sub-services. The process then maps the smart space signature on each service and starts to resort the list by a newly assigned match score. At the end of the process if there is one or more service with a score beyond user or network defined reliability it is setup for execution. If no service is found, individual highest scorers for each periodic table column are short listed and interface matching on their inputs/outputs is checked to see a possible orchestration. If successful newly formed orchestration is temporarily (for the life span of smart space) hosted within the 'Space Execution' module.

If the result is a possibility of more than one service as potential candidate and local SCIM (in the user plane) is not blocking interactivity, the users are presented with a choice to select none or one of the services.

14.4.4 Smart Space Simulator

A simulator called 'Smart Space Simulator (S3)' is under construction to simulate and analyze the performance of the proposed graph-based service and context composition model. Figure 14.7 presents the interface of the simulator. It is designed to enable manual creation of smart spaces and scenarios and dynamic generation by feeding a periodic table of context and running smart space generation algorithms. Early results show that modelling context at atomic level and wrapping it in a model designed to enable dynamic bonding renders richer combinations and compositions of context.

In Figure 14.6 periodic table generation for increasing context elements was examined and it was observed that irrelevant of the number of elements in the table the duration almost stays consistent as compared to the increase in elements. The elements used for this experiment were equally distributed among the six columns. Also, within column reliability values were equally distributed amongst elements. Using the same distribution of context smart

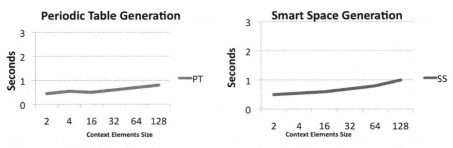

Figure 14.6 Periodic table and smart space generation time for varying context elements.

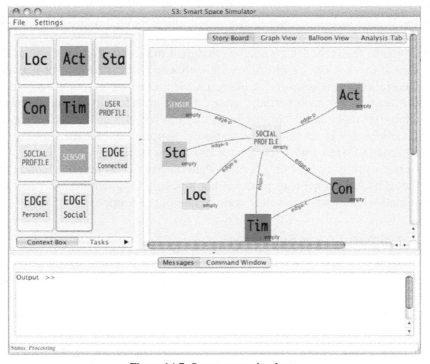

Figure 14.7 Smart space simulator.

space generation time was studied in Figure 14.6. It was observed that time increases with the increase in context elements.

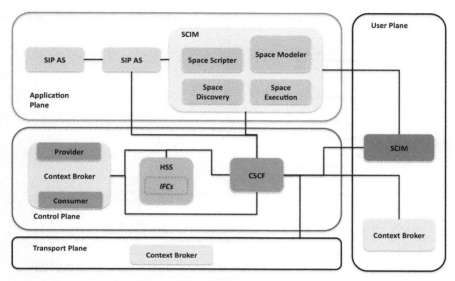

Figure 14.8 IMS smart space architecture overlay.

14.5 Smart Spaces Case Study: IP Multimedia Subsystem (IMS)

IP Multimedia Subsystem (IMS) introduces architecture for offering multi-media services in specific and all applications/services in general, over an IP-based core. IMS being access independent is central towards convergence of services and heterogeneous networks. We have previously discussed in [19] IMS characteristics that support context awareness.

The IMS – Context Awareness convergence happens at each phase of the context management life cycle discussed earlier. The brokering enhances IMS context richness and the context broker functions as a context repository with interfaces to HSS and CSCF. The functioning details of the distributed broker architecture are discussed in detail by Saad et al. [22]. When the user's local SCIM sends a context update via CSCF the filter criteria are preset to redirect messages to the context broker which updates the repository. Similarly, when the user subscribes to 'Smart Spaces' service the service profile is preset with another filter that redirects local SCIM updates to the nearest SCIM in the application plane. Each update sent to SCIM is given a time token and when it expires the SCIM sends the user uri a notification and the user's SCIM can send another update. On receiving each update the SCIM generates the graph

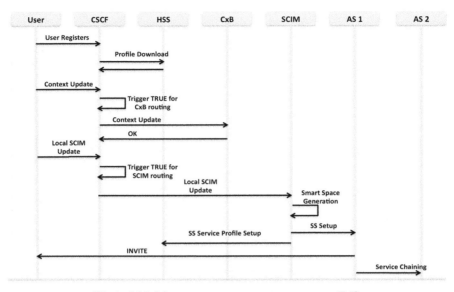

Figure 14.9 Message sequence a smart space over IMS.

based on the rules defined in previous sections and monitors for occurrences of smart spaces. Once a smart space begins to exist, it is sent to the scripting engine to reason about the possible service orchestrations. The reasoning engine is outside the scope of this discussion but for the purpose of this experiment we use a simple rule-based engine which runs a set of rules on the smart space model and generates an output script. For the example scenario of users looking for lunch, the engine examines if there is any multimedia content available in the smart space, if yes, it checks if the content metadata has any tag that matches the activity of smart space, if yes, it decides to share the available content among the participants. It then searches the service directory for a service that can send a given content to a set of users. Later the SCIM hands the control over to the relevant AS which then sends SIP INVITES to the users in the smart space, see Figure 14.9. In more complex scenarios the resulting script might need to orchestrate between different services to achieve the desired service output, e.g., if the device displays in the above scenario are different the service might need an intermediary service that transcodes the available content as per device specifications.

14.6 Smart Spaces Applications

Pervasive smart spaces can be seen as a dynamic flavor of artificial intelligence. Most of the potential applications of smart spaces fall under the category of intelligent, interactive assisting services. Health, environment, gaming, entertainment and advertisement are some of the main domains that can benefit from smart spaces.

Health and environment can benefit from smart spaces in offering intelligent, interactive and context-aware services. The concept of smart homes and smart rooms has existed for sometime now. The proactive European research initiative under the title of 'The Disappearing Computer' [3] and 'Ambient Intelligence' [1] exploring how daily life can be supported and enhanced in such an environment are a few such projects. Similarly, Hristova et al. [16] have discussed the applications of context awareness in assisted living.

Another domain that can benefit from smart spaces is pervasive gaming. On this front various efforts have been made and are in progress, such as, The Sky Remains Project [7], Traces of Hope [4], Comfort of Strangers [2] are some of the examples of real life context-aware games, mainly location based. Some recent European Union projects have also aimed at pervasive gaming, e.g. IPerG [5], IPCity [6]. The aim of IPerG has been the creation of entirely new game experiences, which are tightly interwoven with our everyday lives through the objects, devices and people that surround us and the places we inhabit and IPCity aims on interaction and presence in urban environments.

In short, smart spaces will have applications in almost every domain from daily life shopping to predicting environmental hazards. Smart spaces look promising in engendering an era of services and communication which is highlighted by interactivity and intelligence.

14.7 Conclusion

Evaluating context-aware systems is a complex task. Garzonis [14] discussed different types of evaluation techniques used for context-aware systems and found that there is a balance between findings that support field-based studies and simulation-based studies. As context awareness overlaps with human computer interaction and aims considerably at enhancing usability experience the evaluation process must encompass heuristic walkthroughs. In the scope of this chapter we choose to analyze the modelling proposal by feeding the modelling process with a set of contextual inputs and analyzing the expected accuracy of the output.

Context awareness is often seen as an extension to intelligent systems and most of the artificial intelligence evaluation criteria are considered valid in context-aware systems. Oh et al. [8] proposed 11 heuristics for an end-to-end evaluation of a context-aware system, namely, separation of concerns, flexibility and openness, scalability, reuse, debugging, explanations and accountability, security and privacy, reliability, match with a real world, manual override and reconfiguration.

Future directions of this research include extending S3 simulator to analyze qualitative and quantitative aspects of the context modelling proposal. OpenIMSCore-based field trials are also among the future directions. It is also expected to develop a higher level abstraction layer to the context modelling proposal based on ontologies to enable more specific understanding of a smart space.

References

[1] Ambient intelligence, http://www.eusai.net/, accessed 3 March 2010.

[2] Comfort of strangers, http://www.comeoutandplay.org/, accessed 7 March 2010.

[3] The disappearing computer, http://www.disappearing-computer.net/, accessed 6 March 2010.

[4] Enable interactive, http://www.enableinteractive.co.uk, accessed 7 March 2010.

[5] Ipcity, http://www.ipcity.eu/, accessed 28 September 2010.

[6] Iperg, http://iperg.sics.se/index.php, accessed 28 September 2010.

[7] The sky remains, http://theskyremains.wikidot.com/, accessed 6 March 2010.

[8] Y. Oh, A. Schmidt, W. Woo, and S. Korea. Designing, developing, and evaluating context-aware systems. In *Proceedings of International Conference on Multimedia and Ubiquitous Engineering (MUE'07)*, pages 1158–1163, 2007.

[9] C.A. Licciardi, B. Moltchanov, and M. Knappmeyer. Context-aware content context aware content sharing and casting. In *Proceedings of 12th International Conference on Intelligence in Next Generations Networks*, Bordeaux, France, October 2008.

[10] M. Baldauf, S. Dustdar, and F. Rosenberg. A survey on context-aware systems. *International Journal of Ad Hoc and Ubiquitous Computing*, 2:263–277, 2007.

[11] Claudio Bettini, Oliver Brdiczka, Karen Henricksen, Jadwiga Indulska, Daniela Nicklas, Anand Ranganathan, and Daniele Riboni. A survey of context modelling and reasoning techniques. *Pervasive and Mobile Computing*, 6(2):161–180, 2010.

[12] L. Birkedal, S. Debois, E. Elsborg, T. Hildebrandt, and H. Niss. Bigraphical models of context-aware systems. IT University of Copenhagen (ITU), Denmark, 2005.

[13] H. Chen, T. Finin, and A. Joshi. An intelligent broker for context-aware systems. In *Adjunct Proceedings of Ubicomp*, 2003.

[14] Stavros Garzonis. Usability evaluation of context-aware mobile systems: A review. Technical Report, Department of Computer Science, University of Bath, 2005.

[15] L. Goix et al. Situation inference for mobile users: A rule based approach. In *Proceedings of International Conference on Mobile Data Management*, pages 299–303, 2007.

[16] A. Hristova, A.M. Bernardos, and J.R. Casar. Context-aware services for ambient assisted living: A case-study. In *Proceedings of First International Symposium on Applied Sciences on Biomedical and Communication Technologies (ISABEL'08)*, 2008.

[17] Ahsan Ikram, Saad Liaquat, Madiha Zafar, and Nigel Baker. Experiences in design and development of context-aware IMS-based multimedia services for ubiquitous environments. *Proceedings of International Conference on Next Generation Mobile Applications, Services and Technologies*, pages 15–20, 2009.

[18] L.L. Kelleher and M.B. Cozzens. Dominating sets in social network graphs. *Mathematical Social Sciences*, 16(3):267–279, 1988.

[19] A. Ikram, M. Zafar, N. Baker, and R. Chiang. IMS-MBMS convergence for next generation mobile networks. In Proceedings of the 2007 International Conference on Next Generation Mobile Applications, Services and Technologies (NGMAST'07), pages 49–56, 2007.

[20] S. Dobson, F. Zambonelli, S. Denazis, A. Fernández, D. Gaïti, E. Gelenbe, F. Massacci, P. Nixon, F. Saffre, and N. Schmidt. A survey of autonomic communications. *ACM Transactions on Autonomous and Adaptive Systems*, 1(2):223–259, 2006.

[21] K.E. Kjær. A survey of context-aware middleware. In *Proceedings of 25th Conference on IASTED International Multi-Conference: Software Engineering*, pages 148–155, 2007.

[22] Saad Liaquat Kiani, Michael Knappmeyer, Nigel Baker, and Boris Moltchanov. A federated broker architecture for large scale context dissemination. In *Proceedings of the 2nd International Symposium on Advanced Topics on Scalable Computing in combination with the 10th IEEE International Conference on Scalable Computing and Communications*, Bradford, UK, July 2010.

[23] Address = IEEE, X. Long and G.-S. Kuo. A novel dynamic fuzzy analysis hierarchy model enabling context-aware service selection in IMS for future next-generation networks. In *IEEE Vehicular Technology Conference*, pages 2814–2818, IEEE, 2008.

[24] John O. Connell. Service delivery within an IMS environment. *IEEE Vehicular Technology Magazine*, 2(1):12–19, March 2007.

[25] Seng W. Loke. Building taskable spaces over ubiquitous services. *IEEE Pervasive Computing*, (Ubiquitous Services):72–78, 2009.

[26] H. van Kranenburg et al. A context management framework for supporting a context management framework for supporting context-aware distributed applications. In *IEEE Communications Magazine*, 44(8):67–74, 2006.

[27] P. Zimmer. A calculus for context awareness. Technical Report, BRICS Report Series, 2005.

15

NGN-Based Subscriber Context Processing Architecture

Alberto J. Gonzalez[1], Andre Rios[1], Jesus Alcober[1] and Alejandro Cadenas[2]

[1]*Universitat Politecnica de Catalunya, i2CAT Foundation, C/Esteve Terradas 7, 08860 Castelldefels, Spain; e-mail: alberto.jose.gonzalez@i2cat.net*
[2]*Telefonica R&D, C/ Emilio Vargas 6, 28043 Madrid, Spain*

Abstract

NGN proposes a new architecture to support improved, enriched and innovative service capabilities that engender service customization, personalization, and portability. Specifically, NGN represents a reference model to build network architectures which allows the development of a wide range of IP multimedia services such as VoIP, videoconference, instant messaging, IPTV, etc. reaching all kind of users (both mobile and fixed). Currently, telcos are experiencing a significant decrease of their traditional voice services as they progressively become commodities that do not provide differential services to the subscribers. Accordingly, telcos are observing that users require new services, beyond the traditional ones. Novel solutions are searched, in order to offer new added value and emerging services to attract subscribers' interest and improve quality of experience. In this dynamic environment, personalization of services based on the specific situation (context) of users plays a key role. Personalization is a cornerstone feature for enriching traditional, current and future services. In order to do this, it is possible to take advantage of some NGN capabilities such as ubiquity and QoS provisioning already available. This chapter will analyze how to enable the provisioning of context-aware services from the telcos point of view taking advantage of NGN. In addition,

Anand R. Prasad et al. (Eds.), Advances in Next Generation Services and Service Architectures, 329–353.

it will describe a proposed architecture for the deployment of context-aware services. Based on the real experience of the authors, a prototype giving some recipes for the implementation of a context-aware service, focused on multimedia content adaptation in IMS-based NGN will be provided. Moreover, trends challenges and opportunities will be highlighted.

Keywords: Telco, context-awareness, Next Generation Networks (NGN), IP Multimedia Subsystem (IMS), value added services.

15.1 Introduction

Due to the growing number of players in the telecom arena, telco operators are constantly looking for added value services and capacities to be offered to the end user, in order to be differentiated from the competition. It is quite clear that traditional telco services, like voice calls, are progressively becoming a standard commodity, while the subscriber is looking for personalized services, which can satisfy specific needs that each user may have. This means that the concept of personalization in the service portfolio of telco operators is a must. Such service personalization is extremely expensive to achieve for an operator if it has to be obtained by deploying different types of services, each one being oriented to a subset of users, each one being a small percentage of the total amount of subscribers of the telco operator.

Context can be viewed as a form of high personalization, mainly targeting costumers. Contextual systems are mostly designed to deliver personalized information, suggestions, advertising and services to consumers. It is achieved by exploiting real-time information about location, identity, interests, needs and habits. This is where the concept of context-aware technologies is used [2], in order to deploy common service execution frameworks, but with a centralized users' context monitoring domain. Such a centralized technological option significantly decreases the required capital expenditures from the operator as well as the time to market, while providing fully personalized and context-specific services to the end users [1]. Gartner [3] defines context-aware services as using identity, location, usage, presence, social attributes and other environmental information to anticipate and/or react to end users' immediate needs by offering more sophisticated, situation-aware and usable services.

Significant research has traditionally been done in the area of context-awareness [4, 13], mostly in limited environments like buildings, rooms or university campus areas. In these limited environments, context information is processed by a service platform specific to the type of sensor devices or applications, in such a way that different context-aware service platforms require separate context acquisition elements. A tight coupling currently exists between the application and the sensing devices in such scenarios. Little or no sharing of contextual information will be possible among different context-aware end-user services. This is the main reason why it is necessary to deploy a centralized context processing element, hosted at the telco operator layer that provides all the required functionality to get context-aware behavior from services at a minimum development costs [15].

Within this environment, there are specific context sources, which feed the context broker with external information, for example, the telco operator. Such users' context information is especially relevant, for example, that from social networks, and such information providers are considered as a critical part of this work.

This chapter presents and analyzes the current situation of the context management within the telco operator. The different entities involved in the global system are introduced, as well as a global architecture, which is based on the existing and coming infrastructure deployed by the operators. The proposed architecture will take advantage of these deployed capabilities to address the main issues presented that prevents the context-aware services to be deployed. As a main part of the chapter, a scenario is selected and its implementation described. In this scenario, the advantages of the proposed architecture are analyzed. Finally, some conclusions are drawn.

15.2 Operator Network Architectures

In recent years, remarkable research efforts from the operator service development departments are being dedicated to obtain optimized and fully customized services that behave based on users' context usually in a controlled execution environment, such as the home environment or the enterprise premises environment. This topic has been significantly active due to work in the pure research area related to ubiquitous computing or pervasive computing. However, no commercial deployment at a global level has been carried out by an operator.

Commercial deployments of such research architectures have important drawbacks. The developed context-aware service prototypes are usually implemented in vertical service platforms. That means the service execution platforms locally handle the users' information acquired by the sensors, and applying the specific service logic for the particular context-aware service or set of services. Accordingly, it is not possible for the service platform to reuse in a simple manner the information obtained by other types of sensing devices but only the specific one that the application is designed for. In addition, potentially complex network arrangements will be required in order to give the service platform the possibility to access to the information of sensors located in other types of access networks with no direct connectivity with the service. On the coverage aspect, a huge number of sensors should be deployed in order to increase the coverage area, given that most of the time the application is developed to be integrated with a certain type of sensor. Such specific sensing devices are usually deployed in limited environments and with limited coverage outside of that area, the user will not get such context-aware behaviour.

These issues provide a huge barrier to build a consistent business plan to provide context-aware applications, as the target segment of users would be very limited both on volume and on service coverage. However, telco operators are in a perfect position to take the role of the end users' context aggregator, and open the way for context-aware ubiquitous service development. Telcos already own detailed user information and, additionally, they can deploy enough processing power to run intelligent algorithms [18] over the data and extract higher level end user information.

Therefore, the previous problems can be solved by telcos by hosting a context-aware application at the telco operator network. Given that, one of the main objectives to achieve at an architectural level is the accessibility to the contextual service from any device, location or network. A Control Layer is proposed in order to handle different Access Levels and give connectivity from the information acquisition device to the service layer, hosted in the operator network. Such interworking capabilities are provided by IP Multimedia Subsystem (IMS), specified by 3GPP (Third Generation Partnership Project). One of the objectives of this Control Layer is to provide controlled convergent access to the service layer from any access network technology. If this Control Layer is reused for a context-aware service development and deployment, the immediate consequences are as follows:

- By hosting the Contextual Application at the operator network through an access-agnostic Control Layer, the service can be provided to the user regardless the geographical location and the type of device, as any access network can be integrated in the system. Accordingly, the user can access the service platform and vice versa in a much wider coverage area. This is based on the wide coverage range of telco access networks, typically radio access networks, but some others may apply.

- Based on the previous consequence, a Hosted Contextual Application will see a much higher diversity of users' context information. Based on this, more access networks can be integrated in the system, a much higher number of sensor devices will be able to send the acquired end user information to the hosted contextual application.

- A convergent Service Layer is structured in a Horizontal architecture, in which different elements of the service layer perform the execution of specific service functions in separate optimized modules. The Next Generation Network (NGN) Control Layer will orchestrate the different transactions with different elements of the Horizontal Service Layer during service execution. This service architecture allows information sharing among different elements of the Service Layer, as opposed to a Vertical Architecture. Accordingly, potentially all Service Layer elements may be enriched with the contextual information, including both new services as well as traditional or legacy ones.

This global architecture based on IMS is depicted in Figure 15.1.

15.3 Emerging Customer Demands for Telco

Tendencies in customer demands provide telco operators and players in the telecom industry with ways to differentiate from the competitors and provide added value that means to keep the market share and customers base, or even increase that, reducing subscriber churn.

Customer expectations for newer and better services are forcing telco players to re-evaluate their technology infrastructure in order to capture the growing consumer demands and finding a way to satisfy their requirements. These requirements include an ever-growing list of features and functionality that mainly operate on all three screens: PC, mobile device and television. Additionally, consumers are demanding context-aware connectivity, independent of location and device, as well as an increasing level of flexibility in service portfolio.

In this regard, there is a major trend in telecom users, which is the possibility to be permanently connected with the appropriate people, regardless of the location or even the communication mechanism that would go from traditional voice calls to social network-based communication. Such a consumer trend is complemented with the personalization of the telecom services

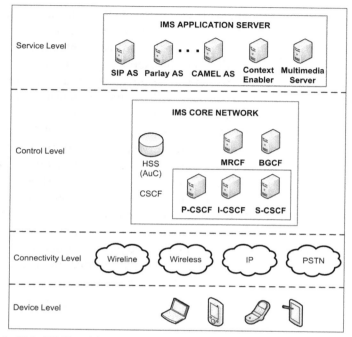

Figure 15.1 Global IMS architecture, including device, access (or connectivity), control and service levels.

and functionalities, as the diversity of available options for the subscriber to communicate different technologies, applications, data, etc. may become a communication barrier rather than an added value for the end user.

This customer trend is associated with a technology trend, which is context-aware personalization of communications. Traditional services need to be enriched, so that telcos are seeking to provide more value-added services to create the required differentiation.

Based on this, service personalization is becoming a key point in developing new and enriched services. Hence, a global personalization architecture is required in order to reduce costs and leverage information reuse among the different service platforms. Moreover, telcos have traditionally been directly impacted by the different privacy issues that the end user may identify. Thus, telcos are uniquely positioned to address such issues, and they are currently doing so by applying different techniques. There are several external references and research works on this particular issue [8], that will not be discussed here as it is outside the scope of this chapter.

15.4 Context-Aware Services in NGN

NGN proposes a new architecture to support improved, enriched and innovative service capabilities that engender service customization, personalization, and portability. Specifically, NGN represents a reference model to build network architectures which allows to develop a wide range of IP multimedia services such as VoIP, videoconference, instant messaging, IPTV, etc., reaching all kind of users (both mobile and fixed). NGN capabilities are tightly related to the options of the design and implementation of reliable, ubiquitous and cost effective context-aware services by the telco operator. This is based on the E2E interoperability capabilities of NGN with potentially any access network. However, context-aware service developments in the operator service layer have been available for quite some time already, and a more detailed analysis will be helpful in order to clarify the notion of context and how it can be leveraged during service design phases to generate added value capabilities.

15.4.1 Context in Operators Networks

Context is a concept so broad that its actual meaning is likely to be different in different situations or fora. Indeed, the definition of context will be completely dependant of the purpose of the service or set of services. Accordingly, there are many research works in which context means physical variables associated to the target user, including location, temperature, noise level, etc., which is the notion of context given by the Ambient Intelligence area (AmI) [5]. With the huge rise of social networks in the Internet worlds, the concept of users' context evolves. In this environment, the context is the situation of friends (or buddies) of a given user, the emotional or mood information that users post on the social profile for other users to see. This information about the user is significantly rich.

In the case of a telco operator, this definition can be: *any information about the situation of a user that may affect the potential ability to request, accept, establish, maintain, reconfigure, terminate, or reject any type of communication session with other entity (subscriber or application).* Such information will include the different user profiles or preferences, stored in service or database nodes owned by the telco operator or by any other third-party provider that may affect the way the telco services are provided.

15.4.2 Types of Context in Operator Networks

In essence, any service that depends on the situation of the user is context-aware. Also, context is a concept that has existed for a long time, but whose formalism is relatively new. Personal communications may have not really considered the context-aware paradigm as it is understood today. However, the situation of the user making use of the service is considered to be critical for the telco service in most of the cases. The inclusion of users' context as a paradigm in service development started in early mobile communication systems, like GSM cellular system, which flourished during the 1990s. This type of context is more related to the physical nature of the situation information:

- Connectivity status of a mobile: stored at the Home Location Register (HLR) database, including also some other information like the Visiting Location Register (VLR) or Visited Network the user is connected to.
- Radio Link situation: coverage, signal-to-noise ratio, macro and microdiversity conditions. All those aspects affect the quality of the signal received at the mobile station and accordingly are considered while the call is ongoing to eventually drop the call, or while the call is to be established, in order to abort its establishment.
- User position information: based on the position of the user, the call can be routed via one specific GSM base station or a different one.
- Situation of the user. Emergency call requests by users in the same cell or area, which may trigger a preemption of an existing call in order to have enough capacity to route the emergency call.

Later evolutions of mobile personal communications, namely 3G/UMTS, in which the user's mobile terminal has access to Packet Switched connectivity as well as Circuit Switched, the nature of users' context information that can be considered for service development starts to increase as the diversity of the applications that the user can get grows. Voice call is not the only application that the user is able to obtain from a mobile terminal or from a telco. Accordingly, an establishment of a videoconferencing session is not requested, if at least one of the involved mobile devices does not support the appropriate radio bearers for that. New context information is considered: the mobile terminal capabilities. Some services can be provided only if the user has the appropriated terminal to support that. Thus, the terminal capabilities need to be monitored or acquired at a given moment before granting access to the service. Up to this point, the users' context, or the early version of that, was fully owned by the telco operator.

Finally, the full data access of the mobile terminals via the telco access network opens a wide range of possibilities for Internet services. This way the user will have access to online agendas (not just the limited capabilities available on the mobile terminals), to social networks, tracking systems, etc. The next step ahead is tremendous both in qualitative and quantitative terms. The users' context moves now to the web application, in such a way that the internet will handle the most complete, robust and diverse set of user information available. However, the context information handled by the web applications is fragmented into different applications, with no information leverage among them. Currently, operators are facing the challenge to take advantage of the access networks, while they are still on time to do that, in order to centralize and aggregate all the fragmented information that the web applications are getting, and build a centralized users' context manager able to develop the most sophisticated context-aware services up to date, or in parallel, open that information to relevant entities that may build the services based on that users' context information.

15.4.3 Context Information Capture and Processing

The design and deployment of context-aware services rely on a set of separate functional levels. Those are the following:

1. Service execution.
2. Context acquisition.
3. Context management.

The first domain is very specific in terms of service functional design, as it depends on the functionality to be provided. The goal in this domain is to implement the global and horizontal mechanisms to obtain the specific context information that is relevant for each service. Some examples will be provided in Sections 15.5 and 15.6 tackling a real case.

The second domain is made up of different elements or context providers, such as sensing devices located in sensor networks with external connectivity to report the results to external elements. However, the elements capable of capturing useful information about the situation of a user (context) can also be telco services, web applications like social networks, etc. In this domain, the main issue is the diversity of context sensor elements, which makes it really complex to obtain and aggregate in a global way all the

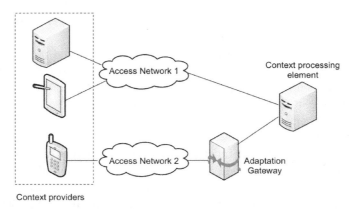

Figure 15.2 Information adaptation for user devices with lower processing capabilities.

context information that is available at different user devices, with different access network connectivities and information formats. This problem is usually solved in lab environments by using specific transport protocol in which the context information is embedded. However, in global telco deployments that option is not valid because the diversity of devices and interfaces makes it impossible to obtain a global transport option. In global telco deployments, multiple transport protocol stacks are defined, with the same contextual application layer protocol embedded in the transport protocols. Accordingly, in the system prototyped in this work, both HTTP and SIP are used as transport and signalling protocols. The HTTP option, either in the form of RESTful or WebServices interfaces, is widely used by context acquisition devices with access to the Internet. The SIP protocol is used by user devices with operator-oriented interfaces.

In any case, the format of the context information must be uniform, so if the acquisition element does not have the processing capacity to work with such format (it may be an infrastructure sensor with no processor), an additional context gateway element should exist to make sure that the information that is processed has a uniform structure. This is depicted in Figure 15.2.

The third domain includes several functions. However, the most relevant one is context processing. Those functions comprise all the related activities to the aggregation of related information received from the context providers, the processing of such information into a meaningful, reliable high-level information that fits the needs of the different services and, finally, handling the subscriptions and interactions with the services or context-consumers that request for context information about specific users.

There are several information mining technologies that may be used for this purpose. The most relevant is the semantic technology [11], which is able to establish relationships among potentially non-correlated concepts, through a conceptual structure called ontology. Based on such conceptual structure, a semantic reasoner is able to infer high-level information from fragmented lower level context reports from the sensors. The drawback of such technology is the amount of processing and memory resources that are required. This drawback makes it impossible to perform telco-level deployments of context-aware services, based only on such technology. There are other options, less resource consuming, like the rule-based processing [17], among others. Fuzzy-logic techniques are still being investigated on the context-aware service development. Thus, they are currently not a consolidated option for telco deployments.

On the contrary, rule-based technologies have been implemented for some time already, and it is a well-known technology that provides excellent results in terms of resource consumption and behavior. The main advantage of such techniques is that, in telco deployments, the different alternatives or scenarios that can be found in contextual monitoring execution are not very diverse, so obtaining a clear set of rules is a viable option.

The acquisition of an optimized context processing technique for a wide-scale deployment, as it is the case of the telco operator deployments, is an area for further research.

15.5 Deployment of Context-Aware Services at Telco Layer: Architecture Proposal

It is important to realize that context information is very useful in a horizontal architecture, as any service at Service Layer can take advantage of it. The only difference among services is the type of context that each service is interested in. In addition, a cornerstone issue to be solved is the acquisition of context, which is going to be used by each service. Scalable mechanisms to be implemented must be found, which will make it possible to request notifications of context updates and to receive them. In order to provide ubiquitous and customized services, communication systems must allow users to transfer their context, even though end user devices or network infrastructure should sense their context and offer enhanced and optimal service provisioning accordingly.

This section proposes a hosted telco architecture for providing advanced context-aware services taking advantage of a horizontal service layer deployment. This architecture is applied to the providing of ubiquitous multimedia services considering the specific conditions of the subscribers.

15.5.1 Proposed Architecture

IMS allows the evolution from a vertical network model, designed for a specific range of services, to a horizontal model of unified network able to support the full range of multimedia services (Figure 15.1).

One of the most relevant aspects of the horizontal Service Level is the concept of Enabler [10]. The Enabler is an entity whose objective is to provide additional information or capabilities to the existing applications at the Service Level. The Enabler is not a service itself, but it provides added value to the service delivery. The contextual processing is understood not as a service, but as a way to significantly enhance the added value that each individual end-user service may provide through contextual service discovery, selection, recommendation, triggering and personalization. Note that once the service is recommended, discovered or personalized, it needs to be executed. Service execution may imply a process of service concatenation or orchestration, as different separated services can be composed to offer an enriched and more complex service. Moreover, the main implications of service discovery, composition and adaptation are the management of all the contextual information received in the system.

Thus, an Enabler that provides information or capabilities of the context of the user is called Context Enabler and provides contextual processing functionalities to the Service Level. Two main Context Enablers are Presence and Location, among others:

- *Location Enabler* is an element that allows the use of location data, determining the position of a person or object. Currently, location services use different location technologies and interfaces, being the cell-based location information from mobile networks (via LBS) and GPS the most representative ones. It is recommended, however, to use a suitable combination of information sources from various positioning technologies in order to determine the most accurate position of a customer.
- *Presence Enabler* is the most common element to deliver context information. In fact, it is already being handled by NGN telco networks

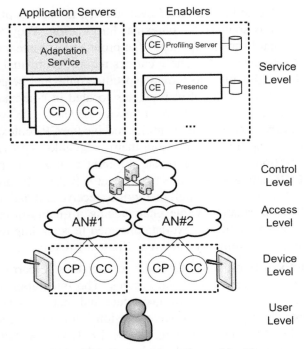

Figure 15.3 Context-aware telco architecture.

without deploying specific context-aware infrastructure. Presence information can be stored and processed in a flexible way at the Presence Server for a given subscriber alias. An alias is any entity associated with a given subscriber, such as a mobile device, or fixed one, PC softphone, etc. Finally, in 3GPP a Presence Server is defined and in OMA (Open Mobile Alliance), there are also several specifications.

To allow the Context Enabler to know the contextual information of each architecture element, a context-aware architecture that allows obtaining this information needs to be implemented. The proposed architecture is shown in Figure 15.3. This architecture provides an optimized way to turn all the traditional services into context-aware applications, by defining a Service Level entity that will centralize the contextual processing. It follows a Server-Side Enabler architecture design. Several contextual entities are identified:

- *Context Providers (CP)*. These entities capture and/or report contextual information from Device Level to Service Level, running at user terminals or even located at Service Level, that is, at service platforms that may act also as sources of contextual information. They are normally software or hardware components that provide raw context data.

- *Context Consumers (CC)*. These entities request contextual information to the Context Enablers to provide specific contextual behaviour. They can be associated with software agents located in contextual services, located either at Device Level or at Service Level.
- *Context Enablers (CE)*. These entities receive the CP contextual information reports and process and store them to be able to provide it to any CC. In addition, CE can aggregate different context information from different CP, acting as a Context Aggregator (CA). CE can implement publish-subscribe, push or pull methods to receive and maintain updated the context information while guaranteeing a quality-of-context data, which ensures that the context data provided fit according to the service goals. They are commonly known as Context Brokers too.

A suitable protocol is proposed for IMS architectures to carry the information among the different contextual entities. The integration of IMS can also provide an easier way to deliver, execute and manage services. In addition, as a signalling protocol, it needs Session Initiation Protocol (SIP) to interact with the IMS platform from the Device Level. At Service Level, communication between deployed services can be done using SOA-based technologies [14], such as Web Services (HTTP/SOAP, REST) or SIP as well. This is advisable due to well known interoperability reasons.

At Device Level, it is suggested to implement the CPs by means of agents installed in the end user terminals to obtain a complete and updated description of the device context. Otherwise, if it is not possible to install an agent at the end device, context can be obtained by means of message exchange. For instance, if a user accesses via a web interface, device capabilities can be obtained by accessing the HTTP User-Agent header or by reading the supported MIME types.

Given this global architectural framework, remarkable research efforts are currently being made to obtain an optimized dimensioning and service provisioning whose behaviour is based on the user context. In this sense, it is proposed a new service responsible of adapting multimedia contents (Figure 15.3).

Three relevant aspects regarding to context for multimedia content delivery applications are: where you are; which kind of device are you using, and what network resources you have nearby. In this environment, the context can include parameters such as:

- User context: the user characteristics, location, preferences, and environmental constraints, e.g. public place where silence is required, working place, home, etc.
- Device context: the type and the capability of the device.
- Service context: service availability, required QoS level, and service performance.
- System resource context: CPU, memory, processor, disk, I/O devices, and storage capability.
- Network context: bandwidth, traffic, topology, and network performance.

Such parameters will be considered by the CEs to enhance multimedia streaming services, enabling personalized and ubiquitous multimedia service provisioning. The proposed architecture will allow telcos to provide enriched and value added services. This is especially important as the competition among different service providers offering a wide spectrum of services, makes subscribers very demanding. Furthermore, it is foreseen that NGN services will involve the provisioning of multimedia contents with certain level of QoS. Providing these services in a transparently manner from the technical and business point of view is still a challenge.

15.6 A Real Case: Context-Aware Multimedia Content Adaptation Service

This section describes the architecture of the Multimedia Content Adaptation Service (MCAS) which was implemented within the i3media project (www.i3media.org). MCAS permits users to consume audiovisual media contents, recognize its context and adapt the requested content in order to be delivered to the user in a personalized manner. Consequently, any device will be able to consume an available multimedia resource anywhere. A profile is defined as the combination of a media container, media codec (this case audio and video) and other coding attributes such as bit rate, frame rate, sample rate or number of channels. These profiles are pre-established in order to accomplish with common used streaming combinations, e.g. H.264 profiles, supported by encoders and decoders. Moreover, the MCAS is able to monitor any possible change of the context,while the user is playing a media resource and to react accordingly. This is called dynamic content adaptation.

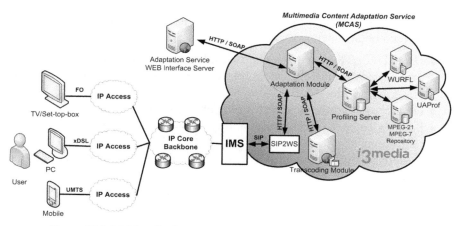

Figure 15.4 Multimedia Content Adaptation Service (MCAS) architecture.

15.6.1 Multimedia Content Adaptation Service (MCAS) Architecture

The MCAS architecture (see Figure 15.4) is composed by the following components:

- The *Adaptation Module* is the core component. It generates the best adaptation recommendation taking into account the user device, network and content characteristics. It acts as a Context Consumer entity.
- The *Transcoding Module* is the element which adapts or transcodes contents, following the recommendation generated by the Adaptation Module and delivers it to the end user.
- The *Profiling Server* contains the multimedia content and user environment (context) descriptors (including: device, network, preferences, location, etc.). It is responsible for maintaining and providing this information to the Adaptation Module. These descriptors follow well-known standards. Concretely, MCAS uses MPEG-21 Usage Environment Description (UED) for describing the user context, and MPEG-7 for representing the multimedia contents. In addition, the Profiling Server contains the transcoding capabilities of the Transcoding Module and the terminal capabilities using WURFL [16] as a basis. This element is considered as a Context Enabler server in the architecture depicted in Figure 15.3. It manages all context information which will be required for service personalization.

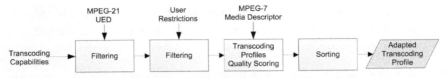

Figure 15.5 Decision process block diagram.

The MCAS architecture defines HTTP/SOAP Web Service interfaces for all the elements described above. In Figure 15.4 its integration with the IMS architecture can be seen. Some well-known implementations of the core components of IMS following the specifications of 3GPP, 3GPP2 and ETSI TISPAN can be used, for instance, the OpenIMS developed by FOKUS or the Software Development Studio (SDS) provided by Ericsson. Note that a SIP to Web Service protocol translator can be used to invoke the corresponding services demanded by end-users. This function is realized by the SIP2WS component. This converged application is composed of a SIP Servlet for processing the incoming and outgoing SIP messages. The SIP Servlet generates a Web service request as a result of the processing of those SIP messages.

It must be noted that the Adaptation Module needs the context information of users. It is obtained from the end user device acting as CP. Morever, it needs the transcoding capabilities of the multimedia transcoder as inputs to recommend the best feasible adaptation. The context information of users is obtained from the Contextual Enabler Architecture (Figure 15.3) and the transcoding capabilities are obtained from the Profiling Server. A tool that performs multimedia quality assessment [12] of each supported coding profile has been incorporated. It guides the decision making process that determines the best profile to be used in an adaptation recommendation. In general terms, Figure 15.5 shows the block diagram of the decision process.

15.6.2 Use Case

This section introduces a use case, which allows one to understand the integration between the MCAS and the proposed architecture. The use case description is as follows:

Margaret is coming back home after a hard day at work. She is going by train and she decides to watch an episode of her favourite TV show on her mobile device. The train journey ends but her favourite program is not finished yet. So, once she arrives at home, she decides to continue watching the show, from the scene she left, on her panoramic TV.

Note that this use case involves heterogeneous devices and access networks. In addition, channel conditions may vary during the multimedia session.

In this situation, there are two CPs: the mobile device of Margaret and her PC connected to the TV. Both devices contain a client application (software agent) that gathers some user context information, thanks to available context sources such as:

- Sensors: integrated GPS module.
- Networks where the device can be connected to: Active Bearers Description (WLAN, GPRS, etc.).
- Device capabilities: User Agent HTTP-header detection.

A good approach to process all this context data to provide customized services is to add a reasoning module, which implements specific rules according to service and business goals. An example of a rule-based system written in Java is JESS [7].

The process starts when Margaret chooses a multimedia content on her mobile device and finishes when she arrives at home and continues watching her favourite TV show on her TV:

1. Margaret executes the mobile client application that works as a CP and gives access to the MCAS.
2. The CP sends a notification to the CE with the contextual information of the device: location, attached network, networks around it and device capabilities. This notification is sent via a SIP PUBLISH message to the IMS domain.
3. While the user is on the train, the CP running on the mobile device periodically tests if there are any changes in the context. If there is a change, a notification is sent to the CE. It generates a notification of the position and sends that information to the Context Enabler through a SIP PUBLISH message with the format presented in Figure 15.6.
4. When Margaret chooses a specific multimedia content, the mobile application sends a notification to the MCAS. A SIP INVITE message is generated, as she wants to start a multimedia session.
5. The MCAS sends a subscription message to the CE (SUBSCRIBE), implemented by the Profiling Server. This way the CE notifies to the MCAS any change in user context.

```
PUBLISH sip:CEnabler@coreims2.hi.inet:6070;transport=tcp SIP/2.0
Call-ID: 16366dbd90fe21f23f80dabf052ebeaa@10.95.29.237
CSeq: 1 PUBLISH
From: <sip:margaret@coreims2.hi.inet:6070>;tag=1
To: <sip: CEnabler@coreims2.hi.inet:6070>;tag=567
Via: SIP/2.0/TCP
10.95.29.237:5060;branch=z9hG4bK2c64853cd3a0457bf2bbcebbfdf8
d448
Max-Forwards: 70
Contact: <sip:margaret@10.95.29.237:5060;transport=tcp>
Event: presence
Content-Type: text/plain
Content-Length: 134
<?xml version="1.0" encoding="UTF-8"?>
<contextual_information>
<userId>1555</userId>
<sessionId>310995000000000</sessionId>
<location>
          <lat>40.869636890588225</lat>
          <lng>-3.822334846993267</lng>
</location>
<network id= "hdspa">
<status>connected</status>
<bandwidth unit= "kbps">1000</bandwidth>
</contextual_information>
```

Figure 15.6 SIP PUBLISH message with mobile context information.

6. The MCAS knows the contextual information of the user, and it adapts the desired content resource according to the user device capabilities, access network characteristics and user preferences. The adapted multimedia content is streamed to the user device (online adaptation).

7. When Margaret arrives at home, the mobile CP notifies the context change to the CE. Once Margaret switches on her TV, this CP (PC) indicates its contextual information to the CE. The CE notices that the two devices of the user are in the same location (home). Then, the CE notifies to MCAS this situation. Thus, MCAS, sends to both device applications a notification indicating that there are two available user devices (mobile and PC). This way the device application offers the user the possibility to switch the multimedia session to other device at the same location.

8. Once Margaret decides to switch from the mobile to the TV station, she selects to start watching the stream through her TV. So, the TV indicates this new situation to the CE. The CE indicates the new user context situation to the MCAS. The MCAS detects the new user context situation and adapts the multimedia content accordingly. Margaret, now

Table 15.1   Traffic loads applied for each scenario and adaptation process delay.

Scenario	Number of users	Request inter-arrival average time	Per-user delay
1	1	1 s	100 ms
2	10	1 s	400 ms
3	20	1 s	1.1 s
4	1000	5 min	125 ms

at home, can switch off her mobile multimedia application as she is able to continue watching her favourite TV show through her TV station.

15.6.3 Evaluation

We studied the performance of MCAS with synthetic workloads (Table 15.1). The purpose of these tests is to evaluate resource consumption and to analyze the average delay for generating context-aware adaptations. This test includes different number of users making adaptation requests to the Adaptation Service at different intervals of time. Jmeter [9] was used for traffic and statistics generation. During test execution, the Transcoding Module and the Profile Module were placed in the same machine (Intel Xeon 2×2.9 GHz, 1 GiB of RAM, Windows XP Pro). The interfaces between them were HTTP/SOAP Web Services. The version of the Java Virtual Machine was 1.6. We used Apache Tomcat 6 as application server. The implementation of the profiling server was based on eXist 1.2 (XML data base). Finally, we have developed a transcoding module based on the well-known ffmpeg software.

In the fourth and more realistic scenario, the average response delay of a request per adaptation is in the millisecond order, so it can be considered an appropriate value in order to attend the user requirements without delay perception. The traffic load follows a Poisson distribution. We measured an average CPU consumption of 25 and 35% of RAM during tests execution. When a user switches from one device to another (session mobility) or vice versa, the required time to signal this situation and transcode the content is significantly higher than the time required for generating the adaptation by the Adaptation Module.

15.6.3.1 Streaming to a Mobile Device

For testing context-aware adaptation, a software agent application was developed for a Symbian platform acting as CP. Another interesting option for developing agent-based applications in Java would be JADE (Java Agent

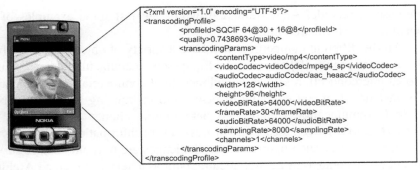

```
<?xml version="1.0" encoding="UTF-8"?>
<transcodingProfile>
        <profileId>SQCIF 64@30 + 16@8</profileId>
        <quality>0.7438693</quality>
        <transcodingParams>
                <contentType>video/mp4</contentType>
                <videoCodec>videoCodec/mpeg4_sp</videoCodec>
                <audioCodec>audioCodec/aac_heaac2</audioCodec>
                <width>128</width>
                <height>96</height>
                <videoBitRate>64000</videoBitRate>
                <frameRate>30</frameRate>
                <audioBitRate>64000</audioBitRate>
                <samplingRate>8000</samplingRate>
                <channels>1</channels>
        </transcodingParams>
</transcodingProfile>
```

Adapted Multimedia Content   Content Adaptation XML Response

Figure 15.7 Adapted multimedia content and content adaptation XML response.

Development Framework) [6], which is in compliance with the FIPA (Foundation for Intelligent Physical Agents) specifications. Figure 15.7 shows how the mobile device receives the stream of the adapted content according to the users' context. It can also be appreciated the XML generated by the MCAS containing the recommended parameters used in the transcoding process.

15.7 Trends, Challenges and Opportunities

Context-awareness is identified by telco operators to be one of the core information mining technologies to enhance the service portfolio in two main areas: added value service functionality, by being able to create new services specific to certain users' situations, and service personalization, by adjusting service triggering, adaptation as well as logic itself to the users' situation. Such trends are being widely investigated by telco operators worldwide, as well as by manufacturers and equipment providers.

This technological trend is based on the assets of additional capabilities of mobile devices, increasing wireless bandwidth technologies and the enhancement of telco service architectures with context information. Currently, the applications and capabilities that are available to the end user in this regard are limited and are based on the use of presence, location or social networking information. But it is just the beginning, as can be seen from emerging players on the telecom field, like Internet companies (Google is a significant example), that are also leaders in user information mining. In the case of Google, AdSense and Adwords are just some examples of the potential of information mining.

All the different competitors will have a variety of objectives to get from the context-aware global services. Equipment manufacturers are pursuing global interconnection standards and their final commercial implementation. For instance, they are adopting diverse solutions ranging from NFCs (Near Field Communications) embedded onto mobile phones to initiatives like SensorML to discover, locate and process sensor information.

In the case of telco operators, the challenges are to mold its core service model to the context-aware architecture, built as a Service Oriented Architecture (SOA) or Event-Driven Architecture (EDA), and not being considered as natural improvements of these architectures. In parallel, telcos are looking to discover which information sources can become context sources, and how this information can be provided in a secure, timely and usable manner to be considered as a context source. Such activities will involve a global analysis of the integration possibilities, including the end user devices, access networks and service layer technologies. The context source enables the context-enriched services, used by the context-aware applications. Currently, all these services are custom-built. There is a lack of standards in this field.

Nowadays, the opportunity of telcos is to become the de-facto context broker of their customers, taking advantage of content aggregators and service providers, which will emerge as an alternative broker context in the coming years. The activity of such a context broker would be to centralize the user's information acquired in any domain or device with connectivity with the telco service layer, either online or offline. In parallel, it will distribute such information, processed, and in the appropriate format, to interested consumers. Those would eventually be the third party companies developing applications that would be able to provide an advanced functionality, like the multimedia content adaptation described in this chapter. Examples of that would be those companies available in the smartphones markets such as Android or Apple Appstore.

In this global entity picture, the main missing point is a business model for each of the involved elements. Such a business model should be able to distribute the associated earnings in an acceptable manner to get a stable ecosystem.

In a somewhat centralized architecture, the centric processing element is in control of the information and business model as well. However, it is quite likely that there will be more than one entity struggling to handle and process users' context for others to use. Thus, in the current extremely diverse technological environment, it is quite likely that several context managers/brokers may appear, probably each one suitable for different domains.

15.8 Conclusions

Currently, telcos are experiencing a significant decrease of their traditional voice services. Users require new services, beyond traditional ones. Novel solutions are searched, in order to offer new added value and emerging services to attract subscribers' interest and improve the quality of their experience. In this dynamic environment, personalization of services based on context plays a key role. Personalization is a cornerstone feature for enriching traditional, current and future services. In order to do this, it is possible to take advantage of some NGN capabilities such as ubiquity and QoS provisioning already available. Based on these deployed capabilities, telcos must play a proactive role in the design and deployment of new capabilities providing public APIs to third parties in controlled and secure environments.

Specifically, context-aware applications are some of the upcoming paradigms in telecommunication commercial environments. Today, it is possible to obtain a great deal of parameters about the context of the user due to the enhanced hardware and software capabilities and functionalities of the communication terminals that is supported by the growing amount of deployed network technologies, as well as the great amount of different users with different preferences and needs. All this contextual information, if processed in an appropriate manner, will become instrumental for providing a variety of added value services.

This topic has been actively researched in recent years, due to the remarkable works related to ubiquitous computing or pervasive computing. Such activities are currently taking place at the research departments of telco operators.

Acknowledgements

This work was supported by the CENIT Programme (Spanish Ministry of Industry) under the i3media project, the MCyT (Spanish Ministry of Science and Technology) under the project TSI2007-66637-C02-01, which is partially funded by FEDER, and by the TARIFA project of the i2CAT Foundation. The authors would like to thank all participants from Telefonica R&D and i2CAT Foundation for their help and support.

References

[1] Stefan Arbanowski, Pieter Ballon, Klaus David, Olaf Droegehorn, Henk Eertink, Wolfgang Kellerer, Herma Van Kranenburg, Kimmo Raatikainen, and Radu Popescu-zeletin. I-centric communications: Personalization, ambient awareness, and adaptability for future mobile services. *IEEE Communications Magazine*, 42(9):63–69, 2004.

[2] Anind K. Dey. Understanding and using context. *Personal and Ubiquitous Computing Journal*, 5(1):5–7, 2001.

[3] Gartner website, WWW page, September 2010.

[4] Hottown Geobots and Theo G. Kanter. Attaching context-aware services to moving locations. *IEEE Internet Computing*, 7(2):43–51, 2003.

[5] Hani Hagras, Victor Callaghan, Martin Colley, Graham Clarke, Anthony Pounds-Cornish, and Hakan Duman. Creating an ambient-intelligence environment using embedded agents. *IEEE Intelligent Systems*, 19(6):12–20, 2004.

[6] Java Agent Development framework website, WWW page, September 2010.

[7] JESS website, the Rule Engine for the Java Platform , WWW page, September 2010.

[8] Xiaodong Jiang and J.A. Landay. Modeling privacy control in context-aware systems. *Pervasive Computing, IEEE*, 1(3):59–63, 2002.

[9] Apache Jmeter website, WWW page, September 2010.

[10] C Konstantinides and C.D. Charalambous. Supervisory and notification aggregator serve enabler in a fixed mobile convergent architecture. *Proceedings of 4th International Symposium on Wireless Communication Systems (ISWCS2007)*, pages 737–741, November 2007.

[11] Jibran Mustafa, Sharifullah Khan, and Khalid Latif. Ontology based semantic information retrieval. *Proceedings of 4th International IEEE Conference Intelligent Systems (IS'08)*, 6–8 September, Vol. 3, pages 22-14–22-19, 2008.

[12] Iain E.G. Richardson. *Video Formats and Quality, Cap. 2 in H.264 and MPEG-4 Video Compression*, Wiley, 2003.

[13] Debashis Saha and Amitava Mukherjee. Pervasive computing: A paradigm for the 21st century. *Computer*, 36(3):25–31, 2003.

[14] Christoph Schroth and Till Janner. Creating web 2.0 and SOA: Converging concepts enabling the internet of services. *IT Professional*, 9:36–41, 2007.

[15] Steffen Staab, Hannes Werthner, Francesco Ricci, Alexander Zipf, Ulrike Gretzel, Daniel R. Fesenmaier, Cecile Paris, and Craig Knoblock. Intelligent systems for tourism. *IEEE Intelligent Systems*, 17(6):53–66, 2002.

[16] WURFL website, WWW page, September 2010.

[17] Suyun Zhao, Eric C.C. Tsang, Degang Chen, and XiZhao Wang. Building a rule-based classifier – A fuzzy-rough set approach. *IEEE Transactions on Knowledge and Data Engineering*, 22(5):624–638, 2010.

[18] He Zhen, Tao Tang, and Lu Qirong. Research on personalized application recommendation system in the ota system based on mining users' interests. *Proceedings of 4th International Conference on Wireless Communications, Networking and Mobile Computing (WiCOM'08)*, 12–14 October, pages 1–5, 2008.

16

Classifying Users Profiles and Content in a Content-Zapping IPTV Service

João Rodrigues, António Nogueira and Paulo Salvador

Instituto de Telecomunicações, University of Aveiro, 3810-193 Aveiro, Portugal; e-mail: joaoffr@ua.pt

Abstract

The television was not only one of the most remarkable inventions of all times, but also its evolution and all the related opportunities revolutionized and continue to revolutionize the world. The multimedia content sources are becoming more and more abundant and diversified. The high number of contents and its dispersion can complicate the content dissemination and access by the target audience. Besides, users have increasing difficulty to obtain multimedia contents that best fit their preferences.

In order to try to overcome the previous issues in multimedia content dissemination, we propose an approach that, in a few words, try to index all the available multimedia contents and dynamically select the best contents available for each user at each time. This approach defines our content-zapping methodology, which uses the user's profile information, users clustering/correlation and the user's dynamic context information. Our aim is to create a transversal and global classification infrastructure for multimedia contents and users, which allow the proposed service to classify and distribute contents in an intelligent and dynamic way accordingly to the user's needs, regardless the multimedia provider, the end-user device/application or the access network.

After analyzing the problem specification, we conclude that the best way to implement a content-zapping service is to dynamically classify all the

Anand R. Prasad et al. (Eds.), Advances in Next Generation Services and Service Architectures, 355–382.

involved entities in the system. In terms of profile definition and classification, we used a dynamic multidimensional hierarchical classification. As is known, in chemistry each compound can be described using a finite formula which mixes several simpler elements, quantified by numerical values in a continuous range. Using this premise as inspiration, we decided to describe each entity in our system as a multidimensional variable, where each dimension is a specific weighting factor related to a specific parameter (classifier), available to all entities in the system.

The proposed service tries to conjugate several aspects from different recommendation systems and techniques in a very flexible and dynamic system architecture. After and during the system implementation, we conducted preliminary intensive simulations using a controlled environment. The tests show us good results not only in terms of the service's computational performance but also in terms of our recommendation algorithms accuracy. At the end, we conclude that the implemented algorithms can be greatly improved, since they do not explore all the potentialities of our architecture, in terms of available data and profiles correlation possibility.

Keywords: multimedia content-zapping, recommender systems, adaptive classification, profile classification, multidimensional hierarchical classification, IPTV, multimedia aggregation.

16.1 A Content-Zapping IPTV Service

Nowadays, the Internet is a place with an extensive and increasing number of multimedia contents that are located in a highly distributed way. The sources of multimedia contents are becoming more and more abundant and diversified, but also increasingly dispersed, which makes them more difficult to be found and accessed by the target audience. Besides, users have an increasing difficulty to obtain multimedia contents that best fit their preferences, finding most of the times contents that they do not want, do not like and do not need or are inappropriate to their current context. Thus, there is a increasing need to develop a transversal and global classification infrastructure for multimedia contents and users that is able to classify and distribute contents in an intelligent and dynamic way. This infrastructure must have not only knowledge about the users and contents profiles, which are the central entities in the system, but also provide context-awareness mechanisms in order to improve non-linearity in the suggestion algorithms. The adaptive classification of users and contents will allow the system to recommend the most suitable contents for a particular user based both on his adaptive profile and context.

By definition, the result of a recommender system is not necessarily a list of recommendations, that is, a list of items that the user will probably find interesting based on their profiles. A recommender system can also provide other means of guiding a user to choose items that are potentially interesting to him, like highlighting interesting items [2]. Nevertheless, the core task of a recommender system should be providing "users with a ranked list of recommended items, along with predictions for how much the user would like them" [4]. There are several recommendation techniques: *collaborative*, *content-based*, *demographic*, *utility-based* and *knowledge-based* [2].

All learning-based techniques (*collaborative*, *content-based* and *demographic*) suffer from problems like the *cold start* problem [8], also known as *bootstrapping* problem in the literature [6], which leads to a high *ramp-up* in the users and items profile construction. The vulnerability to obfuscate attacks that try to manipulate the internal knowledge in order to control the system's recommendations depending on the attacker behavior [7], the system stability versus plasticity (that is, the difficulty to change a well-established user profile in the system versus rapid profile updates) and user privacy [1] are other problems generally associated to recommender systems.

The *collaborative* recommendation is probably the most well-known and widely implemented recommendation method [2,11]. The typical user profile in a *collaborative* recommender system is defined by a vector of rated items and corresponding users ratings that is continuously updated, using a specific algorithm, by the user-system interactions over time. The rates can be binary ("like-it!" or "hate-it!") or numerical values in a well-defined continuous scale, indicating the degree of preference related to one specific item for the target user. Burke refers that "the greatest strength of collaborative techniques is that they are completely independent of any machine-readable representation of the objects being recommended, and work well for complex objects such as music and movies where variations in taste are responsible for much of the variation in preferences" [2].

In a *content-based* recommender system, the objects of interest are defined by their associated features [2]. In other words, the items of interest are defined by a set of simpler nuclear elements that parameterize the target items, acting as its basic components. Therefore, a *content-based* recommender system learns a profile of the user interests based on the nuclear elements ("features") present in items that the user has rated. In the literature, this technique is also called "item-to-item correlation" [9]. The *content-based* approach can use learning methods such as decision trees, neural networks or vector-based representations. As in the *collaborative* recommender system, *content-based* user profiles are *long-term* models [2].

Demographic recommender systems aim to characterize the user based on its personal information and make recommendations based on demographic classes, creating clusters of users with similar demographic attributes. The most salient benefit of a *demographic* recommender system is that it may not require a history of the user ratings [2].

Utility-based and *knowledge-based* recommendation techniques do not attempt to build *long-term* generalizations about the user, basing their advice on an evaluation of the match between the user needs and the set of available options. In other words, the system calculates the utility of each item for the target user and suggests the set of items that most fit the user needs. This type of recommendation is generally achieved through inferences about the user needs and preferences. *Utility-based* approaches compute the utility coefficient based on functional knowledge [2].

From our perspective, a recommender system is a platform that can learn with the user (using direct and implicit feedbacks) and is able to suggest to him specific contents, based not only on his current profile but also on the specific user context. The proposed system introduces the concept of content-zapping, in contrast to the traditional channel-zapping approach. Each user receives a multimedia stream that is automatically composed by the system and the user can simply interact with the system by requesting a content change, marking the content as a favorite or finding contents based on some criteria (Figure 16.1). The system will incorporate adaptive classification of both contents and users based on the users behaviors about the received contents.

The adaptive classification of users and contents will allow the system to recommend the most suitable contents for a particular user, based not only on his adaptive profile but also in his dynamic context. Due to this feature, the proposed system is not only a typical recommender system using typical prediction techniques, but also a context-aware system that uses the available user context information (such as the current time, GPS (Global Positioning System) coordinates, temperature, device capabilities, network capabilities, etc.) to improve and personalize the composed multimedia stream. So, innovative and integrated adaptive content recommendation techniques based on multidimensional data analysis and statistical clustering were investigated, developed and tested. The proposed system can classify a new user and/or a new multimedia content based on the well-classified contents and users that

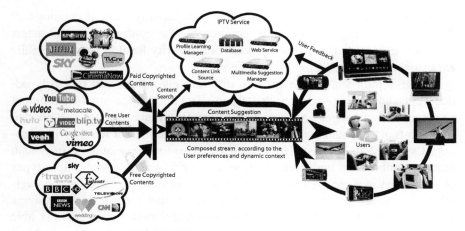

Figure 16.1 Conceptual diagram of the proposed service.

are already stored in the system. Using all the well-classified contents and users, we can easily detect "misbehaving" users and contents that are very far from the admissible behavior that is typical for its profile. The trustable entities are crucial for the system convergence.

Each stored item (multimedia content) is suggested only one time to each user, unless there are no more distinct items available in the persistent unit (real-time multimedia contents can be suggested more that one time). User profiles are dynamically built and updated in order to be flexible over time. In order to overcome the insufficient and inaccurate profile classifiers that are available at the beginning, the well-known *cold start* problem, users should give some information about their personal, demographic and multimedia interests profiles when they initially register in the system. Due to the dynamic update of the user profile, classifiers follow the global profile evolution tendency. In terms of user profile updates, the oldest feedback data, stored in the persistent unit, will be less significant than the most recent information. The content profiles are built in a very similar way when compared to the user profiles. Initially the system uses the tags and metadata that are attached to the multimedia contents to initialize the dynamic classification of each content profile.

In the proposed system, users can explicitly stop the default recommendations, allowing them to take full control of the multimedia stream and/or make a request for a specific multimedia content. The recommendation process is fully configurable, but the standard configurations generally fit the user needs and allow the user to use the service transparently without noticing that the stream is being automatically composed. The configurable parameters allow the user to define his own privacy and security level in terms of profile exposure and data collected by the client.

Using a plug-and-play approach, different algorithms can be easily added, replaced or removed from the system. The knowledge modules (content recommender manager and user profile learning manager) are sub-divided into sub-functions and sub-categories based on each target feature. The proposed system can be classified as a *mixed hybrid recommender system* with feature combination and augmentation, since it is able to use multiple types of prediction and recommendation techniques in order to improve the overall recommendation accuracy [10]. It is known that recommender systems which combine two or more recommendation techniques are generally able to achieve higher performances and avoid problems like the *ramp-up* problem [2].

The knowledge that is stored in the system persistent layer is organized in a hierarchical way, taking advantage of the intrinsic data relationships, forming a *taxonomy* [3]. All the knowledge belonging to the multiple entities stored in the system (users, multimedia contents, contents providers, content authors, etc.) uses the same *taxonomy*, that is, is described by the same nuclear descriptors, thus improving generalization and allowing a direct relationship between all stored knowledge.

Modularity, flexibility and scalability were some of the most important project requirements. Structurally, the system is composed of five main modules: the client-server communication manager, the multimedia sources manager, the multimedia suggestion manager, the profile learning manager and the data persistence unit. The first four modules can work separately and asynchronously, using the data persistence unit as the common point for data interchange and synchronization. Finally, a transversal management platform was also built to monitor and manage in real-time the status and events in each module and its multiple instances.

16.2 IPTV Server Architecture

The multimedia system must maintain a list of multimedia contents residing in other systems (media servers) and keep a dynamic classification of both the core entities (users and multimedia contents) and the secondary entities (content providers, content authors, users groups, etc.). This classification is initially established and gradually refined based on the interactions between

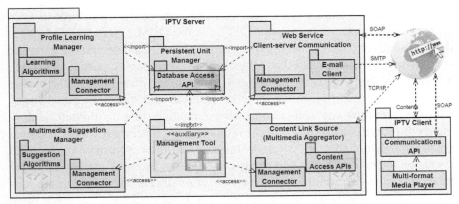

Figure 16.2 General diagram of the service architecture.

the users and the requested multimedia contents. The system should store all the users' explicit reactions (direct feedback) and their behaviors (implicit feedback) for all watched multimedia contents, in order to build and update both the users preferences and the multimedia contents profiles. The implicit feedback data details and complements the context data, since the user behavior is conditioned by his current context.

The user interaction with the system will be made through a client software that allows access to all multimedia contents and includes interaction capabilities with the proposed service. The service features will be available through a web service (HTTP (Hypertext Transfer Protocol) or HTTPS (HTTP Secure)), allowing public access using a key that identifies each client software.

As mentioned above, the basic architecture of the proposed system is based on five distinct modules with specific functionalities, connected through a TCP/IP (Transmission Control Protocol/Internet Protocol) network. The persistence unit has a nuclear role on the correct system operation, since it enables the global system synchronization and crucial data storage, acting as the common point for data interchange. Due to this fact, it is considered the critical point of the system and should be replicated over multiple servers. Since this kind of system must have a high level of availability, accessibility and reliability, we decided to build each module in such a way that it should support service replication through multiple physical (or virtual) computational systems. Figure 16.2 depicts the general system architecture, in a very simplified way.

Regarding the technologies that were used, the system was built in Java, the current persistence unit uses MySQL as the DBMS (DataBase Management System) and the connection between each Java module and the database is made using the Hibernate Java object-relational mapping framework. Each module can be separately configured through its own XML (eXtensible Markup Language) file. Since the configuration and monitoring tasks may increase in complexity and become very repetitive on a distributed deployment with service replication over several different machines, we use a simple management tool that allows us to receive real-time information about all events on the entire system and easily configure and control each module and all its instances. The entire system is platform independent thanks to the Java platform. Each module manages its own computational load, so, when there are enough free resources available, the system will run some operations/tasks that are not essential/required at that moment in order to anticipate future needs and use the available resources in a better way. The next subsections will describe each system module in more detail.

16.2.1 Client-Server Communication Manager

The client-server communication manager follows the web service model that allows the system to expose its functionalities to any client. All client requests have to be addressed to this module or to the transparent communications manager gateway (if it exists). Each request/operation is executed concurrently with all others that may exist on the system. Each request unequivocally corresponds to one server response, except if the system does not receive the request, the network is down or the system is offline. Each available method returns a well-defined XML message that describes the requested transaction end-state and result. The client-server communication manager also controls the process of sending emails to users and all user-related periodical procedures, like account lock/unlock, long inactivity logout, among others. The client-server communication module is stateless, that is, each request/operation is treated as an independent transaction that is unrelated to any previous request. The system, as a whole, maintains the necessary data corresponding to each operation in the persistence unit. The stateless approach greatly improves the system performance and simplicity, because there is no need to dynamically allocate storage to deal with conversations in progress and, if any client "dies" in the middle of a transaction, no part of the system needs to be responsible for cleaning or restoring the current state of the server. However, this approach has some little disadvantages, like the need for some extra data in each message related to the specificities of the request context.

The mutual exclusion between each client request is assured using one session per request and multiple transactions inside each section, creating an isolation between nuclear operations, specially in write or update tasks. The DBMS is the core element responsible for managing all the sessions and transactions. The generic database access API (Application Program Interface), created based on the Java Hibernate framework, separately manages the first level of synchronization inside each module.

In order to make the proposed system context-sensitive, the client software should send the initial user context (after user login) and should update the entire context or some context parameters when the data changes. Due to the modular approach, the system allows the introduction of new features and their easy availability by adding new methods to the web service that uses them. This module has an internal connector that allows it to be monitored and controlled by the system management tool.

16.2.2 Multimedia Source Manager

One of the objectives of the proposed system is to develop a multimedia content aggregator, allowing easy and intelligent content access. To realize this purpose, we built a simple recursive multimedia content aggregator, that can be partially described as a simple *web crawl*. The aggregator module searches for multimedia contents based on an updatable list of providers (based on absolute and relative URIs (Uniform Resource Identifiers)) that acts as base locations. For each provider in the list, the system gets the target file type. If the target URI corresponds to an allowed multimedia content, the aggregator tries to get its metadata and, before adding it to the database, the system compares it with the stored contents in order to avoid replicated entities. If the content already exists in the system, the aggregator searches for updates based on the most recent content metadata. Otherwise, if the target file that is pointed out by the URI is a text-based file (e.g. HTML (HyperText Markup Language), XML, RSS (Really Simple Syndication), etc.) or belongs to an unknown file format, the aggregator invokes the respective parser that will try to extract all URIs and the corresponding metadata from it. After parsing the data and creating a list of new URI-metadata pairs, the system executes the crawling process from the beginning in order to follow the extracted URIs. Each recursive branch stops when the data pointed out by a URI is a valid multimedia content, is unavailable, cannot be parsed, or when the maximum search depth is achieved (in order to avoid infinite loops).

Apart the *web crawl* functionality, the modular structure of the multimedia source manager allows it to handle almost any type of interface between the aggregator and the multimedia providers, like specific/proprietary APIs. In fact, the use of specific APIs is preferred rather than other generic sources. An API provides a more structured access to the contents and their corresponding metadata, allowing high level features like content search based on some criteria or content retrieval using internal similarity relationships. The aggregator also allows users to insert content links directly on the system. In this last case, the system prompts the user for information related to the added content, in order to improve its classification. If users want to share their own contents, they must first upload them to a multimedia content provider (e.g., Google YouTube, Vimeo, etc.) and then publish the new links and the respective metadata/description in the proposed system.

Since the proposed service does not store the contents but only indexes them, it is not possible to guarantee QoS (Quality of Service) requirements between the service providers and the clients at this level. The multimedia source manager tries to reduce this limitation by periodically checking each indexed content in order to verify if it is available and computing some statistics related to the provider access network. Moreover, each time a certain content is requested, the system stores some metrics related to the provider network and QoS. When a client (user) requests that content, all available statistical information about the content and its respective provider/author is sent to the client together with the requested content link and metadata.

16.2.3 Multimedia Suggestion Manager

This module allows the system to handle the multimedia content recommendation process. The implemented recommendation process can be divided into two main categories: profile-based and context-based. As will be explained in Section 16.4.2, in this early development phase, we are only using profile-based recommendation. However, we are conducting performance tests in order to evaluate context-based recommendation algorithms and, despite the promising results that have been obtained so far, we think that it is possible to greatly improve them and establish a parallel approach where contexts can be composed and shared among a group of users in similar contexts in a transparent way.

The multimedia suggestion manager module was developed using a modular approach, so it is very easy to add, replace or remove any of its internal components, such as recommendation algorithms. Like all other modules of the proposed system, this module also allows service replication and has an internal connector that allows the management tool to collect relevant events and fully control the module operations (e.g., start, stop, restart, configure, etc.).

The current suggestion algorithm is based on Euclidean distances in a multidimensional environment and takes advantage of the global classification methodology to simplify the suggestion process. Since all contents and users are described by the same nuclear structure (classifiers tree), calculating the multidimensional distance between the target user and all the available contents (real-time contents or recorded contents not yet viewed) in the system is an easy and intuitive start. Considering that the user prefers to watch contents which are related to his profile, the suggestion manager selects the multimedia contents having the smallest distance between the content profile and the target user profile. This approach should not be strictly linear because, if the system does not try to suggest items that are less related to the user profile, the learning process will be stuck in the initial user profile and will not be able to quickly evolve in order to generate knowledge about the largest possible number of permutations of the classifiers. In other words, once in a while, we should extrapolate the user profile in order to confirm if the extrapolation was correct (user profile ramifications) or deny the extrapolation (increasing the certainty over the contents that the user does not like). A simple extrapolation module and simple algorithms are currently under development and test.

16.2.4 Profile Learning Manager

This module is one of the nuclear components of the proposed multimedia system, since it aims to learn the interests profile from the target entities (multimedia contents, users, multimedia providers, multimedia authors and user friends groups). In each interaction, the learning manager tries to refine the entity profile based on direct and implicit feedback. The system correlates profile updates from multiple entities in order to discover possible implicit relationships. An example of this occurs when a user gives some feedback (implicit or explicit) about a specific content, making this new information to be used as input not only in the target user profile learning algorithms but also in other related entity profiles, such as target content, author, provider and current user group profile, weighted by metrics that translate user credibility.

In a perfect system with perfect users, all of them would express their multimedia preferences in a consistent and correct way, without noise, errors and inconsistencies (voluntary or involuntary generated). But, in a real operation, the learning system needs to compute, in some way, a trust factor that translates the entity profile certainty and consistency. In the proposed system, we use a statistical certainty value updated in every interaction between the target entity and the system. This approach allows the system to weigh each profile update according to the source profile certainty. The profile learning methodologies that were adopted will be presented in Section 16.4.1.

16.2.5 Data Persistence Unit

This is the most critical module of the proposed system, since it allows the global system synchronization using the stored data. The persistence unit was implemented using the MySQL 5.1.36 database manager and the Java Hibernate 3 object-relational mapping framework. The use of the Hibernate framework gives a high portability degree to the system, in terms of the used database, by creating an abstraction layer between the application and the persistence unit: we built a transversal data access API that is used by each module in its interaction with the database. The current use of MySQL DBMS is only an initial approach to the final system that we want to build: we plan to replace the existing MySQL database by an Oracle database, allowing us to set up a distributed and reliable database architecture, ideal for the proposed service. Using the abstraction layer, a nuclear change can be easily made, without implying any changes in the server modules and only residual changes in the access API.

16.3 IPTV Client

A web-based IPTV client prototype was built in order to communicate with the proposed system, not in a simulated environment but using real users and real multimedia contents. To improve flexibility, scalability and portability, the client software must be platform independent, both in terms of its operating system and hardware, thus improving the universality and convergence of the users access to the content-zapping service. Due to this requirement, the Adobe Flash technology was chosen as the platform that best achieves our purposes.

The client must guarantee access to the multimedia contents that are stored in the multimedia providers and allow the correct interaction between the user and the proposed recommender system. The software that was built uses a set of Flash multimedia players that can handle the different multimedia streams from different providers. The IPTV client software was developed based on a modular architecture design: the client is composed by a set of independent modules that implement each control bar, feature and each multimedia player. When the client starts, it loads the most common players from the providers and a generic multimedia player named Flowplayer. The client software manages the different players and anticipates the multimedia contents by loading according to the current user playlist, that is dynamically generated and updated by the IPTV service according to the authenticated user profile. This pre-loading approach improves fast and smooth transitions between multimedia contents from the same and/or from different providers, and also attenuates some effects of the connection delay, poor bandwidth or high jitter between the client and the multimedia providers. The client was also designed to integrate other services besides the multimedia players: local information, weather forecast, schedule of local movies that are currently in exhibition, RSS feeds reader, short dynamic footnotes and almost any other services can be easily integrated in the client software, using (or not) the available user profile information. The proposed IPTV client software implements all features that are currently provided by the proposed content-zapping service.

By analyzing all the possible targets of our IPTV client, such as Personal Computers (PCs), mobile devices, Set-Top Boxes (STBs) and TV equipment, we concluded that the software should keep a simple interface that needs to be very robust, customizable and highly adaptable to the target device. Some devices have reduced interaction and control features, low memory and low processing capabilities, so the client software needs to be scaled according to the capabilities of each specific device not only at installation time but also during its execution. Figure 16.3 depicts the main IPTV client interface with all menu bars simultaneously active, due to space constraints. The menu bars are active in a mutual exclusive fashion, which means that when the user activates one of the four menus (using a mouse, a keyboard, remote-control or even touch-screen capabilities) and there is another already active menu bar, this will be automatically closed. The user is allowed to adjust all the software and multimedia service settings at any time, including its own dynamic profile or privacy parameters. The client can take advantage of all the device capabilities in terms of features and sensors (accelerometers,

Figure 16.3 IPTV client interface.

temperature sensors, GPS receiver, etc.). The web-based version of the client software was successfully tested on several devices with browser and Adobe Flash Player capabilities, like PCs (Windows, Ubuntu and Mac OS X) and SmartPhones (Android).

16.4 Classification Methodology

Our approach to the intelligent content recommendation problem uses some concepts that were already proposed in the literature and an innovative concept associated with the autonomic classifier evolution. Besides using concepts from usual classifying methods, our approach introduces a classification methodology that can update the classifiers structure by itself in order to discover new classifiers (new dimensions or ramifications) or compress some segments of the current taxonomy, thus decreasing redundant ramification. In a very simplistic way, if our algorithms detects that, in a significant number of situations, the current set of classifiers related to a segment of the current taxonomy is not enough to correctly distinguish between different profile segments that have similar classifiers but a different correlated overall behavior, the system can conclude that it is necessary to create at least one more classifier in order to handle the unknown user profile descriptor. On the other hand, regardless of the user and item in question, if one or more classifiers assume the same behavior and high correlated values in a sufficient number of situations, the system will compress the taxonomy in order to eliminate redundant classifiers that do not correspond to an overall information gain. The articulation between all these concepts and features is currently under research and development.

In the proposed system, the user preferences are described by a finite set of interest areas (classifiers), defined *a priori*, whose elements maintain inheritance relationships between them, in order to extract more knowledge from the structure. The basic principle is that with a finite number of base elements it is possible to create an infinite set of configurations (with new and more complex elements) by conjugating them and varying, in a continuous range, at least one of the parameters that constitute their weights. Besides the classifiers that are initially defined, the system allows the dynamic manipulation and construction of new classifiers based on the previous ones. In order to do that, multiple "base" classifiers are weighted by real values in the standard interval $[0.0, 1.0]$, thus defining new classifying elements that can serve as "base" classifiers for new future elements. All these elements can be dynamically used to describe each entity profile. Each entity is represented by an ordered set of classifiers (tuples), where each one is limited by a minimum and a maximum bound calculated using the classifier certainty values and other dynamic information from the target user. The classifiers are hierarchically organized based on generalization relationships that are, in the simplest case, simple inheritance relationships (simple tree structure) but can also correspond to multiple inheritance relationships (oriented graph).

This approach uses a multidimensional (multivariable) strategy of generic order n. The idea is to define a n-dimensional point that is specific to each entity that has to be classified and, using each classifier certainty, define a probabilistic area (or "cloud") around it. The probability that each user is interested in a content whose cloud intersects its own depends on the intersection degree. This degree should be large enough if we want to suggest that content to this user with a high probability of success. The profile learning approach can use the vector defined by the difference between two multidimensional points (e.g., a user profile and a content profile), to create a controlled displacement for one entity profile depending on other entity profile. The magnitude of the entity profile upgrade should be controlled by parameters such as the classifiers certainty, the context, the previous feedback, etc.

From a mathematical point of view, the areas of interest (classifiers) are organized in generalization relationships that reflect the "social" reality and are unequivocally identified by variables a_i, $i \in \mathbb{N}^{0+} \wedge a_i \in [0.0, 0.1]$. Each user can thus be identified by the tuple $(a_0, a_1, a_2, \ldots, a_{n-1}) \in [0.0, 0.1]^n$,

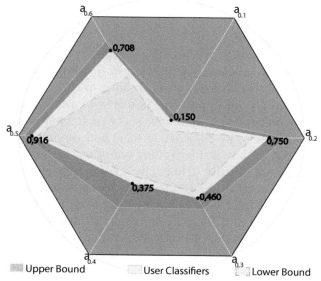

Figure 16.4 Profile representation.

thus defining a point in a n-dimensional positive unitary space. We also generate two different tuples with equal cardinality n, one that serves as the upper
additive bound and the other as the lower subtractive bound of a_i, both of
them in the interval [0.0, 1.0]. The tuples are used to define lower and upper
bounds for each variable associated to an entity. These bounds are calculated
using the respective classifiers, the classification certainty and other dynamic
parameters. After each user interaction, all parameters can be readjusted in
order to reflect the entity dynamic profile, weighing the newly obtained information with all past information that was cumulatively stored. Note that
these weights can also be dynamically adjusted, depending on the relative
weight of the past and present data corresponding to each entity. As an example, Figure 16.4 represents the classifiers with values (0.150, 0.750, 0.460,
0.375, 0.916, 0.708) corresponding to one level of the classifiers tree, for one
specific user at a specific moment in time. Classifiers values near to 0.5 are
not strongly defined, while classifiers near to 1.0 or 0.0 are strongly defined.
Labels "$a_{0.1}$", "$a_{0.2}$", "$a_{0.3}$", ..., unequivocally identify the classifiers, using
their hierarchy composition path.

16.4.1 Profile Learning Methodology

The profile learning is based on two main components: direct knowledge acquisition and implicit knowledge discovery by analyzing the user behavior, which is called implicit feedback [5]. The first component is a direct process, using several kinds of explicit data, like initial profile description, feedback, metadata, direct rating, favorite contents or tagged contents. The second uses deductive methods to infer information from the user interaction with the system, such as the playing percentage for each content, the number of times that a content was replayed, user-to-user content recommendation, among others, in other to obtain more information about the user profile and, consequently, to achieve a better classification accuracy for users and contents. If we have a sufficiently large collection of users with a well-defined profile, based on a sufficient set of contents also with a well-defined profile, it is possible to completely leave contents classification in the hands of trusted users. Using the same approach, if we have a sufficiently large content set with a well-defined profile, we can classify the profiles of new users that joint the system based on the profiles of the trusted contents. Thus, by using the correct parameters to weigh the influence of each stored user and content profile over new users and contents, it should be possible to obtain a stable system that allows a natural and automatic clustering of the previous entities.

16.4.1.1 Direct Knowledge Acquisition

While watching a specific multimedia content, each user can give zero or more direct feedbacks about it at any time, related to the entire content or associated to a specific timestamp. The direct feedback is cumulative and the most recent feedback should always have more relevance than older feedbacks. User feedback is discrete ("4 stars", "love-it!", "hate-it!", etc.) and corresponds to a specific value in the range [0.0, 1.0]. The mapping between the discrete values and the continuous range should not be linear, due to some entropy that can exist around some of the discrete feedback options.

The initial content classification is obtained from the content metadata that is received from the multimedia content provider. This metadata (tags, keywords, etc.) needs to be translated to the internal classifiers using the similarity degree between the metadata and the meta-description of each classifier. In this case, the classifier value (rate) is the similarity degree between the metadata and the classifier tags. The classifier certainty is a constant value in the middle of the admissible range (0.5 by default, in order to allow quick changes in the initial profile).

Algorithm 1 Apply user direct feedback.

Inputs: U is the user structure, C is the content structure and F is the new feedback structure from U to C

procedure APPLYUSERFEEDBACK(U, C, F)

    SAVEFEEDBACKINDB(U, F);

    $C_{ip} \leftarrow$ FINDCONTENTINTERESTSPROFILE(C);

    $U_{ip} \leftarrow$ FINDUSERINTERESTSPROFILE(U);

    if ($\#C_{ip} \neq \#U_{ip}$) **then return** inconsistencyError;

    end if

    $r \leftarrow \{0.0, 0.0, ..., 0.0\}$;

    for $i \leftarrow 0, \#C_{ip} - 1$ **do**

        $r[i] = C_{ip}[i].rate - U_{ip}[i].rate$;

    end for

    $f \leftarrow$ FEEDBACKAGEFACTOR(U, C); $\alpha \leftarrow 1/4$;

    if $f \geq 0.5$ **then**

        if $f < 0.7$ **then** $f \leftarrow f * \alpha$;

        end if

    else

        if $f > 0.3$ **then** $f \leftarrow f * (-\alpha)$;

        else

            $f \leftarrow -f$;

        end if

    end if

    $\tau \leftarrow$ GETCURRENTTIME;

    for $i \leftarrow 0, \#U_{ip} - 1$ **do**

        $U_{ip}[i].rate \leftarrow U_{ip}[i].rate + r[i] * f$;

        $U_{ip}[i].updatedate \leftarrow \tau$;

        SAVEUSERINTERESTSPROFILETOBD($U, U_{ip}(i)$);

    end for

end procedure

function FEEDBACKAGEFACTOR(U, C)

    $F_l \leftarrow$ FINDCFEEDBACKS($U, C,$ *"order by date desc"*);

    $j \leftarrow 0; \kappa \leftarrow 0; \Gamma \leftarrow 0; \rho \leftarrow 1.2$;

    for all $f \in F_l$ **do**

        $\gamma \leftarrow 1/\rho^j; \Gamma \leftarrow \Gamma + \gamma$;

        $\kappa \leftarrow \kappa + \gamma * f.rate; j \leftarrow j + 1$;

    end for

    return κ / Γ;

end function

The system saves all users' feedbacks and associated information. Due to that, it is possible to rollback a "bad" cumulative learning step at any moment. During a content reproduction, our approach allows the user to tag specific sections of the content, in order to improve the granularity and accuracy of the content feedback. Using the information contained in each tag, the system generates an automatic feedback related to the requested tag with two timestamps: one in the target content time referential and another in the time-zone referential.

Algorithm 1 shows the general technique that describes the update of the user interests profile after a user direct feedback related to a visited content, considering direct knowledge acquisition. The feedback is always given in the continuous interval [0.0, 1.0], as well as the classifier rate and certainty. The algorithms that are used to compute other direct knowledge sources are similar to this one, being only distinct in their input parameters and their corresponding meanings. Algorithm 1 is very simple and based on direction vectors that are computed using the difference between the content profile and the user profile. Other more complex and accurate approaches are currently being under development and test.

The first task of Algorithm 1 corresponds to saving the feedback F of the new user on the database; then, the algorithm fills variables C_{ip} and U_{ip} with the content and user interest profile, respectively. If there is an inconsistency in the previously loaded lists of classifiers, the algorithm returns an error; next, variable r (a vector with the same dimension of C_{ip} and U_{ip}) is initialized; then, the algorithm computes the vector direction that approximates user U to the content C and saves it in variable r; next, the algorithm computes the feedback age factor. This value is a weighted average of all feedbacks from user U for content C, where the most recent feedbacks have more impact (see sub-algorithm *feedbackAgeFactor*); next, due to the entropy around the median value of the considered range ([0.0, 1.0]), the algorithm gives more importance to feedback values that are located around the ends of the scale; finally, the algorithm applies the previously computed direction vector r to the current user profile, modulating the profile shifting intensity by the last feedback and the feedback age factor. Note that all the α and ρ values were empirically tuned in order to balance the user profile update gain and the feedback age, respectively.

16.4.1.2 Implicit Knowledge Discovery

A significant number of the implemented recommender systems uses only direct knowledge acquisition. This approach produces good results but can

Table 16.1 Parameters used in the indirect knowledge discovery.

Parameter	Transfer function	Description
ω factor	$\omega = \frac{\upsilon}{\tau}$	relationship between the content duration (τ) and the watching time (υ)
ψ factor	$\psi = 1 - \frac{1}{\xi^{\tau}}$	distinction factor for the length of time content; τ is the content duration and ξ is the gain (]0, 1[)
r factor	$r = \frac{1}{\upsilon^{n}}$	replay watched contents factor. n is the number of replays and υ is the gain (]0, 1[)
s factor	$s = \frac{1}{\chi^{n}}$	seek in the content factor; n is the number of consecutive seeks and χ is the gain (]0, 1[)
f factor	$f = 0.9$	a constant value that give more importance to the contents that was been searched by the user
p factor	$p = 1.0 \vee p = 0.0$	a binary factor that is 1.0 if it is a paid content or 0.0 if it is a free content

be significantly improved by adding an indirect knowledge discovery component, taking advantage of small user-system interaction details, without overcharging the user with feedback duties. In this work, we gave more emphasis to the indirect knowledge acquisition using implicit feedbacks, that is, non directly-qualifying user behaviors, which allows the user to face the multimedia system in a natural way, without having too much concern about giving his opinion regarding the watched content.

This type of knowledge can be intrinsically subjective. All implicit information used for building the profile is weighted using parameters that are empirically fine-tuned. These parameters are global for all entities in the system. The parameters that were identified in order to obtain more information about the user profile are presented in Table 16.1. Each parameter has a transfer function, with a specific domain and co-domain in the interval [0.0, 1.0], that translates the weight of the parameter to the overall profile update. Due to some limitations in terms of available space, we will not describe the algorithm that handle the impact of an implicit feedback in the target entities profiles, because it is very similar to Algorithm 1. The most significant differences lie in the gain parameters, α and ρ, that were adjusted to give less importance to implicit feedback when compared to direct feedback. In the implicit feedback case, parameter F of Algorithm 1 is now the result of the transfer function of each implicit knowledge factor described in Table 16.1. Due to the subjective connotation of the implicit feedback, the system gives a lower weight to those factors when updating the entities profiles.

16.4.2 Recommendation Methodology

Parallel to the user profile learning mechanisms, we need a methodology that is able to suggest at each moment the most appropriate content to a specific user, taking into account not only the current user profile but also the dynamic context/environment. Due to the usually fast transitions between different contexts, suggestions related to the user context must be *short-term* recommendations. On the other hand, suggestions based on the user profile should be *mid-term* recommendations, since this profile cannot change abruptly.

The context needs to be parameterized in order to be mathematically described. Information such as the user spatial location, the content spatial location, the weather at the user location, the place, the watching mode, the noise level, the light intensity, the user movements, the client device capabilities, and so on, need to be translated into parameters that will be inputted to the context-dependent algorithms. At this time, we are describing and parameterizing the context data in order to build context-dependent algorithms that will improve the multimedia content suggestion.

Contents that were suggested by other users ("friends" of the target user) must have a higher initial priority, when compared to the recommendations triggered by the system, in order to initially increase the user-to-user suggestion priority. For each pair or users, the user-to-user recommendations priority will be iteratively readjusted according to the behavior of each user. Considering only the user profile-based recommendations, the system implements a simple algorithm that selects the best contents for each user according to the Euclidean distance between the user profile and each content profile. Considering the profile definition described in Section 16.4 and assuming that U is the target user profile given by $U = (u_0, u_1, u_2, \ldots, u_{n-1})$ and C is the target content profile given by $C = (c_0, c_1, c_2, \ldots, c_{n-1})$, the distance between U and C is given by

$$d_{U,C} = \sqrt{\sum_{i=0}^{n-1} (u_i - c_i)^2}.$$

A small $d_{U,C}$ Euclidean distance corresponds to closer profiles in terms of classification and, therefore, content C should be more interesting to user U. Algorithm 2 implements this approach by selecting the N most appropriate contents, sorted in ascending order of their Euclidean distance, for target user U from the available contents. User profile-based suggestions can be generated both in real-time and pre-processing mode. In other words, the

---

Algorithm 2 Select the best not viewed contents.

Inputs: U is the user, N the maximum number of selected contents
procedure TOPN(U, N)
    $\varepsilon \leftarrow$ MINCURRENTSUGGESTEDDISTANCE(U);
    $C_l \leftarrow$ NEARESTCONTENTSNVNS(U, ε, N);
    if ($\#C_l = 0$) **then**
        if NUMOFSUGGESTEDCONTENTSINDB(U) $= 0$ **then**
            $C_l \leftarrow$ BESTRATEDCONTENTS($U, +\infty, N$);
        end if
    else
        return
    end if
    $\tau \leftarrow$ GETCURRENTTIME;
    for $i \leftarrow 0, eta - 1$ **do**
        SAVESUGGESTEDCONTENTTOBD($U, C_l(i), \tau$);
    end for
end procedure

---

system can recommend contents based on the user profile even if the user is not online and, consequently, not requesting suggestions.

The first task of Algorithm 2 consists of calculating the minimum distance (ε) between user U and the current list of suggested and not viewed contents for user U. If there are no suggested contents for user U, the minimum distance (ε) is equal to ∞; next, the algorithm gets a list of the contents that were not yet viewed and suggested to the current user U, sorted in ascending order of their Euclidean distances, and have a distance that is smaller than the previously calculated minimum distance ε; next, it evaluates if the previously calculated list of contents C_l is empty or not. If C_l is empty but there are suggested contents to user U in the system, the algorithm ends, because the system already has the best available contents suggested to user U and just needs to wait for the user to view the currently suggested contents in order to capture more feedback about them or wait for new multimedia contents that will fit even better the current user profile U. If C_l is empty and the current suggested content list is also empty for user U, this means that user U has already watched all the contents in the system. In this case, we cannot give him new contents for now, so the algorithm will suggest the best rated contents (not necessarily the nearest contents for the target user profile) that are indexed in the system, in other to capture again a feedback for these contents until the system has new contents available; finally, the algorithm saves on the database the list of contents C_l that was previously selected for user U.

16.5 Performance Tests

The system simulation proceeds through phases, where each phase is composed by several tasks that have to be completed before entering in the following phase. In terms of hardware, we used two *Asus* rack servers (named *iptvserver1* and *iptvserver2*), both equipped with an Intel Xeon 3.0GHz 64bit dual-core processor, 4GByte of RAM, two 500GB SATA (Serial Advanced Technology Attachment) hard discs using RAID1 (Redundant Array of Independent Disks) and a 100Mbps/s Ethernet connection. The operating system was Ubuntu 8.04.4 LTS. The Java SE Runtime Environment, version 1.6.0_17 (64bit version), was installed and MySQL version 5.1.36 was used. The complete system (all modules without service replication) was installed in the *iptvserver1* machine, while the simulator was executed in *iptvserver2*. In order to simulate the network delay and jitter, the servers are not directly connected, although the simulation runs inside the same private network.

The first simulation step (step 0) was the database initialization with basic pseudo-static data. In this step, the system also generates the XML file of the classifiers tree and inserts it in the database. The classifiers tree structure is randomly generated according to some parameters, like the maximum allowed number of children at each node and the maximum depth, creating random relationships between nodes (classifiers). For now, we use a flat list of classifiers, in which the maximum number of elements is calculated by $e = c^d$, where c is the maximum allowed number of children at each node and d is the maximum tree depth. For this simulation we only considered 20 different classifiers.

Next (step 1), the simulator generates one XML file per each simulated user, containing its complete, accurate and objective profile. The maximum and minimum number of simulated users simultaneously authenticated on the system are 555 and 5, respectively. In step 3, the system will insert some users (55) from the complete pool of users: those that remain will be added progressively by the simulator during the simulation in order to recreate the continuous user registration process in real-world applications. The XML file of each user profile is composed by the typical personal data (name, gender, etc.) and a list of all classifiers that are stored in the system. Each classifier value (rate $\in$ [0.0, 1.0]) is randomly generated according to a normal distribution ($\mu = 0; \sigma = 0.1$) in its domain. A Bernoulli random variable ($p = 0.35$)

is generated in order to randomly set each user profile classifier as "visible" or "not visible" for the multimedia system. This means that when the simulator loads the XML file of each user profile (user registration process emulation), only a set of classifiers will be "visible".

In the next procedure (step 2), one XML file is generated for each simulated multimedia content that we want to consider. Each content profile is also generated in a complete, accurate and objective way, without any noise or error. Initially, the system only inserts 6553 items from the total set of 65536 randomly generated contents profiles. During the simulation, the content source manager module progressively inserts the remaining generated contents. Each classifier value (rate $\in$ [0.0, 1.0]) corresponding to each content profile is randomly generated according to a normal distribution ($\mu = 0$; $\sigma = 0.1$) and all content profile classifiers can be "visible" or "not visible" to the multimedia system according to a Bernoulli random variable ($p = 0.35$). The next step (step 3) inserts the initially required number of users and contents in the multimedia system. All the "visible" profile information, in both cases, can be disturbed using statistical noise, in order to emulate the real-world incomplete and inaccurate data.

Step 4 launches all the initial users life-cycle and starts the users life-cycle logging. The simulator generates new user authentications or waits for the end of some of the current authenticated users, according to a normal distribution ($\mu = 55$; $\sigma = 5$) in the admissible range [5, 555]. Using the definition file of each user and a list of well-defined random variables, the system generates specific random parameters that will deeply influence each simulated user life-cycle, changing it according to the respective user profile and adding a small random effect. The last step is optional: we implemented a learning analyzer module (step 5) and a GUI (Graphical User Interface) that can give us a better feedback about the simulation process in real-time and allow us to have a better control over the simulator core (step 6). In this initial approach, only the user profile will be adapted after each implicit or explicit feedback. The contents are fixed points in our multidimensional world defined by the classifiers.

In this simulation, we only used 20 different classifiers in the range [0.0, 1.0]. Due to that, the maximum distance between any profile is $\sqrt{20} \approx$ 4.47. In order to compare any two profiles, we always used percentages relative to the maximum allowed distance between any profile. So, a percentage close to 100% means a poor classification improvement (Euclidean distance is high), while a small percentage means a good classification improvement (Euclidean distance is low). Two distinct test sets were conducted:

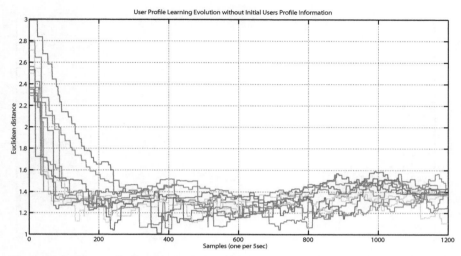

Figure 16.5 User profile learning evolution during the simulation, considering the *cold-start* problem.

user registration (initialization) without any initial information about the user preferences (simulation of the *cold-start* problem) and user registration using some information (perturbed) about the user interests profile (initial classification). Figures 16.5 and 16.6 show the real-time evolution, in terms of Euclidean distance, between the current profile (located in the system) and the real profile (located in the XML files) for a group of 10 users randomly chosen from the pool of users initially inserted in the simulator. For both situations, the users represented in Figures 16.5 and 16.6 are the same.

Comparing the real, the initial and the final user profiles in the system, we can express the results by dividing them in two segments: without the *cold-start* problem and with the *cold-start* problem. In the first case, the average distance between the initial users profiles and the real profiles was equal to 1.68 (38% of the maximum distance) and the final distance between the refined users profiles and the real profiles was 1.16 (12% of the maximum distance). With the cold-start problem, the average distance between the initial users profiles and the real users profiles was equal to 2.97 (66% of the maximum distance) and the final distance between the refined users profiles and the real profiles was 1.20 (27% of the maximum distance).

If we analyze both Figures 16.5 and 16.6, it is possible to observe that the proposed system would be able to converge even more to the real user profile in an inverse exponential style (decreasing the Euclidean distance),

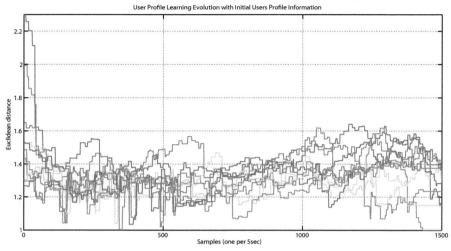

Figure 16.6 User profile learning evolution during the simulation, without considering the *cold-start* problem.

if the system had more contents that could better fit the user profile. Due to the limitation we have imposed, the last iterations between the user and the system were not very constructive in terms of user profile learning. On the other side, this limitation was useful to study the impact of bad suggestions in the user profile learning process. Due to the fine tuning of Algorithm 1, the learning process does not suffer a major degradation when the remaining contents did not best fit the target user. We are currently conducting other types of tests in order to evaluate other aspects of the system performance, like the contents aggregation performance and the initial classification accuracy.

16.6 Conclusions

The proposed system is a work in progress in terms of the used and tested algorithms and in terms of the techniques for profile learning and content recommendation. We have already built a robust system that can be used to virtually apply and test any algorithm and technique both for profile learning and item recommendation. Knowing that there are periods of intense message exchange, database access and processing during the simulation and testing operations, the system behaved in a very stable way, without compromising the algorithms simulation and without outputting any unexpected error.

The simulation results follow our expectations. We conclude that, even using simple learning and suggestion mechanisms, we were able to achieve a good user profile classification improvement (in terms of Euclidean distance), of approximately 26% (scenario without *cold-start* problem), when comparing the user profile that was initially inserted into the system and the final profile after the simulation. For the other scenario (with *cold-start* problem), the system took more time to reach an equivalent improvement degree, but we were also able to achieve an improvement near 30% when comparing the initial and the final user profiles.

When we gave the system some information about the initial user profile, the proposed system was able to rapidly converge to the best possible user profile representation. Due to the learning algorithm limitations, we cannot improve the final user profile representation, even with the initial user profile data. To archive a better profile convergence, we need to use each classifier certainty, the classifiers tree structure meta-information, the user context and other features that the proposed system offers, trying to correlate the classifiers corresponding to all entities in the system.

References

[1] Shlomo Berkovsky, Tsvi Kuflik, and Francesco Ricci. Enhancing privacy while preserving the accuracy of collaborative filtering. In *Proceedings of ECAI 2006 – Workshop on Recommender Systems*, pages 49–53, August 2004.

[2] Robin Burke. Hybrid recommender systems: Survey and experiments. *User Modeling and User-Adapted Interaction*, 12(4):331–370, November 2002.

[3] Davide Grossi, Frank Dignum, and John-Jules Meyer. Contextual taxonomies. *Computational Logic in Multi-Agent Systems*, 3487:33–51, August 2005.

[4] Jonathan Herlocker, Joseph Konstan, Loren Terveen, and John Riedl. Evaluating collaborative filtering recommender systems. *Transactions on Information Systems*, 22(1):5–53, January 2004.

[5] Henry Lieberman. Letizia: An agent that assists web browsing. In *Proceedings of 14th International Conference on AI*, pages 924–929, Montreal, Canada, 1995.

[6] Paolo Massa and Paolo Avesani. Trust-aware bootstrapping of recommender systems. In *Proceedings of ECAI 2006 – Workshop on Recommender Systems*, pages 29–33, August 2004.

[7] Michael O'Mahony, Neil Hurley, and Guenole Silvestre. Attacking recommender systems: The cost of promotion. In *Proceedings of ECAI 2006 – Workshop on Recommender Systems*, pages 24–28, August 2004.

[8] Seung-Taek Park and Wei Chu. Pairwise preference regression for cold-start recommendation. In *Proceedings of RecSys '09*, pages 21–28, October 2009.

[9] Ben Schafer, Joseph Konstan, and John Riedl. Recommender systems in e-commerce. In *Proceedings of the 1st ACM Conference on Electronic Commerce*, pages 158–166, ACM, 1999.

[10] Janusz Sobecki. Implementations of web-based recommender systems using hybrid methods. *International Journal of Computer Science & Applications*, 3(3):52–64, 2006.

[11] Xiaoyuan Su and Taghi M. Khoshgoftaar. A survey of collaborative filtering techniques. *Advances in Artificial Intelligence*, 1–19, January 2009.

PART 4

SECURITY

17

Security Challenges in Emerging Service Architectures

Masayuki Terada

Research Laboratories, NTT DOCOMO, Inc., Yokosuka, Kanagawa 239-8536, Japan; e-mail: teradam@nttdocomo.com

Abstract

The recent proliferation of wireless and wired broadband Internet access environments and the evolution of computing technologies have induced the emergence of many innovative services based upon new service platform architectures. These service architectures offer exciting business opportunities and new lifestyles that are beyond our experience. However, for every head there is a tail – they also pose new kinds of problems that can damage society. Mainly focusing upon *user-to-user* digital content distribution and the utilization of so-called *big data*, this chapter overviews the characteristics of the anticipated service architectures and their negative impacts, and discusses the security challenges raised by the emerging services.

Keywords: user-generated content, big data, lemon problem, privacy.

17.1 Introduction

The recent proliferation of wireless and wired broadband Internet access environments and the evolution of computing technologies have induced the emergence of many innovative services based upon new service platform architectures. These service architectures will likely offer exciting business opportunities and new lifestyles that are beyond our current experience. But

Anand R. Prasad et al. (Eds.), Advances in Next Generation Services and Service Architectures, 385–405.

unfortunately, there are two sides to every coin. These emerging services also pose new kinds of problems that can damage our society.

For example, the recent spread of broadband access and the evolution of digital content creation instruments are redirecting the Internet towards being more *user-to-user* oriented. This trend dynamically reforms the service architecture of content distribution. In particular, user-generated content (UGC), self-made content created and distributed by ordinary people outside of professional practices, greatly impacts the content distribution architecture – both positively and negatively.

The diversity and cost-efficiency of UGC indubitably enhance our cultural life. In addition, the utilization of UGC by businesses, namely crowd-sourcing, is expected to enable firms to inexpensively mobilize the collective intelligence of thousands of experts in the world. On the other hand, the creators of UGC are usually not paid for their creative efforts. This may prevent UGC from experiencing sound and sustainable growth, and may even damage the labor market for creative jobs; that is to say, user-generated un-employment. A possible solution is to promote a UGC market where creative work is sold at fair prices, but such a market will be doomed to collapse by Akerlof's *lemon market* effects [5], caused by the uneven quality of UGC and the consequent mutual distrust between a user and a creator.

Another example is utilization of so-called *Big Data*, such as log data of business activities and customer interactions, which potentially contains valuable knowledge to increase customer loyalties and business opportunities. Big data was considered too large to work with using traditional systems, but the recent evolution of data mining and cloud computing technologies has made big data easier to deal with. A good statistician with large-scale parallel computing techniques is now able to mine useful knowledge from piles of data at reasonable cost by using cloud computing systems. Big data is said to be "changing the rule" [2] of conducting business activities.

As the flip side of the coin, skeptics are concerned that big data can lead to "Big Brother"-like situations. The utilization of big data can degrade our privacy since big data also contains a lot of information that relates to privacy of customers or users. Privacy breach can happen even if the data miner who extracted some knowledge from big data was not willing to disclose someone's privacy. The extracted knowledge itself may involve privacy information that can be linked to individuals by further data mining processes.

These negative impacts of emerging services should be mitigated by security technologies. Resolving mutual distrust between creators and users is required to establish a fair and sustainable market for UGC, and assurance of preserving privacy to the respondents in datasets must be provided to utilize big data dispensing with concern about "Big Brother"-effects. However, they cannot be fully addressed by the traditional security approach, providing a solution to *prevent an adversary from doing something undesirable*, and pose new security challenges involving difficult but interesting trade-offs that should be addressed with considering the social and economic background of the problems: a trade-off between content quality risks of users and the risks of creators not being paid, and a trade-off between privacy preservation and data utility.

Mainly focusing on these two emerging trends in service architectures, user-to-user and big data utilization, the remainder of this chapter overviews the characteristics of emerging services and their negative impacts, and discusses the security challenges raised by these services.

17.2 Characteristics of New Service Architectures

17.2.1 User-to-User

The notion of "user-to-user" can be roughly divided into two parts; one at the network level and the other at the content level. The former is better known as peer-to-peer (P2P) networking, while the latter is exemplified by users who are not only consumers of content but also generators of content, also known as user-generated content (UGC). This chapter mainly focuses on the latter, which is likely to yield stronger social impacts.

User-generated content (UGC), also known as user-created content (UCC) or consumer-generated media (CGM), has been rapidly proliferating due to the recent penetration of wired or wireless broadband Internet access. According to surveys by OECD, 26% of Internet users in the US had published self-made "artwork, photos, stories or videos" as of 2006 [28, 29]. eMarketer also reported that 82.5 million people (43% of Internet users in the US) had published content as of 2008 and they estimate that this number will reach 114.5 million in 2013 [10, 27].

There is no widely accepted definition of UGC (or UCC/CGM), but the definition in OECD reports [28, 29] well captures the distinctive features of UGC,[1] which is defined as content that

1. is made publicly available over the Internet,
2. reflects a "certain amount of creative effort", and;

---

[1] To be exact, Vickery and Wunsch-Vincent in [28, 29] refer to UGC as UCC.

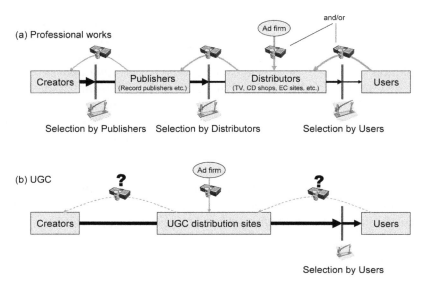

Figure 17.1 A comparison of (a) the conventional value-chain for professional works and (b) the value-chain for UGC.

3. is "created outside of professional routines and practices".

While the first and second conditions simply exclude private (i.e., not publicized) content and non-creative works, the last condition is the most important characteristic that distinguishes UGC from professional works.

Here *professional works* do not mean the content created by professional artists, but any content that becomes available through "professional routines and practices" such as the selection and review processes performed by publishers, who are responsible for paying creators for their content and who guarantee the quality of published content to the buyers (users). This is realized by filtering out the content that is considered to have little mass appeal (also called "long tail" content) and thus few sales. It is an intrinsic and important characteristic that UGC is published outside these conventional processes because the absence of such hurdles in the publication process of UGC offers much more diversity and lower distribution costs (Figure 17.1).

The utility and impact of UGC have lately attracted not only Internet users but also more traditional content holders, who initially treated UGC as a menace to their businesses mainly because of concerns about literary piracy. This initial fear has abated somewhat and examples of a more cooperative approach include the British Broadcasting Corporation (BBC), which is providing content under the Creative Commons (CC) license, and Kadokawa-shoten, which is acknowledging superior user-created secondary works (a.k.a. mash-ups) as "official" secondary works. As these examples show, content holders of professional content, the opposite of UGC, have started to seek ways to utilize UGC for the promotion of their businesses, rather than remaining hostile to it.

Another trend in the utilization of UGC by businesses is *crowdsourcing* [13], a portmanteau created from *crowd* and *outsourcing*, which means the act of taking a job traditionally performed by a designated agent (usually an employee) and outsourcing it to an undefined, generally large group of people in the form of an open call [14]. Crowdsourcing is said to enable firms to inexpensively mobilize the creative works of sometimes highly skilled people as a resource for the generation of value and profits [15].

17.2.2 Big Data

More and more of our daily personal and business activities are being conducted on computer systems. These activities are continuously producing huge amounts of data, namely *Big Data*, which must contain a lot of valuable information. An interesting chart titled *Data inflation* in [1] comprehensibly illustrates that the amount of data surrounding us is rapidly escaping all forms of human understanding. According to this chart, the sum of all the cataloged books in America's Library of Congress, a great compilation of the wisdom of mankind, totals 15 TB.[2] This can be stored on eight 2 TB hard disk drives, whose current street price is no more than about $100 each. In contrast, the amount of data on the Internet is mind boggling. According to the chart again, Google processes around 1 PB (Petabyte = 1,000 TB) *every hour*, and the total amount of information in existence this year is forecast to be around 1.2 ZB (Zettabyte = 1,000,000 PB).

Most of this "big data" is either archived or just intermittently deleted since it has been difficult and expensive to successfully mine useful knowledge from them. Recent advances in the fields of data mining and scaling-out[3] parallel computing technologies have been changing this situation. The most famous company to utilize big data must be Google, whose success largely depends on the scaling-out technologies to efficiently analyze and rank a huge amount of web pages. Other precedents include on-line

---

[2] The prefixes here (T, P, and Z) are SI prefixes; e.g., 1 KB = 1,000 bytes ($\neq$ 1,024 bytes).

[3] An approach to employ multiprocessor systems for computing. To *scale-out* (or *scale horizontally*) means the deployment of applications on multiple small servers [17].

shopping sites such as Amazon.com and Netflix, which provide personalized recommendations to each customer typically calculated from collaborative filtering of their massive sales data.

The tools required to work with big data, those for data mining and scaling-out parallel computing, are rapidly being commoditized. A good example is the open-source combination of Hadoop and R, a scalable distributed computing framework for processing large data sets on clusters of computers and a programming environment for statistical computing and graphics. This combination and emerging cloud computing platforms enable almost anyone who has moderate skills in programming and statistics to resolve large-scale business problems at reasonable cost – basically just the fee for renting cloud time. A quick check in July 2010 found that the cheapest plan cost only $0.029 per hour per CPU instance.[4] These tools and situation have started to attract the attention of not only IT engineers but also business people. The Economist, a famous weekly news magazine mainly focusing on international economic and politic affairs, recently published a special report [2] consisting of nine articles. They discussed the business trends in and the impact of the age of big data, while introducing a number of advanced business practices as examples.

17.3 Negative Impacts

The innovation of these emerging services will most likely make our daily life more convenient and more efficient, but they are also likely to pose severe challenges to society.

The elimination from the value-chain of the intermediaries who were responsible for paying creators as well as ensuring the quality of distributed content raises the specter of *unfair remuneration to creators*, which is thought likely to damage the current market for creative works and consequently deprive creators of decent jobs. *No quality assurance of content* is another risk; it will make the quality of content reaching the user extremely uneven thus obstructing the establishment of a sound market for UGC.

Big data contains sensitive information and so poses *privacy disclosure risks*. This problem is exceedingly difficult to resolve since privacy disclosure occurs not only by the direct disclosure of sensitive data, but also by the

---

[4] Amazon EC2 for *standard spot instances*; price "fluctuates periodically depending on the supply of and demand for spot instance capacity" [6].

mining of insufficiently "anonymized" data. This is discussed in detail in a later section.

17.3.1 Unfair Remuneration Risks to UGC Creators

The direct delivery of content in the UGC value-chain, dispensing with the selection processes by the publishers and distributors, enables the users to enjoy a wide variety of content. However, the trade-off is that the creators[5] have difficulty in receiving compensations for their creative efforts.

If the situation where creators are not fairly remunerated in the UGC value-chain is not changed, the future growth of UGC and its corruption of the conventional value-chain will severely challenge creators, no matter whether they provide UGC or not. Upon its first mentioning in [13], crowdsourcing was introduced as "the new pool of cheap labor: everyday people using their spare cycles to create content, solve problems, even do corporate R&D (research and development)", and introduced several stories of professionals who had lost their jobs as a result of crowdsourcing. Some critics claim that crowdsourcing is "the equivalent of slave labor" and will severely damage the job market as a result [18].

Other critics have a different opinion; it is a natural consequence of the free market principle to substitute the employed or contracted creators with crowds, and there is little problem since the substitution will likely create a new world-wide job market for user-generated works. This discussion postulates the existence of a sound market where the creators in the crowds can be fairly remunerated, but it is likely to differ from reality, at least currently. There are several large UGC distribution sites such as YouTube, which are attracting the participation of users from around the world. A number of excellent user-generated videos in these sites entertain great number of people in the world,[6] which may often exceed the viewership of typical TV programs. However, their creators are rarely remunerated, unlike the creators of professional works broadcast on TV.

---

[5] UGC creators are also "users" by definition (i.e., UGC stands for *user*-generated content), but this chapter refers to users who created content as *creators* so as to avoid confusion.

[6] According to YouTube official charts in October 2010, the most viewed video in YouTube has approximately 360 million views and about 100 videos have more than 50 million views.

17.3.2 Content Quality Risks to UGC Users

Another penalty of dispensing with the selection processes is that the consumers of UGC have no guarantees of content quality. The absence of the traditional gatekeepers yields the diversity and low distribution costs of UGC as mentioned before, but as a natural consequence, the quality of the published content, UGC, is highly uneven; that is to say, UGC is inherently a mixed bag.

Professional works are distributed via conventional value-chains and the "brand names" of the creators and/or the publishers provide some implicit guarantee of quality, since established publishers have kept their brand name by continuously filtering out inferior content so as not to damage their brand. In contrast, no one selects which UGC is published. The consumer has very little guidance in choosing which content to view. Most UGC sites provide reputation information such as "ratings by viewers", the quality of which is also uneven. This is currently not a big problem since most UGC users pay nothing for content, but it is most likely to become one of the key obstacles in establishing a market for UGC, where UGC creators can be fairly remunerated, as discussed in the next section.

17.3.3 Privacy Disclosure Risks Posed by Big Data

While big data is a rich source of valuable information for improving the quality of our daily life, it also contains a lot of information that we want to keep private. Access to this data must, therefore, be carefully controlled so as not to permit any privacy breach. This is not so easy to achieve mainly because of the difficulty of removing all information that (potentially) could cause a privacy breach, without spoiling the usefulness of the data.

Narayanan and Shmatikov [19] applied their de-anonymization methodology to the Netflix Prize dataset, which contains the anonymous movie ratings of 500,000 Netflix subscribers, and concluded that they had successfully identified the Netflix records of known users, uncovering their apparent political preferences, and other potentially sensitive information, by using the Internet Movie Database as the source of background knowledge. Of course, Netflix would never intentionally publish its "anonymous" data, but the attack demonstrated that sophisticated data mining techniques can be effective in extracting private information from data.

While the Netflix case is an example of a privacy breach due to (insufficient) anonymization, similar threats can be made to public data such as profile and friendship information in SNS (social networking services) sites.

If a user does not want to make his or her account on an SNS site identifiable, i.e., the user's name or address cannot be linked to the account, the user may keep such information confidential. However, this is not enough to preserve the privacy of the user; an adversary can mine the public data in SNS, information willingly provided by users. For example, given two social graphs, one is anonymous while the other is autonymous, their collation can re-identify people who are involved in both graphs. Another work of Narayanan and Shmatikov [20] showed that "linking" the data from SNS sites can successfully identify a fairly large portion (about 30% in their experiment) of the users. This result means that just keeping identifiers (e.g., the name and address of the user) confidential is not sufficient to protect privacy, which can be mined from public data especially if the data contain information linkable to another database.

17.4 Security Challenges

The problems derived from the emerging services introduced in the previous sections show that they have the potential to negatively impact our economy and society. These problems are not so easy to resolve since they involve significant and difficult trade-offs.

The unfair remuneration problem for UGC creators may be addressed by providing a market platform for UGC where creative works can be sold at reasonable prices as professional works, but the unevenness in content quality makes the consumer unwilling to buy anything, and accordingly a sound market can never by established. Relying on voluntary donations from consumers reduces the user's risk but increases creator's of not receiving proper payment for valuable works.

The privacy problem raised by big data involves another trade-off; a trade-off of privacy and data utility. This trade-off arises from the difference between anonymization and encryption. Content of an encrypted message has to be readable only for the authorized viewers in possession of the corresponding decryption key, and its confidentiality must be kept against third-parties. On the other hand, access to anonymized data is shared by others who want to extract some useful knowledge from the data, and the privacy must be preserved against them. That is, anonymized data must provide enough information to extract useful knowledge to the recipients, while it must not provide them with data that have been obtained by breach of privacy legislation. These somewhat contradictory requirements lead to a seemingly impossible result: a notion of "perfect privacy", nothing about individuals is learnable from a database that cannot be learned without access to the database. This is impossible to achieve.

17.4.1 Resolving Risk Trade-off between Creators and Consumers

To fairly remunerate creators for their creative efforts, one solution is to monetize UGC. Given the characteristics of the UGC value-chain described earlier, however, it is not so easy to fairly remunerate creators; the uneven quality of UGC becomes an impediment.

While UGC is rarely charged, professional works are widely sold in digital format by digital content distribution services; a few examples include iTunes Store for iPod portable players, Amazon MP3 for PCs, and "Chaku-uta" ring tone services for Japanese 3G mobile phones. In these services, users pay to acquire (download) content like purchasing physical media such as CDs or DVDs. Why not sell UGC in a similar manner and share the profits with its creators? If it succeeds, the creators of excellent works may have major sales and be remunerated well.

This direct monetization of UGC, known as promoting the "UGC store", is also known as the *charging for download* model. This model is simple and natural, however, applying this model straightforwardly to UGC burdens the consumer with too much risk with regard to the quality of the content as mentioned before.

Unlike professional works, whose quality is assured by publishers and distributors, the quality of UGC solely depends on the skill and motivation of the creators who include anonymous professionals, rising indie producers, amateur creators and, maybe, malicious fraudsters. This inherent unevenness in UGC quality yields brilliant and unprecedented content, but users are forced to run the risk of paying for undesirable content since they cannot judge the quality of the content beforehand; this is quite unfair to the user.

Such an unfair market, where the quality of merchandise (i.e., content) is highly uneven and unknown to the buyer (i.e., consumer) before making the purchase, is known to yield *adverse selection* by consumers and creates a *lemon market* [5]. In such a market, the buyer has to estimate the quality of merchandise, but his best guess is the average of the merchandise because of the uncertainty of quality. The average quality as estimated by users might be rather inferior in the UGC market, since the few brilliant works are surrounded by many more crude ones.

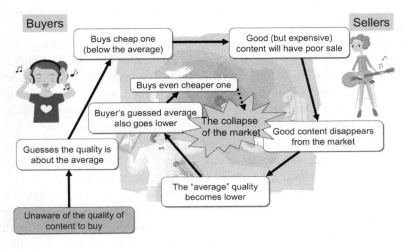

Figure 17.2 The death spiral of the lemon market.

A consumer will, accordingly, tend to purchase cheap content to avoid the quality risk regardless of its real quality (adverse selection), and therefore superior (but rather expensive) content provided by "good" creators will disappear from the market. As a result, the market will be full of cheap but inferior content, i.e., the market becomes a market for lemons (Figure 17.2). This result is, ironically, contrary to the motivation of monetizing UGC, i.e., facilitating the sustainable and sound growth of UGC by appropriately compensating the creators.

The lemon problem is a consequence of placing too much risk on the consumer, who has to pay for content without knowing its quality. Then, how about allowing users to pay for content after discovering its quality? That is, content is available for free and a user voluntarily pays for content if he feels that the content is good enough, just like donations to street performers.

This model, often called *voluntary donation*, is rarely used for professional works, but is common in free-software distribution sites with a "donate" button. A user who enjoyed the content and decides to donate to the site owner pushes the button and pays the site owner for the content via an online P2P payment scheme such as PayPal. A significant advantage of this model is that the user can completely avoid any risk of paying for useless content since payment is made only if the user enjoyed the content. However, this model obviously creates many "free riders" and is unfair for creators (and users who made donations); it is unlikely to resolve the problem of too little remuneration to UGC creators.[7]

A famous example of this approach is the album "In Rainbows" by the English rock band Radiohead. It was first released on October 10, 2007 from the band's official website, and was downloadable from there until November 3, 2007. Fans were asked to pay what they wanted to by credit card when downloading the album, and those who were not willing to pay were asked to give their email addresses. This case is somewhat an extreme case, since the creator, Radiohead, was an established band with a very good brand reputation, and the album, which was released over four years after their previous album, was one that their loyal fans had been eagerly waiting for.

According to a study by comScore [7], however, 62% of those who downloaded the album from Radiohead's official website chose to pay *nothing* even for this album, and the average amount was no more than $2.26. Furthermore, there were many more freeloaders who used illegal downloads, even though they were allowed to legally download the album for free from the official site. Page and Garland [23] estimate that there were 2.3 million downloads via BitTorrent during the period that the album was officially downloadable, which far exceeds the estimated download total from the official site; thus the correct average price is extremely low.

Given that even a well-established band could receive so little from their brilliant work, we are pessimistic that the majority of UGC creators, most of whom are obscure, can receive fair compensation for their works in this model.

The charging for download scheme is too risky for consumers and raises the lemon problem, while the voluntary donation scheme is too risky for creators and triggers the freeloader problem. We need to resolve the trade-offs, and find the solution that removes all risks.

17.4.2 Resolving Trade-off between Privacy and Usefulness of Data

Disclosing data that contains information that can be linked to someone obviously violates privacy. Masking the information that would allow a person

---

[7] Nevertheless, voluntary donations (of time or money) mainly from the creator side continue to support the *gift economy* (in contrast to market economy), and the Internet and freeware (or open source software) movements. A discussion about whether the gift economy or the market economy will provide a better future for UGC might be an interesting theme, but is beyond the scope of this chapter.

to be identified such as name and address, often (inappropriately) called anonymization, may not be enough to prevent privacy disclosure as mentioned before. Then, how can we correctly "anonymize" big data and extract valuable information from the data without violating privacy?

This is a very difficult question to answer, mainly because we have to consider the external data, the so-called *background knowledge*, which is information available to any adversary trying to extract private information. For example, consider a dataset that contains the record (age: 60, sex: male, disease: colon cancer). Publishing this record does not seem to disclose any private information because we cannot identify the person from this record. However, if the dataset contains only one 60 year old male and the adversary knows that "Taro Yamada", a 60 year old male, is included in the dataset as background knowledge, the adversary can know that Taro is suffering from colon cancer; his privacy has been breached.

The above example describes a privacy breach from raw data (often called *microdata* in statistics literature), but even statistically processed data can break privacy. Consider a set of tabulated data that shows how many people suffer from colon cancer, classified by age and sex, made from the dataset in the previous example. If the cell value of "60 year old male" is not zero, the same adversary can learn that Taro suffers from colon cancer from the tabulated data.

These examples show that publishing or sharing too fine-grained data may break the privacy of the people in the data, whether it is raw data or statistical data. So the dataset must be (partly) suppressed or perturbed to reduce the privacy risks, but such processes will also reduce the usefulness of the data, the value of the knowledge that can be extracted from the data.

We therefore need criteria that tell us how far a dataset has to be distorted to preserve privacy, and methodologies to achieve the criteria while minimizing the value lost by the distortion. However, there have not been found feasible criteria for "perfect" privacy, where a satisfied dataset is guaranteed not to disclose any information of individuals.

There is an interesting but rather pessimistic impossibility result about absolute disclosure prevention [9]. An intuitive goal of preserving privacy in statistical data analysis, "access to a statistical database should not enable one to learn anything about an individual that could not be learned without access", cannot be achieved, while the seemingly similar goal with regard to secrecy, "nothing can be learned about a plaintext from the ciphertext that could not be learned without seeing the ciphertext", namely *semantic security*, is a widely accepted definition for security and is achieved by a number of cryptosystems.

Dwork [9] provides an intuitive proof of impossibility by offering the following example. Suppose an adversary knows that Terry Gross is two inches shorter than the average Lithuanian woman as background knowledge,[8] and there is no public data about the height of Lithuanian women. The exact height of Terry is uncertain since the height of Lithuanian women is unknown at this point. A statistical database of the height of Lithuanian women from which their average height can be derived allows the adversary to discover the exact height of Terry Gross, which "could not be learned without access" the database.

17.5 Related Works

17.5.1 Reducing Risks in Monetizing UGC

As measures to avoid the lemon market effect, two possible solutions are mentioned in [5]: to guarantee the quality of distributed content (screening), and to provide means to inform the quality of the content to consumers (signaling).

These approaches just focus on mitigating the consumer's risks, but the recent evolution of Internet and security protocol technologies makes it possible to consider an unexpected approach, where the notion of fair exchange [24] is introduced to minimize both creator's and consumer's risks.

17.5.1.1 Screening Approach

The first approach, known as "screening", is equivalent to resurrecting the gatekeepers of the conventional value-chains for professional works; it may work, but it means that we have to give up content in the "long tail", which may include many striking and unusual contents that characterize UGC. The market based upon this approach therefore contradicts the definition of UGC and makes it questionable that it can be called a "market for UGC". However, it does provide a clue to involve consumer-generated "screening" in the value-chain for UGC. An example (partly) of this approach is the technology-oriented news site Slashdot, where discussions are moderated by a consumer-generated moderation system. A comment (to some news article) is assigned points by moderators randomly selected from the consumers

---

[8] In [9], background knowledge is referred to as *auxiliary information*.

having accounts on the site, and is shown if the total score is positive (in the default setting). The comments on this site do not generate any remuneration to the poster or moderators, but Noaki et al. [21] proposed to apply this approach to the UGC market, where the moderators of content can also receive some share of profit from the content so as to let them provide honest evaluations.

17.5.1.2 Signaling Approach

The second approach, known as "signaling", includes free previews and reputation management systems.

Free previews, which are common in digital content shops, enable users to access partial content (typically the first 30 seconds) for free, to allow users to assess if the content is valuable enough to justify purchase. This may ease the consumer's quality risk provided that the quality of the rest, the greater part of the content, matches the preview part, however, there is no one to guarantee this assumption in the UGC value-chain; in an extreme case, the rest of content could be just noise if its creator is malicious. Previews are certainly helpful for consumers, but fail to fully eliminate the quality risk.

The reputation management systems collect comments from consumers who have accessed the content, and provides a summary of them. The simplest examples, which are common in UGC distribution sites, are "rating by stars" and "posting comments" to enable viewers to post comments on the content.

It is doubtful that these simple solutions can adequately reduce the quality risk to consumers, but further advances in reputation management schemes may save the day. In particular, Yamagishi et al. [30] described experiments that showed that providing reputation information of merchandise providers is effective in preventing the C2C market from being lemonized.

17.5.1.3 Gradual Release/Payment Approach

This new approach reduces both the consumer's risk with regard to content quality and the creator's risk about freeloading by introducing the "bit by bit" play and pay of content; i.e., dividing the content into a number of microblocks, which the consumer can purchase and view one by one [26].

The idea behind this approach is similar to that of gradual exchange protocols, which are fair exchange protocols where two parties exchange information gradually. In these protocols, intuitively, when two parties Alice and Bob are willing to fairly exchange their information with each other, each divides his or her information into n pieces,[9] and both repeat the exchange of pieces n times. Since if Alice stops sending pieces, Bob can also stop, the unfairness for Bob is at most n times less than exchanging information by simply sending the complete set of information to each other (and vice versa).

Adapting the above idea to content payment, this approach divides each content into a number of micro-blocks, and the consumer pays per bit for playing each micro-block. Since the consumer can stop accessing the content whenever desired and is not liable for the unused portion of the content, the consumer's risk of paying for worthless content is reduced. From the creator's point of view, freeloading is prevented provided that the security of the payment scheme is not broken. This process can be considered as a series of (micro-)trades. Such an iterative trade process is known to encourage the trading parties to behave cooperatively (cooperative strategies become better than betrayals); e.g., since brilliant works will be viewed to the end by many users and so return a lot of profit to their authors, creators will be further motivated to provide better content.

An implementation example of this approach is a micro-billing scheme for UGC [26]. This scheme utilizes a smart card, e.g. a SIM or UIM card in the mobile phone, to enforce the correspondence between micro-content access and micro-billing receipt, i.e., a smart card guarantees no content use without the corresponding payment and vice versa.

17.5.2 Preserving Privacy in Data Utilization

The privacy problems in processing data that (potentially) contains sensitive information can be roughly divided into *respondent privacy*, *data owner privacy*, and *data user privacy*, in terms of whose privacy is to be protected [8].

Respondent privacy concerns the people whose private information is used to populate the dataset records, and is mainly discussed in the statistical community, which refers to the measures to address this privacy problem as statistical disclosure control (SDC). In recent years, the technologies in this area have come to be called *privacy-preserving data publishing* (PPDP) [12], mainly in the database community.

Data owner privacy focuses on protecting the privacy of dataset owners, who are willing to extract some valuable knowledge by combining datasets possessed by different owners, while they do not want to disclose the data-

---

[9] To be more exact, each piece includes some cryptographic information that gradually increases the probability of restoring the original information to be exchanged.

sets themselves to each other (or one another). This problem usually implies the respondent privacy problem because the dataset might contain data of individuals and therefore access to the dataset would involve breaching the privacy of the respondents. Of course, a data owner can only know a part of the data to be processed while the other part is kept secret by other owner(s). Measures to address this problem are called *privacy-preserving data mining* (PPDM),[10] a term independently coined in 2000 by Agrawal and Srikant [4] in the database community, and Lindell and Pinkas [16] in the cryptographic community.

Data user privacy is to keep user queries to some database private, e.g., the queries to web search engines, which could identify the individual if enough information was collected by the database owner. Attempts to guarantee this sort of privacy, which allow a user to retrieve an item from a database without revealing which item the user is retrieving, originated in the cryptographic community, namely private information retrieval (PIR).

As Domingo-Ferrer and Torra [8] stated in 2008, the technologies to deal with the above three privacy areas had evolved "in a fairly independent way within research communities with surprisingly little interaction". However, recent developments that combine the techniques from different research fields suggest new vistas in these fields.

One prominent example is *differential privacy* [9], which introduces the simulatability approach from the cryptographic literature to PPDP. Intuitively, a database that satisfies ε-differential privacy assures that "any given disclosure will be, within a small multiplicative factor (ε), just as likely whether or not the individual participates in the database", by perturbing the result of a database query with Laplace noise whose strength is determined by ε. Since differential privacy does not assume anything about the background knowledge of the adversary, the privacy of the database respondents can be protected against any adversary while other well-known privacy measures such as k-anonymity, which assures that there exist at least k records sharing

---

[10] The term PPDM in [4] originally meant a methodology to modify original data so as to "mask" sensitive information contained in the data while enabling retrieval of data mining results from the modified data, and that in [16] means a cryptographic multi-party computation methodology to compute some query results across different databases without revealing the content of the databases to each other. However, its meanings have expanded to cover many other techniques including measures to address respondent privacy and data user privacy [3].

the same combination of key attributes,[11] may be vulnerable to adversaries in possession of some specific background knowledge.

This "ad omnia" characteristic and its very light-weight procedure, just adding some Laplace noise to query results, may potentially make differential privacy the standard security measure for preserving privacy in data utilization. However, these promises are currently under inspection. Sarathy and Muralidhar [25] state that the Laplace-based noise addition procedure in [9] does not satisfy the requirements of differential privacy. They accordingly conclude that the whole concept of differential privacy fails as a security standard since satisfying differential privacy does not guarantee the promised level of security, or else, differential privacy is a security standard which like Dalenius' definition of privacy is unachievable in practice.

From a practical point of view, one of the biggest problems with differential privacy is to explain "what ε means" to non-expert people, and therefore it is not clear how lowering ε can be accepted by society for publishing statistical results from sensitive data. Conventional privacy measures such as k-anonymity are ad hoc but intuitively comprehensible. This problem is not only a burden on differential privacy, but may be also shared by a number of more sophisticated enhancements of k-anonymity proposed in recent years.

17.6 Concluding Remarks

This chapter gives an overview of the characteristics of emerging service architectures and their negative impacts, and discussed the security challenges raised by these emerging services, while emphasis is placed on the services of "user-to-user" and the utilization of big data.

There are several important trends not mentioned in this chapter. One of them is *mobility*. In Japan, the success of the "i-mode" service, launched in 1999, has made tens of millions of mobile phone users able to access not only Internet services such as web and e-mail but also various (charged and free) mobile phone-based services including music, games, regional weather forecasts, GPS-based navigation, ticket booking, and mobile e-money services. The recent evolution and rapid deployment of smart phones such as Google's Android-based mobile phones and Apple's iPhone are now rapidly increasing smart phone users around the world; a study by Nielsen [11] pre-

---

[11] Key attributes are also known as *pseudo-identifiers*, which themselves cannot identify respondents, but which may be linkable with external information to re-identify respondents in combination.

dicts that smart phones will overtake feature phones in the US by 2011. The enriched functionality of smart phones makes them more like PCs, but the mobile phone-based services are not mere subsets of the PC-based Internet services, mainly thanks to the location-awareness and the capability of tight combination with billing. This means that mobile phone-based services must face not only threats similar to those faced by PC-based services, but new privacy and financial risks; if an adversary could successfully corrupt a phone by using some malware, the user's movements in his or her daily life may be easily disclosed, as well as allowing the adversary to financially damage the user by way of unintentional payments. There is another difficult trade-off between phone functionality and the threat from malware (and malicious users). Fine-grained access control techniques are helpful to address this problem, but, unfortunately, they may not be the ideal solution. It is difficult even for security experts to correctly configure permissions in systems with fine-grained access control, such as SELinux. Such access control is likely to be indispensable, but some more user-friendly measures should be provided.

As the OECD guidelines [22] stated in 1992, a "secure" system is required to offer the properties of CIA, which stands for confidentiality, integrity, and availability. Accordingly, security technologies have mainly focused on "(perfectly) protecting something" related to these properties, in particular, the confidentiality property. To provide these properties and develop methodologies for them are, of course, important and remain necessary, but it must not be assumed to be sufficient for systems behind recent emerging services as discussed in this chapter.

The emerging services pose difficult but interesting trade-offs to the security research field; future advances in security technology will be required to pay more attention to social and economical aspects, such as the security challenges mentioned in this chapter.

Advances in security technologies should, and can, optimally resolve these trade-offs and thus create the bright future suggested by the emerging services, while minimizing the corresponding negative effects.

References

[1] All too much: Monstrous amount of data. A special report on managing information. *The Economist Newspaper*, February 2010.
[2] Data, data everywhere: A special report on managing information. *The Economist*, 27 February 2010.

[3] Charu C. Aggarwal and Philip S. Yu (Eds.). *Privacy-Preserving Data Mining: Models and Algorithms*, Advances in Database Systems, Vol. 34, Springer, 2008.

[4] Rakesh Agrawal and Ramakrishnan Srikant. Privacy-preserving data mining. In *ACM SIGMOD Record*, Vol. 29, pages 439–450, ACM, June 2000.

[5] George A. Akerlof. The market for 'lemons': Quality uncertainty and the market mechanism. *Quarterly Journal of Economics*, 84(3):488–500, 1970.

[6] Amazon web services. *Amazon EC2 Pricing.*

[7] comScore. For Radiohead fans, does "free" + "download" = "freeload"?, November 2007.

[8] Josep Domingo-Ferrer and Vicenç Torra. A critique of *k*-anonymity and some of its enhancements. In *Proceedings of 3rd Intl. Conf. Availability, Reliability and Security, (ARES08)*, pages 990–993, IEEE, IEEE Computer Society Press, March 2008.

[9] Cynthia Dwork. Differential privacy. In *Proceedings of 33rd Intl. Colloquium on Automata, Languages and Programming, Part II (ICALP 2006)*, Springer-Verlag, July 2006.

[10] eMarketer. Can user-generated content generate revenue?, April 2008.

[11] Roger Entner. Smartphones to overtake feature phones in U.S. by 2011. Nielsen, March 2010.

[12] Benjamin C. M. Fung, Ke Wang, Rui Chen, and Philip S. Yu. Privacy-preserving data publishing: A survey of recent developments. *ACM Computing Surveys*, 42(4):14:1–14:53, June 2010.

[13] Jeff Howe. The rise of crowdsourcing. In *Wired Magazine*, 14, June 2006.

[14] Jeff Howe. *Crowdsourcing: Why the Power of the Crowd Is Driving the Future of Business*. Random House, August 2008.

[15] Frank Kleemann, G. Günter Voß, and Kerstin Rieder. Un(der)paid innovators: The commercial utilization of consumer work through crowdsourcing. *Science, Technology & Innovation Studies*, 4(1):5–26, July 2008.

[16] Yahuda Lindell and Benny Pinkas. Privacy preserving data mining. In *Advances in Cryptology – CRYPTO 2000*, volume 1880 of *LNCS*, pages 36–53, Springer-Verlag, 2000.

[17] Maged Michael, José E. Moreira, Doron Shiloach, and Robert W. Wisniewski. Scale-up x scale-out: A case study using nutch/lucene. In *Proc. 21st IEEE Intl. Parallel and Distributed Processing Symposium (IPDPS 2007)*, pages 1–8. IEEE, IEEE Computer Society Press, March 2007.

[18] Erika Morphy, The dark side of crowdsourcing. LinuxInsider, April 2009.

[19] Arvind Narayanan and Vitaly Shmatikov. Robust de-anonymization of large sparse datasets. In *Proceedings of 29th IEEE Symposium on Security and Privacy*, pages 111–125, IEEE Computer Society, May 2008.

[20] Arvind Narayanan and Vitaly Shmatikov. De-anonymizing social networks. In *Proceedings of 30th IEEE Symposium on Security and Privacy*, pages 173–187, IEEE Computer Society, May 2009.

[21] Kozo Noaki, Masayuki Terada, and Kimihiko Sekino. Why the UGM market does not emerge? In *Proceedings of Computer Security Symposium 2008*. IPSJ, October 2008 [in Japanese].

[22] Organization for Economic Co-operation and Development (OECD). *Guidelines for the Security of Information Systems*, April 1996. Adopted by the Council of the OECD on November 1992.

[23] Will Page and Eric Garland. In Rainbows, on Torrents. Economic Insight 10, MCPS-PRS Alliance, September 2008.

[24] Henning Pagnia, Holger Vogt, and Felix C. Gärtner. Fair exchange. *The Computer Journal*, 46(1):55–75, January 2003.

[25] Rathindra Sarathy and Krish Muralidhar. Differential privacy for numeric data. In *Proceedings of Joint 2009 UNECE/Eurostat Work Session on Statistical Data Confidentiality*, No. CE/CES/GE.46/2009/WP.9. UNECE Statistical Division, December 2009.

[26] Masayuki Terada, Kozo Noaki, and Kimihiko Sekino. Smartcard-based micro-billing scheme to activate the market for user-generated content. In *Proceedings of 4th International Conference Ubiquitous Information Management and Communication (ICUIMC2010)*, January 2010.

[27] Paul Verna. User-generated content: More popular than profitable. eMarketer Reports, January 2009.

[28] Graham Vickery and Sacha Wunsch-Vincent. *Participative Web and User-Created Content: Web 2.0, Wikis and Social Networking*. OECD Publications, October 2007.

[29] Sacha Wunsch-Vincent and Graham Vickery. Participative web: User-created content. Technical Report DSTI/ICCP/IE(2006)7/FINAL, Committee for Information, Computer and Communications Policy, OECD, April 2007.

[30] Toshio Yamagishi and Masafumu Matsuda. Improving the lemons market with a reputation system: An experimental study of internet auctioning. Hokkaido University, May 2002.

18

Reputation-Based Service Management and Reward Mechanisms in Distributed Cooperative Personal Environments

Malohat Ibrohimovna[1], Sonia Heemstra de Groot[1,2], Vijay S. Rao[1] and Venkatesha Prasad[1]

[1] *Faculty of Electrical Engineering, Mathematics and Computer Science, Delft University of Technology, 2600 GA Delft, The Netherlands; e-mail: k.m.ibrohimovna@ewi.tudelft.nl*
[2] *Twente Institute of Wireless and Mobile Communications, 7521 PK Enschede, The Netherlands*

Abstract

Personal Network (PN) is a person-centric, distributed environment of a person's devices that provides access to personal resources and services regardless of the location of the person. A Federation of Personal Networks (Fednet) is a group-oriented network of PNs. A Fednet is a pervasive and ubiquitous computing technology that enables the users to enjoy cooperation and promises exciting opportunities for different applications in various fields, such as education, healthcare, entertainment, business and emergency.

Since each PN is associated with a person, i.e., the PN owner, the cooperation of the PNs reflects the social behavior of the PN owners, and therefore a Fednet can be seen as a social network of PNs. Trust and reputation influence the real-world interactions; similarly, using reputation as a metric for interactions between PNs is an interesting topic. In this Chapter, we look at the Fednets from the 'social' angle and discuss how the Fednets can benefit from using reputation. We propose a reputation-based framework for Fednets and present ideas on applying reputation information for service management and

Anand R. Prasad et al. (Eds.), Advances in Next Generation Services and Service Architectures, 407–430.

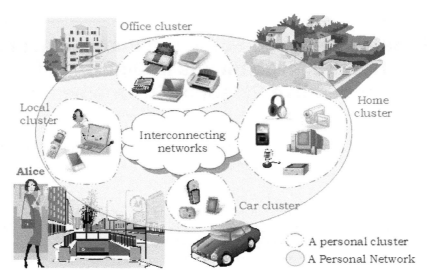

Figure 18.1 Example of a personal network.

reward mechanisms in Fednets to improve the quality of cooperation between the PNs.

Keywords: personal networks, Fednet, sharing personal services, dynamic access control, stimulating cooperation, reputation, reputation-based systems.

18.1 Fednets: Federation of Personal Networks

Almost everyone today uses one or more personal devices in their daily work, entertainment, communication and social activities. Most of these devices also have networking capabilities. Examples of such personal devices are mobile phones, PDAs, digital cameras, handheld game consoles, laptops, desktops, personal navigation systems, MP3 players, printers, home appliances, gadgets, etc. It would be useful if they could communicate with each other and provide added-value and meaningful services to their owners independent of their geographic location. This is the idea behind the concept of Personal Networks. A Personal Network (PN) [13, 14] is a person-centric, distributed environment of a person's devices, and provides access to personal resources and services regardless the location of the person. This is illustrated in Figure 18.1.

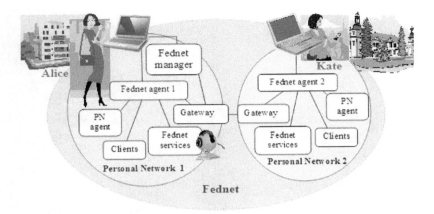

Figure 18.2 Basic architecture of a Fednet.

Having a personal network with a variety of personal services and resources, one can benefit from sharing them with others in order to reach a
common objective, for example sharing sensor information from different
sources for rescue of people in disaster relief, getting real-time information
from devices that belong to other persons in healthcare applications, sharing
digital media for business or entertainment. A Fednet is a secure group-
oriented network composed of PNs, in which PNs collaborate with each other
and share resources and services in a peer to peer (P2P) manner, i.e., they may
be the producers and consumers of the services in the Fednet [4, 6, 7, 12].

Figure 18.2 illustrates the basic architecture of a Fednet. In this Fednet,
the colleagues Alice and Kate staying in different locations federate their PNs
to share camera view, videos and pictures related to their business affairs.
Types of resources and services shared in Fednets vary depending on the
application area. Each Fednet is tailored for a specific goal; therefore each
Fednet has different characteristics.

A Fednet is a dynamic entity, because it evolves incrementally as more
PNs join, and it ceases to exist when all PNs leave it. Each PN has the
Fednet agent functionality, which is responsible for joining/leaving a Fednet
and controls access to personal resources and services of a PN. Furthermore,
each Fednet has the Fednet manager functionality, which is responsible for
management and control of the Fednet, such as creating, dissolving a Fednet
and accepting/removing Fednet members. The Fednet manager functionality
can be hosted by one of the PNs, as is illustrated in Figure 18.2, or can be
provided externally by a third party.

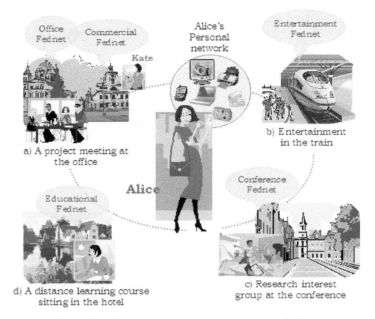

Figure 18.3 Fednets support various activities of Alice.

When joining a Fednet, a PN obtains a membership credential signed by the Fednet manager. It contains the PN's name/pseudonym and its contact information. The membership credential has a validity period. Consequently, the credential should be periodically refreshed. This credential is used to authenticate the PNs within this Fednet and for accessing the Fednet services. There are many possible situations in which the user can take advantage of using a PN and a Fednet. Figure 18.3 illustrates Alice's various activities that can be supported by the Fednets that her PN is engaged in.

(a) Business (Office) Fednet: At work Alice attends a project meeting. In the office colleagues federate their PNs to prepare a project plan efficiently, by sharing important documents, photos, video clips, screen and presentation slides with each other. Furthermore, the PNs of Alice and her colleague Kate, from another branch of the company, are engaged in a commercial Fednet. The commercial Fednet is established between the company's various branches from different cities to provide 'a virtual shopping mall' service to multiple customers and clients of the company from different cities.

(b) Entertainment Fednet: Alice goes to a conference scheduled in another city by train. In the train, Alice federates her PN with another person's PN, who has the same hobby: collecting choreography of dances from around the world. She shares the pictures and videos of her collection using her laptop with this person, who shares with Alice fabulous photos and videos of Asian dance fragments.

(c) Business (Conference) Fednet: Later at the conference, people with close research interests (e.g., marketing and selling products and goods) federate their PNs to make contact, to exchange documents, contacts and references, photos of the conference and other material, to conceive a business.

(d) Educational Fednet. After the conference Alice wants to attend her distant learning course on modern art. Sitting in the hotel room, Alice federates her PN with her course-mates and attends the online course.

Here we can see how Alice's PN and various Fednets support her social life and daily activities. Note that while the essence of a PN is providing the user with personal ubiquitous services (e.g., Alice can access her documents, course materials stored in her home PC from anywhere, in this case from the hotel), the essence of a Fednet is sharing these ubiquitous services with others for a common goal (e.g., Alice can share some of her documents with course-mates of her distance learning course to prepare an online assignment).

In this chapter, we address the issues of a dynamic access control to the services in a Fednet and stimulating the cooperation in Fednets. For this purpose we look at Fednets from a 'social' angle. Since behind each PN there is a person, i.e., the PN owner, the cooperation of the PNs reflects the social behavior of the PN owners, and therefore a Fednet can be seen as a social network of PNs. Reputation information is one of the driving forces in social networks. To enable a dynamic access control and a dynamic cooperation of PNs, we propose reputation-based framework for Fednets. We introduce the concept of a Federated reputation identity for Fednets/PNs, which enables them to collect and use reputation information across multiple Fednet domains.

The rest of this chapter is organized as follows. In Section 18.2, we present the reputation-based framework for Fednets. In Section 18.3, we describe the operation of the reputation framework in Fednets. In Section 18.4, we discuss the results of the simulation which is being carried out to support the ongoing research. In Section 18.5, we briefly discuss the related work on reputation-based systems. Finally, in Section 18.6, we draw conclusions and discuss future trends.

18.2 Reputation Framework for Fednets

The core of the reputation-based system is the reputation information. Reputation built based on the experience of a single entity is called local reputation [11], also referred to as subjective reputation or first-hand information. Reputation built in cooperation, by all participants in the network, as a combination of all local reputations is called global reputation [11], also referred to as objective reputation.

18.2.1 Requirements to the System Design

A reputation based system must meet the following minimum requirements to benefit from using reputation information [5]:

1. The participants should collect feedback from their interaction experience and optionally can exchange or distribute their opinions to each other.
2. The source for observations must be identifiable, so that the reputation information can be accumulated for this source. The identities of the participants should be traceable, so that it should be possible to recognize them in the future interactions.
3. There should be a sufficient number of events/interactions to acquire the experience and to learn the behavior.
4. Experience and observations must be factors in future decisions.
5. There should be incentives for the group members to collect and exchange their experiences during their interactions.

18.2.2 Reputation Information in Fednets

Reputation is a measure of quality of interactions between the PNs. We define reputation as a previous interaction quality and trust as a belief in future interaction quality. Building reputation information requires monitoring and observing the behavior of the participants. Reputation information is collected by all parties after every interaction. This local experience information is stored as credit points (or reputation value). These credit points are incremented when there is a positive experience and decremented when there is a negative experience. We define the reputation for a PN, a PN's service and a Fednet in the context of Fednets.

- The *reputation of a PN* is based on its service to other PNs, service consuming and cooperating behavior.

- The *reputation of a PN's service* is based on the service's quality, content, availability, performance, price, etc.
- The *reputation of a Fednet* is the reputation that a PN concludes about the Fednet, based on the quality of cooperation experienced while being a member of this Fednet. It is a reputation that the Fednet has obtained from its members.

18.2.3 Building Reputation in Fednets

In our framework, the PNs monitor each other and log their experiences. They report the local reputation value about other PNs to the reputation manager of the Fednet. As criteria for monitoring, evaluating the behavior and building a reputation for a PN in Fednets, we consider the interactions between the PNs and the contributions of the PNs to this Fednet. Interactions between the PNs are observed at the PNs. The primary information measured in any reputation-based system is the first-hand observation, i.e., the local reputation value. During the operation of the Fednet, the local reputation value is updated based on the number of successful interactions of the PNs and the satisfaction level of the PN owners. There are different ways to consider the interaction as successful, for example, system-level and user-level observations.

System-level observation is logging and counting successful interactions by the Fednet. For example, the reputation of a service can be incremented, if the service response time is acceptable by the application. Furthermore, the local reputation of a PN can be increased, if the requested service was made available as it was promised. User-level observations are based on the quality of the service and satisfaction level of the user. This information is created by the user (PN owner)'s feedback on whether or not the quality of the service (provisioning) was satisfactory, after the service was delivered/consumed.

Observations on the contributions of the PNs are made at the Fednet manager which is aware of the amount and type of contribution of every PN in the Fednet. Initial reputation values can be assigned based on this contribution. If the contribution increases, the global reputation value increases as well.

18.2.4 Architecture of the Reputation Framework for Fednets

We motivate our choice for an architecture of the reputation-based framework based on the usability of the reputation information in long-term and short-term Fednets. For a reputation-based system, the duration of the cooperation of PNs is important, according to Requirement 3 stated in Section 18.2.1.

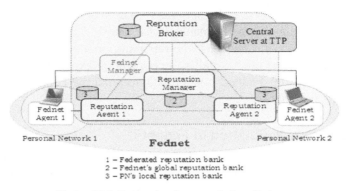

Figure 18.4 Reputation framework for a Fednet.

Let us consider long term and persistent cooperation of PNs in a Fednet. Long-term and persistent cooperation of the Fednet members experiences several life-cycles before the Fednet's disbanding. Such cooperation allows the building of reputation information. Collecting and building up reputation information is more challenging in Fednets with a short-term cooperation between its members, especially, when the environment is dynamic, in which the PNs are anonymous and dynamically joining/leaving. In this case, we look beyond Fednets.

If the reputation information is collected beyond a single Fednet at some commonly known entity, such as a reputation broker, then several Fednets can benefit from this reputation service. This will act as a federated reputation identity service provided by the reputation broker at a Trusted Third Party (TTP). As a result, a PN can use its federated reputation account in other Fednets as well and can benefit from its collected reputation value in various short-term or long-term instances of Fednets. Based on these considerations, we propose a hierarchical reputation framework for Fednets.

Figure 18.4 illustrates the location of the reputation framework components in the Fednet architecture. The reputation broker and its Federated Reputation Bank (1) are located at the TTP, an example of which can be the PN Directory Service proposed in [10]. The reputation manager can be located at the Fednet Manager (FM). The reputation manager maintains the Fednet's global reputation bank (2), which stores all information related to the reputation of the PNs and their services. At the bottom of the hierarchy are the reputation agents of the PNs. They can be located at the Fednet agent (FA) of the PN and maintain their own Local Reputation Banks (3). The local reputation bank of the PN contains the reputation information collected by the PN during the participation of this PN in various Fednets. Participants of the system are the Fednets and the Fednet members and each of them have their reputation accounts in this system.

Reputation Agent is responsible for the following tasks:

- Monitoring, collecting experience information from the participation of this PN in a Fednet. Moreover, it calculates the local reputation values for peer PNs, their services and the Fednets in which they took part and stores in its local reputation bank.
- Periodically, on demand or immediately after update, the reputation agent sends reports to the reputation manager of its Fednet on the reputation values collected about the peer PNs and their services based on the experience.
- Furthermore, when there is insufficient information, the reputation agent can request the reputation manager of its Fednet to provide additional information on the reputation of a particular service or a peer PN.

To implement the idea of reputation information, we define types of reputation as follows: local reputation (LR) and global reputation (GR). Local reputation is based on the local observations of a single PN (Figure 18.5a), and the global reputation is based on the observations of the group of PNs participating in the Fednet (Figure 18.5b).

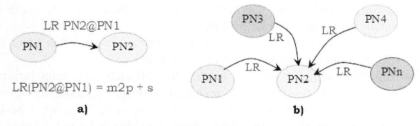

Figure 18.5 Local and global reputation of PN2.

LR2@1 is the reputation value of the PN2 calculated at the PN1 as is illustrated in Figure 18.5a. It is the opinion of the PN1 about the PN2, built up from the experience with the PN2.

p is the number of the PN1's positive (successful) interactions with the PN2. If the number of successful interactions increases, the reputation value also increases.

m2 is a membership class of PN2 and it is indicated in the membership credential of PN2. The higher it is, the higher the resulting reputation value will be, and in addition, the quicker the reputation grows.

S is a satisfaction level of the PN owners from the interaction: positive (1), neutral (0) and negative (-1).

The local reputation value for a Fednet at the RA:

$$LR\ (Fednet) = LRo + s\ (positive,\ neutral,\ negative)$$

Here LRo is previous or initial reputation value if available.
    The local reputation value for a service at the RA:

$$LR\ (Service) = LRo + s\ (positive,\ neutral,\ negative)$$

Reputation Manager collects and stores the reputation information received from the Fednet participants about the PNs and their services. The reputation manager is responsible for:

- Calculating the global reputation in the Fednet domain based on the local reputation values provided by the reputation agents of the participating PNs.
- Maintaining the reputation information bank of the Fednet and storing the global reputation information.
- Making decisions on the access control to the Fednet by using the reputation value of the requesting PNs.
- Retrieving the reputation information from its bank, when there is a request from the reputation agent of some PN.
- Reporting to the reputation broker the collected reputation information during its operation. The reputation manager periodically uploads the reputation information of its participants to the reputation broker. Moreover, the global reputation information should be uploaded to the reputation broker when dissolving the Fednet.

The RM calculates the global reputation of the PN2 as follows:

$$GR\ (PN) = Contributions + Average\ SUM(LR) + Feedback\ index$$

The average summation of the reported LR values for the PN2:

$$Average\ SUM(LR\ for\ PN2) = 1/n\ (LR2@1 + LR2@3 + LR2@4 + \cdots + LR2@n)$$

The feedback index is a reward given to a PN by the RM for providing a feedback.

The RM calculates the global reputation of a service as an average summation of reported LR values for this service:

$$GR \ (Service) = Average \ SUM(LR)$$

The global reputation information is stored in the global reputation account, which contains the combination of the PN's pseudonym and its global reputation value, as for the service, it is a combination of the service's ID and its global reputation value.

Reputation Broker, an entity located outside the Fednet, is responsible for collecting the reputation information about the PNs and Fednets. It is a centralized functionality; therefore multiple Fednets and PNs can use the services of the same reputation broker. The reputation information at the reputation broker is federated reputation information accumulated for the PNs and the Fednets registered with this reputation broker. The reputation broker requires the PNs to register with their true identity. Here, we assume that the PNs and the Fednets register themselves with their true identity. This ensures that the identity of a PN can be traced by the reputation broker and this will allow building up the reputation scores for the PNs. However, the PNs can use different pseudonyms while joining different Fednets, so that they can preserve their anonymity within different Fednets. To prevent ambiguity, the reputation broker should control the uniqueness of the pseudonyms when registering PNs and Fednets. Although they can have multiple pseudonyms and their true identities are not revealed to others, they are still identifiable by the reputation broker. The reputation value the PNs obtain from their each cooperation and interaction in different Fednets will be accumulated at their single federated reputation account, since multiple (anonymous) pseudonyms of a PN are linked to the same federated reputation account. The federated reputation (FR) account contains the Fednet/PN name, pseudonyms and earned reputation values in different instances of Fednets. This account is dynamically updated based on the reports of the Fednets and the PNs.

The Fednet's FR value at the RB:

$$FR \ (Fednet) = Average \ SUM(LR) + Feedback \ index$$

The PN's FR value at the RB:

$$FR \ (PN) = Average \ SUM(GR) + Feedback \ index$$

The feedback index is a reward assigned to this Fednet/PN by the RB for providing a feedback. Reputation information at the reputation broker is collected not only during long-term cooperation but even in short-time co-operations of the PNs in various Fednets. Obtained credits in one Fednet can be utilized in other Fednets in the future. This is owing to the Federated Reputation IDentity (FRID) that contains the unique federated reputation account number and the federated reputation value.

$$FRID = (account\ number,\ reputation\ value)$$

18.3 Working of the Reputation Framework

We propose to use the reputation information in Fednets for service management and for rewarding of cooperative PNs.

18.3.1 Reputation-Based Service Management

In our context, service management denotes admission control to the services, service discovery, selection and provisioning.

Reputation of PNs in the admission control. Reputation information can enable a dynamic admission control, which means that the access to the Fednet and its services is periodically adjusted based on the reputation values of the PNs. When joining the Fednet a PN presents its membership profile to the FM, which includes the PN's services list contributed to this Fednet. Once the PN joins the Fednet, the member list at the Fednet manager is updated with a new entry: PN name (ID), Contact info (IP of the FA), Service list, Membership class, Global reputation value. The membership credential of this PN, once created at joining the Fednet, is periodically refreshed and a PN will have new values for its global reputation in its subsequent membership credentials. Since the FM signs this membership credential, the reputation values can be verified by all members of the Fednet. The service offering PN can grant access to its personal services based on the reputation of the service requesting/consuming PN stated in their membership credentials. Since the membership credential is adjusted from time to time based on the reported reputation values, eventually, cooperative PNs collect higher values of reputation; non-cooperative PNs get less reputation; and misbehaving PNs get black-listed and isolated from the cooperation. As a result, the access rights also will be adjusted based on the reputation, i.e., the higher the reputation values, the higher the access rights granted to the PN.

Reputation-based service discovery. The reputation-based service discovery is realized in the lookup service provided by the Fednet manager. The lookup response from the Fednet manager is a sorted list of services containing the information: Service name; Contact information; Popularity index of the service based on how often this service was requested; Reputation of the service based on the feedback of the other PNs.

In our design, a Fednet service requires a certain reputation value (similar to the price) from its clients in order to allow access. The lower the reputation of the client-PN (e.g., number of contributions, number of positive feedback from peers), the fewer services are visible to this PN in the lookup response list.

Reputation-based service selection. The services in the Fednets are rated based on their reputation value. When there are multiple instances of the same service in the Fednet, service consuming PNs can use the reputation of the service for choosing the best option. The reputation of the service is included in the lookup response list provided by the Fednet manager. When the user gets the list, it is sorted based on popularity and reputation. The user can also give preferences, so that the FA automatically sorts and chooses the service based on them. In addition, the reputation of the server-PN can also be used. The reputation of the server-PN can be retrieved during the authentication phase from the membership credential of the server-PN.

Reputation of PNs in service provisioning. Based on the reputation value of the PN, the policy evaluation produces a decision on service provisioning: either proxy-based or overlay-based service provisioning will be set up between the PNs. We consider two application environments: an anonymous environment and a friendly environment for service provisioning in Fednets. In anonymous, privacy-sensitive environment, the participants do not have sufficient reputation information about each other. In such an environment, we consider keeping the privacy of the users as a priority. In this case, we use reputation information for choosing the way of service provisioning in the Fednet. When the reputation values are high among the community members (above some threshold), then the focus of the service provisioning can be shifted towards better quality (less delay, less overhead, etc.). In this case, the service provisioning can be done from any point in the PN that has the best link quality connection with the client PN.

18.3.2 Reward Mechanisms to Improve the Quality of Cooperation

The most effective way to stimulate the participants in group cooperation is reward [15]. Some examples of reward and incentive mechanisms reported in the literature are nuglets [1], virtual credits [16], traffic credits [17] and team based rewards [2]. To stimulate cooperation in Fednets we propose a reputation-based reward mechanisms for Fednets. Reward mechanisms in Fednets can be implemented in a distributed fashion, between the participants of the Fednet in a P2P manner without a trusted third party and in a centralized fashion, involving a third party (a reputation manager/broker).

Mutual Reward – Reputation tokens as a virtual currency. An example of a distributed reward mechanism is self-signed reputation tokens exchanged between the Fednet members. This approach is based on the reciprocity between the PNs, in which the reputation tokens play a role of non-monetary reward. The client PN1 gives a self-signed reputation token to the PN2 if the PN2's service was satisfactory. Eventually, while requesting a service from the PN1, the PN2 can present this token, signed by the PN1, to prove its reputation earned from the PN1. This mutual reward mechanism fits in ad hoc spontaneous 'Business Fednets' (Scenario c, Figure 18.3), created by special interest groups that share access to the resources and services such as internet access, printers, storage, processing, screens and various types of data. However, this approach does not scale to multiple Fednet domains, since the reputation tokens signed by one PN can not be presented to a PN from another Fednet.

Promotion – Multiple domain service access. The usage of these collected reputation tokens can be extended to various PNs and multiple Fednet domains, when a trusted third party comes into the play. In this case, the reputation credits are collected by the third party and can be used as virtual payments even between anonymous PNs that participate in different Fednets. This allows earning (reputation-based) virtual currency in one Fednet and spending it in various other Fednets, supported by the same trusted third party or a chain of trusted parties. This is the idea behind the concept of the Federated reputation identity presented in Section 18.2.4. An example application for this case can be 'Commercial Fednet' in Figure 18.3, created to enable virtual shopping malls in different cities, for selling personal items, community goods and company products. Suppose that Alice and Kate are the managers of the virtual shopping malls in different cities. The reputation managers of their Fednets keep the reputation scores for their members based on the quality of their services, and the satisfaction level of the customers. The reputation scores are reported by the reputation managers to the reputation broker, which creates Federated reputation identity for the participant PNs of these Fednets. The members, who earned high reputation values by advertising and providing the good quality services, will have their reputation class increased. As a reward, they will have the rights to access the services and to advertise their services in more areas and cities, i.e., will obtain broader range of discovery for their services in various virtual shopping malls.

18.4 Simulation

18.4.1 Scenario

We consider a scenario of communities, in which a large number of people want to sell their goods and also want to buy some products. The community members use their PNs to share their files and images of their products, which are various types of furniture, second hand sporting goods, home made handicrafts, unique pieces of art, etc. Each PN has a personal folder with a number of multimedia files describing particular products they are selling. They use file sharing to share multimedia files about the products for sale. The PNs based on their interest and experience can form Fednets, and can participate in multiple Fednets using different pseudonyms. Since anonymity is allowed, some people advertise bad quality goods. Therefore, the communities want to impose the admission control measures while forming Fednets and joining Fednets to increase the quality of goods that they offer to each other. In addition, the communities want to stimulate their members to behave in a cooperative way.

We will address these issues with our reputation framework for Fednets. To test the reputation framework we will need a large number of participant PNs, having long or short term cooperation for building the reputation between them. We simulate various scenarios. The goal of the simulation is to prove the usability of the reputation framework in (a) admission control which can improve the quality of services, and (b) enforcing the cooperative behavior of the PN owners.

Behavioral strategies of PNs. Similar to the participants of the social networks, PNs in a Fednet might behave with different patterns to maximize the profit from the cooperation. We define three behavioral strategies of the PNs in a Fednet.

- Random behavior, may be caused by unreliable communication medium or by selection by the user.
- Cooperative behavior, always behaving honest and cooperating with others.
- Malicious behavior, we generalize here the behaviors such as non-cooperative, selfish, providing bad quality service intentionally, giving incorrect feedback about others.

To validate our design of the reputation-based framework, we focus on building reputations for the PNs with different behavior strategies, reputation in admission control and stimulating the member to cooperate.

Simulation parameters. We implemented the simulator using Matlab with parameters:

- Behavioral strategies of PNs (random, cooperative, malicious);
- Membership class of the PN (unknown, familiar, trusted);
- Reputation class of the PN (bronze, silver, gold);
- Total number of interactions for each PN is 100;
- The number of positive interactions with other PNs is simulated;
- Satisfaction levels of the user are derived based on the interactions (positive, negative);
- Local, global and federated reputation values are calculated.

We suppose that there are 100 PNs, which are clients and servers in Fednets. There are three behavioral strategies for the PNs, random, cooperative and malicious. Cooperative PNs are 75, random behavior-PNs are 15 and malicious PNs are 10. PNs participate in different Fednets. The simulator each time chooses a random number of PNs to federate. The number of Fednets in the simulation setting is 100. Out of 100, 40 Fednets require silver and gold reputation class, which in the simulation settings has a minimum of 300 for the reputation value. Another 40 Fednets require a minimum of 100 for the reputation value, which can be reached with bronze reputation class. The other 20 Fednets can accept anyone. The PNs with low reputation might have chances to join these Fednets only.

Additional settings are:

- Adaptive malicious strategy, a PN that gets fewer opportunities to federate, hence access to fewer services, due to its behavior, changes its behavior to be cooperative. The reputation threshold value that forces the PNs to change the strategy to get admitted to the Fednets is taken as (-60).

- The fading of the reputation is 2%, which means if the PN does not contribute to the Fednet its reputation value will decrease by 2%. This is to make all PNs to participate in the Fednets and also to reduce the effect of reputation in the distant past.

After receiving a service the feedback is generated at the Reputation agent and a local reputation value is calculated. This local reputation value is reported to the Reputation manager and eventually to the Reputation broker, which also updates the global and federated reputation values, respectively. Local, global and federated reputation values are calculated using the formulas presented in Section 18.2.4. In these simulation settings, we assume that the reputation values are reported truthfully. The maximum reputation value for this simulation setting is 400 and the minimum is -100. Admission to the Fednet is based on federated reputation information collected in previous Fednets. Higher the reputation means higher the chances to federate. We can see the reward here as the chance for admittance into the Fednets.

18.4.2 Simulation Results and Analysis

We analyze the results from the simulation run with respect to the following:

1. The reaction of the reputation system to different behavior strategies;
2. Admission into the future Fednets based on reputation that is built over a period;
3. Influence of the system to the change in the behavioral strategy.

Reputation for different behavioral strategies. Figure 18.6 illustrates the reaction of the reputation system to different behavioral strategies of the PNs. The reputation value is collected across multiple Fednets, so that it is accumulated as a federated reputation value for the PNs. Figure 18.6a shows that the most profitable strategy to build the reputation is the cooperative strategy. Cooperative PNs build the highest federated reputation, while the malicious PNs will have their federated reputation reduced. When the PNs report truthfully, reputation information accurately estimates the real quality of the services delivered by the PNs in the Fednet. Furthermore, Figure 18.6b demonstrates the influence of the reputation system to the behavioral patterns of the PNs. Malicious PNs, if it is adaptive, changes its strategy when its reputation reduces dismally. This is depicted in Figure 18.6b with the increasing reputation value for the adaptive malicious PN. An important observation is that, even with changing the strategy, the adaptive malicious PNs never reach the same level as cooperative-only PNs. While cooperative PNs

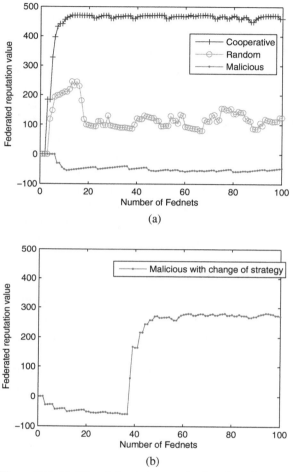

(a)

(b)

Figure 18.6 Building federated reputation based on the strategy.

build up their reputation, malicious PNs decrease their reputation and the random strategy-PNs can expect unstable reputation within Fednets. Service failure or unsatisfactory service provisioning caused by the changing network conditions and circumstances is a typical situation for Fednet environment. Additional mechanisms will be required to distinguish the real cause of this random behavior, to prevent reduction of the reputation of an innocent PN.

Admission to the Fednet based on reputation. Federated reputation identity enables the mobility of reputation information across Fednets. When a

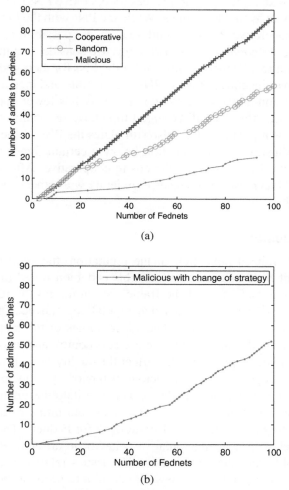

Figure 18.7 Admission control based on the behavioral strategies of PNs.

member moves from one Fednet to another, or joins multiple Fednets, the PN's reputation earned in various Fednets is stored at the reputation broker, so that the PNs can accumulate and benefit from its reputation from multiple Fednets. Fednets admission control policy based on reputation information allows acceptance of a PN if the federated reputation of the PN is within the required range to join this Fednet. Figure 18.7 illustrates the admission to the Fednet based on the reputation built due to the behavioral strategies of the

PNs. The PNs with high reputation values (cooperative PNs) have a chance to join the maximum available Fednets. While the PNs with the lower reputation values (non-cooperative PNs) will end up with less possibilities.

Stimulating the good behavioral strategy. We consider that the reputation-based admission to a Fednet is a reward for the PNs for their good reputation. This reward gives incentives for the PNs to cooperate and increase their reputation. When a non-cooperative PN continuously has fewer opportunities to join other Fednets, the PN will be urged to behave rationally, so that it will try to increase its reputation. This also stimulates the PNs to cooperate better. Figure 18.6b illustrates that reputation below a certain threshold forces the PN to change its strategy from malicious to cooperative. After changing the strategy, the PN increases its chances to federate with others, as is illustrated in Figure 18.7b.

18.4.3 Discussion

Due to the centralized collection of the reputation, the short-term Fednets can also benefit from using reputation. This is demonstrated in Figure 18.6, which proves the usefulness of the framework in the run of 100 short-term Fednets. Figure 18.7 demonstrates that reputations can be efficiently used for admission control in Fednets. The added values of the reputation-based framework here are dynamic and flexible access control and incentives for the PNs to obtain higher reputations that affect the quality of their cooperation.

The system copes with the malicious behavior by reducing chances to federate with others or to join Fednets. This can help the Fednets to reduce the selfish behavior of the PN owners. These are our initial results in designing and simulating reputation-based framework for Fednets. The work tracks inconsistent behavior, i.e., manipulative behavior in different Fednets to get maximum profit from the cooperation. The issues related to overhead and complexity of the reputation framework taking into account the issues from untrustful feedback of the PNs need to be looked into.

18.5 Related Work

A lot of interesting works have been reported in the literature on reputation-based systems. These systems tackle different problems in different layers, starting from the networking and routing to the services and applications [5]. In this section, we discuss some of the reputation-based systems and compare their approaches with ours. Hwang et al. [3] propose a dynamic incentive mechanism to motivate the personal network nodes in participating in cooperative relaying. As a reward for cooperative behavior, the nodes receive additional throughput based on reputation calculations for individual contributions. Dynamic assignment of the reward prevents the nodes from adversely manipulating their behavior after getting the reward.

In [16] the reputation information is used in improving the quality of service provided by the ISPs in wireless hotspot environments. After receiving a service from ISP, the mobile node sends to the Trusted Certification Authority (TCA) its feedback on the service received from ISP. Based on the feedback the TCA issues a new certificate with the updated reputation value of the ISP. While the incentives for the ISPs are to get more clients, the incentives for the mobile nodes are to receive a better service and some amount of credit from their home ISP for feedback.

Another interesting approach discussed in [9] is based on penalty. An independent reputation mechanism requests binary feedback about interactions, 1 for high quality service and 0 for low quality service. The reputation of a provider is computed as an average satisfaction rate of the clients for a given period of time. At the end of each period, the reputation mechanism publishes the reputation of every provider, and service providers are expected to refund every client the monetary penalty specified in the SLA. When the penalty is large enough, Jurca and Faltings [9] prove that rational service providers keep their promises.

While Hwang et al. [3] focus on the single service, i.e., the forwarding service between the PNs, our goal is to assist the PNs and Fednets in selecting the reliable PNs and Fednets to federate, based on the satisfaction level from previous interactions. In [3] reputation mechanisms should be implemented in each node participating in cooperative relay process, i.e., user terminals, relays and base stations. In [9, 16] reputation mechanisms should be implemented in user terminals and in the infrastructure. Our approach has less complexity, since the reputation mechanisms can be co-located with the existing Fednet functionalities, the Fednet Agent and the Fednet Manager. The only additional functionality that is required for the centralized reputation collection is the reputation broker functionality at a trusted third party. All systems [3, 9, 16] use centralized mechanisms for reward and punishment, while our approach is hierarchical: distributed in a Fednet domain and centralized across multiple Fednet domains. This makes our system scalable, since the system can be used within single Fednets independently and between multiple Fednets cooperatively.

18.6 Conclusion and Future Trends

In Fednets, motivating the members to cooperate is a challenge. We address this challenge with a reputation-based approach. A reputation-based system motivates the participants to obtain a good reputation in order to benefit from the cooperation. Using reputation can help to bring the system into balance. For example, by using reputation in the system, the good members get rewarded for their good behavior and the bad members get punished. As a consequence, the members are motivated to behave well, e.g., to cooperate and share their resources. With time, the effect of this reward and punishment will be seen as self-healing, i.e., getting rid of malicious, non-cooperative behavior, since this behavior will become irrational.

In this chapter, we presented a reputation-based framework for Fednets and showed how the Fednets can benefit from using reputations in service management and improving the quality of cooperation between the PNs. The framework provides a dynamic access control in Fednets, which means that the access control to the Fednet membership and resources, is periodically adjusted based on the reputation values of the PNs, i.e., higher the reputation value higher the access rights of these PNs.

Although the requirements to the reputation-based systems state that the Fednets should have a long-term and stateful cooperation of the PNs, we argue that even short-term and stateless cooperation of PNs in a Fednet can benefit from using reputation and social control. We introduced a reputation broker functionality that operates beyond the Fednets and serves multiple Fednets and PNs as a trusted reputation authority. We introduced the federated reputation identity for the PNs and the Fednets, which is a service provided by the reputation broker located at the TTP. The reputation broker functionality makes the system scalable and applicable for multiple Fednets linked with a common third party (or a chain of TTP). The simulation results proved the usability of the reputation framework for admission control and for stimulating cooperation in Fednets. The results presented in this chapter reflect our initial research on applying reputation in Fednets, which is a promising and interesting avenue for further research.

In the future, managing a large number of personal devices will be a challenge and an important issue for the users. The users are becoming providers and consumers of the services. In this sense, Fednets will become attractive to the users, since they are Personal Networks, can be operated independently from the operators, can be suited to various scenarios, can support various types of group-oriented applications tailored for infotainment, communication and collaboration, distributed computation, internet service support and content distribution, remote healthcare and social networking and, finally, can support the social life and business activities of PN owners.

The advantages of Personal Networks and their federations will become more obvious as the number of personal devices increases. Although nowadays we have just a few devices, in the near future this number is expected to increase tremendously. It is predicted by the Wireless World Research Forum [8] that 10 years from now, there will be 7 trillion wireless devices serving 7 billion people, that is, an average of 1000 wireless devices per person. The proliferation of all types of sensors and portable devices with networking facilities will result in new paradigms for service delivery models and service architectures. Fednets are one example. Personal networks and their federations can be seen as a next generation networking concept with service-orientation and personalization features: the concept that allows users to share their personal services in a seamless, secure and flexible way; the concept that allows organizing a big part of personal devices in order to make them cooperate in an effective way and provide next generation services.

References

[1] Levente Buttyan and Jean Pierre Hubaux. Stimulating cooperation in self-organizing mobile ad hoc networks. *ACM/Kluwer Mobile Networks and Appl.*, 8:579–592, 2003.

[2] E.C. Efstathiou, P.A. Frangoudis, and G.C. Polyzos. Stimulating participation in wireless community networks. In *Proceedings of 25th IEEE International Conference on Computer Communications (INFOCOM2006)*, pages 1–13, 2006.

[3] Junseok Hwang, Andrei Shin, and Hyenyoung Yoon. Dynamic reputation-based incentive mechanism considering heterogeneous networks. In *Proceedings of 3rd ACM Workshop on Performance Monitoring and Measurement of Heterogeneous Wireless and Wired Networks*, pages 137–144, ACM, New York, 2008.

[4] M. Ibrohimovna and S. Heemstra De Groot. Policy-pased hybrid approach to service provisioning in federations of personal networks. In *Proceedings of 3rd International Conference on Mobile Ubiquitous Computing, Systems, Services and Technologies (UBICOMM'09)*, pages 311–317, 2009.

[5] M. Ibrohimovna and S.M. Heemstra de Groot. Reputation-based systems within computer networks. In *Proceedings of 5th International Conference on Internet and Web Applications and Services (ICIW'10)*, pages 96–101, 2010.

[6] M.I. Ibrohimovna and S. Heemstra de Groot. Proxy-based Fednets for sharing personal services in distributed environments. In *Proceedings of 4th International Conference on Wireless and Mobile Communications (ICWMC'08)*, pages 150–157, 2008.

[7] M.I. Ibrohimovna and S.M. Heemstra de Groot. Fednets: P2P cooperation of personal networks, access control and management framework. In *Handbook of Research on*

P2P and Grid Systems for Service-Oriented Computing: Models, Methodologies and Applications, IGI Global, January 2010.

[8] N. Jefferies. Global vision for a wireless world. Wireless World Research Forum, 18th WWRF Meeting, Helsinki, Finland, June 2007.

[9] Radu Jurca and Boi Faltings. Reputation-based service level agreements for web services. In *Proceedings of 3rd International Conference on Service Oriented Computing*, pages 396–409, 2005.

[10] MAGNET. IST 6FP project my adaptive global network. www.ist-magnet.org, 2006–2008.

[11] H. Massum and Y. Zhang. Manifesto for the reputation society. *Internet First Monday*, 9, 2004.

[12] I.G. Niemegeers and S. Heemstra de Groot. Fednets: Context-aware ad-hoc network federations. *Wireless Personal Communications*, 33(3):305–318, 2005.

[13] I.G. Niemegeers and S.M. Heemstra de Groot. Research issues in ad-hoc distributed personal networking. *Wireless Personal Communications*, 26(2):149–167, 2003.

[14] PNP2008. The Dutch freeband communications project personal network pilot 2008, www.freeband.nl, 2004–2008.

[15] David G. Rand, Anna Dreber, Tore Ellingsen, Drew Fudenberg, and Martin A. Nowak. Positive interactions promote public cooperation. *Science*, 325:1272–1275, 2009.

[16] Naouel Ben Salem, Jean Pierre Hubaux, and Markus Jakobsson. Reputation-based Wi-Fi deployment. *Mobile Computing and Communications Review*, 9:69–81, 2005.

[17] Attila Weyland, Thomas Staub, and Torsten Braun. Comparison of incentive-based co-operation strategies for hybrid networks. In *Proceedings of 3rd International Conference on Wired/Wireless Internet Communications (WWIC)*, pages 169–180, 2005.

Author Index

Subject Index

About the Editors

Anand R. Prasad, Ph.D. & Ir., Delft University of Technology, The Netherlands, CISSP, Senior Member IEEE and Member ACM, is a NEC Certified Professional (NCP) and works as a Senior Expert at NEC Corporation, Japan, where he leads the security activity in 3GPP. Anand is Member of the Governing Council of Global ICT Standardisation Forum for India (GISFI) where he also chairs the Green ICT group and founded the Security SIG. He has several years of professional experience in all aspects of wireless networking industry. Anand has applied for over 30 patents, has co-authored 3 books and authored over 50 peer reviewed papers in international journals and conferences. He is also active in several conferences as program committee member.

John Buford is a Research Scientist with Avaya Labs Research. He is co-author of the book *P2P Networking and Applications* (Morgan Kaufman, 2008), co-editor of the *Handbook of Peer-to-Peer Networking* (Springer-Verlag, 2009) and has co-authored more than 120 refereed publications. He has been TPC chair or TPC co-Chair of more than ten conferences and workshops. He is a member of the editorial board of the *Journal of Peer-to-Peer Networking and Applications*, the *Journal of Communications*, and the *International Journal of Digital Multimedia Broadcasting*. He is an IEEE Senior Member and is co-chair of the IRTF Scalable Adaptive Multicast Research Group. Dr. Buford holds the PhD degree from Graz University of Technology, Austria, and MS and BS degrees from MIT.

Vijay K. Gurbani works for the Security Technology Research Group at Bell Laboratories, the research arm of Alcatel-Lucent. He holds a B.Sc. in Computer Science with a minor in Mathematics, a M.Sc. in Computer Science, both from Bradley University; and a Ph.D. in Computer Science from Illinois Institute of Technology. Vijay's current work focuses on security aspects of Internet multimedia session protocols and peer-to-peer (P2P) networks. He is the author of over 45 journal papers and conference proceedings, 5 books, and 11 Internet Engineering Task Force (IETF) RFCs. He is currently the co-chair of the Application Layer Traffic Optimization (ALTO) Working Group in the IETF, which is designing a protocol to enable efficient communications between peers in a peer-to-peer system. Vijay's research interests are Internet telephony services, Internet telephony signaling protocols, security of Internet telephony protocols and services, and P2P networks and their application to various domains. Vijay holds three patent and has nine applications pending with the US Patent Office. He is a senior member of the ACM and a member of the IEEE Computer Society.

RIVER PUBLISHERS SERIES IN COMMUNICATIONS

Volume 1
4G Mobile & Wireless Communications Technologies
Sofoklis Kyriazakos, Ioannis Soldatos, George Karetsos
September 2008
ISBN: 978-87-92329-02-8

Volume 2
Advances in Broadband Communication and Networks
Johnson I. Agbinya, Oya Sevimli, Sara All, Selvakennedy Selvadurai, Adel Al-Jumaily,
Yonghui Li, Sam Reisenfeld
October 2008
ISBN: 978-87-92329-00-4

Volume 3
Aerospace Technologies and Applications for Dual Use A New World of Defense and
Commercial in 21st Century Security
General Pietro Finocchio, Ramjee Prasad, Marina Ruggieri
November 2008
ISBN: 978-87-92329-04-2

Volume 4
Ultra Wideband Demystified Technologies, Applications, and System Design
Considerations
Sunil Jogi, Manoj Choudhary
January 2009
ISBN: 978-87-92329-14-1

Volume 5
Single- and Multi-Carrier MIMO Transmission for Broadband Wireless Systems
Ramjee Prasad, Muhammad Imadur Rahman, Suvra Sekhar Das, Nicola Marchetti
April 2009
ISBN: 978-87-92329-06-6

Volume 6
Principles of Communications: A First Course in Communications
Kwang-Cheng Chen
June 2009
ISBN: 978-87-92329-10-3

For Product Safety Concerns and Information please contact our EU
representative GPSR@taylorandfrancis.com
Taylor & Francis Verlag GmbH, Kaufingerstraße 24, 80331 München, Germany

www.ingramcontent.com/pod-product-compliance
Ingram Content Group UK Ltd.
Pitfield, Milton Keynes, MK11 3LW, UK
UKHW010834250425
457613UK00026BB/91